'Focusing on the relationship between Asia and Europe, this is an important and fascinating account of the development of military practice and systems. The concept of convergence and divergence serves Professor Roy well in a study that deserves widespread attention.'

Jeremy Black, *University of Exeter, UK*

'This outstanding work is a hugely ambitious and scholarly analysis of the diversity of the human experience of war over millennia. Kaushik Roy eschews simplifications and slogans to come to grips with one of the most important – and most horrible – aspects of human behaviour and reveals patterns of change and development across the millennia of human existence.'

John France, *Swansea University, UK*

'Military historiography has long favored the narrative of Western exceptionalism, rising out of Greece and Rome, and carried forward by knights against Islamdom until European gunpowder armies and navies conquered the world as part of the supposed "Military Revolution." Kaushik Roy's present book is a tour de force in its reassessment of Eurasian military history and its demonstration that many of the elements of war in different periods attributed to a Western lineage beginning with the Romans and the Greeks were actually born in and shared across Eurasia, that there were in fact many military revolutions across Eurasia that marked the rise of Asian military prowess, and that will into the eighteenth century, European armies could still have been outmatched by the Mongol armies of the thirteenth century. This book with its more complex distillation of the findings of the growing literature on nonwestern military developments, its timely relocation of historical military developments to multiple Eurasian nodes, and its erudite analysis of the factors that kept Asian armies competitive with those of Europe is a significant contribution to our understanding of the world of warfare prior to 1500.'

Michael W. Charney, *SOAS, University of London, UK*

'This sweeping new survey persuasively challenges many of the hoary axioms that have long dominated Western historians' views on premodern warfare. Kaushik Roy takes aim at earlier analyses that turned on over-simplified notions of "Eastern" and "Western" Ways of War, and the dominance of technological change. In their place, he offers a more textured model of recurring convergence, divergence and amalgamation in the organization of military forces and the conduct of war within and among the diverse civilizations of premodern Eurasia, driven by a matrix of cultural, political, geographic, and technological factors. This is an essential read for all students of world military history.'

Karl Friday, *University of Georgia, USA*

A GLOBAL HISTORY OF PRE-MODERN WARFARE

This book examines the military histories of the regions beyond Western Europe in the pre-modern era.

Existing works on global military history mainly focus on warfare in the western part of Eurasia after 1500 CE. As regards the ancient period, such works concentrate exclusively on Greece and Rome. So, 'global' military history is actually the triumphal story of the West from Classical Greece onwards. This volume focuses not only on the eastern part of Eurasia but also on South America, Africa and Australasia and seeks to explain the history and varied trajectories of warfare in non-Western regions in the pre-modern era. Further, it evaluates whether warfare in non-Western regions should be considered primitive or inferior when compared with Western warfare. The book notes that Western Europe became militarily significant only in the early modern era and argues that the military divergence that occurred during the early modern era is not unique – it had also occurred in the Bronze Age, the Classical era and in the medieval period. This was due to the dynamism and innovativeness of non-Western militaries and the interconnectedness that existed in parts of the Eurasian landmass. Further, those polities which were able to construct a balanced military force by synthesising diverse elements were not only able to survive but also became capable of projecting power across continents.

This book will be of much interest to students of military history, strategic studies and world history.

Kaushik Roy is Guru Nanak Professor in the Department of History, Jadavpur University, India. He is a Global Fellow at the Peace Research Institute Oslo, Norway.

Cass Military Studies

Understanding the Impact of Social Research on the Military
Reflections and Critiques
Eyal Ben-Ari, Helena Carreiras, and Celso Castro

Managing Security
Concepts and Challenges
Laura R. Cleary and Roger Darby

Military Strategy of Great Powers
Managing Power Asymmetry and Structural Change in the 21st Century
Håkan Edström and Jacob Westberg

A Global History of Pre-Modern Warfare
Before the Rise of the West, 10,000 BCE–1500 CE
Kaushik Roy

Counterinsurgency Warfare and Brutalisation
The Second Russian-Chechen War
Roberto Colombo and Emil Aslan Souleimanov

War and International Relations
A Critical Analysis
Balazs Szanto

Russia's Armed Forces
A Modern Military for 21st Century Warfare?
Rod Thornton

French Defence Policy Since the End of the Cold War
Alice Pannier and Olivier Schmitt

For more information about this series, please visit: www.routledge.com/Cass-Military-Studies/book-series/CMS

A GLOBAL HISTORY OF PRE-MODERN WARFARE

Before the Rise of the West, 10,000 BCE–1500 CE

Kaushik Roy

Routledge
Taylor & Francis Group
LONDON AND NEW YORK

First published 2022
by Routledge
2 Park Square, Milton Park, Abingdon, Oxon OX14 4RN

and by Routledge
605 Third Avenue, New York, NY 10158

Routledge is an imprint of the Taylor & Francis Group, an informa business

British Library Cataloguing-in-Publication Data
A catalogue record for this book is available from the British Library

Library of Congress Cataloging-in-Publication Data
Names: Roy, Kaushik, 1971– author.
Title: A global history of pre-modern warfare : before the rise of the West, 10,000 BCE-1500 CE / Kaushik Roy.
Description: Abingdon, Oxon ; New York : Routledge, [2022] | Series: Cass military studies | Includes bibliographical references and index.
Identifiers: LCCN 2021010150 (print) | LCCN 2021010151 (ebook) | ISBN 9780367247164 (hardback) | ISBN 9780367247171 (paperback) | ISBN 9780429284069 (ebook)
Subjects: LCSH: Military history, Ancient. | Military history, Medieval. | War—History—To 1500. | Military art and science—History—To 1500.
Classification: LCC U29 .R68 2022 (print) | LCC U29 (ebook) | DDC 355.0209/01—dc23
LC record available at https://lccn.loc.gov/2021010150
LC ebook record available at https://lccn.loc.gov/2021010151

ISBN: 978-0-367-24716-4 (hbk)
ISBN: 978-0-367-24717-1 (pbk)
ISBN: 978-0-429-28406-9 (ebk)

DOI: 10.4324/9780429284069

Typeset in Bembo
by Apex CoVantage, LLC

CONTENTS

PREFACE AND ACKNOWLEDGEMENTS

It always struck me how the military history of the regions beyond Western Europe, especially before the rise of the West from around *c.* 1500 CE, is neglected. Western scholars have extensively studied the Greco-Roman military affairs, but in such accounts the non-Western regions always appear as 'victims'. There have been specialist studies of some parts of the pre-modern non-Western world, but a connected narrative is missing. Moreover, what has been accepted as global or world military history is actually the triumphant story of the West from Classical Greece onwards. Therefore, I always wanted to study and write about the rich and divergent military activities of the regions beyond Western Europe, especially in the pre-modern era.

Frankly, this volume does not have any competitors within the academic circle. We have some serious research work dealing with ancient, medieval and early modern Eurasian military history, but there are no specialised or quasi-academic popular books dealing with precolonial global military history available in the market. The existing works on global military history deal mostly with the western part of Eurasia, and their focus remains the post-1500 CE period. This volume focuses not only on the eastern part of Eurasia but also on South America, Africa and Australasia. Further, in this book, an attempt has been made to explain the history of the non-Western regions from their own perspectives.

This volume will appeal to students (both undergraduate and postgraduate) and researchers of history, political sciences and international relations. In addition, it will be useful for students pursuing courses on ancient and medieval history, military history, Asian empires, precolonial Africa and the Americas. Serious researchers dealing with different aspects of precolonial military history will also find this volume useful as a starting point. And the general public will get an idea about the dynamics of warfare beyond Western Europe in the pre-modern era. Finally, to an extent this book will also function as a handy volume for policymakers and military officers studying in the various academies. For instance, US and British military officers facing future deployment in a foreign region might find this volume useful to acquaint themselves with the military practices that existed long ago in that particular area.

Credit is due to Andrew Humphrys, my editor at Routledge, who encouraged me to write a work on the history of different military organisations that existed in various parts of the globe before the rise of the 'West'. Special thanks also to Bethany Lund-Yates of Routledge for patiently bearing with me. I am also indebted to the three external referees for their hard-nosed criticism.

Professor Jeremy Black's writings inspired me to think about the military affairs of the non-Western world. Furthermore, I am grateful to my friend Professor Peter Lorge for pointing out certain limitations in my original framework. This work would not have been possible without the support of my research scholars, who not only functioned as a sounding board for various ideas over innumerable cups of tea and coffee but also supplied me with various sources. Their interest kept me going. Moumita, Arka, Sohini and Aryama, my dear research scholars, it is thanks to you that this monograph is a reality. While working on this book, I was also encouraged to provide optional courses on global history of warfare both at the undergraduate and postgraduate levels at Jadavpur University, Kolkata. I sincerely hope that this work will encourage both young and senior historians to write a more nuanced global military history of the pre-modern world.

Last but not the lease, credit is also due to my wife, who besides her own teaching tasks took up the bulk of the household chores so that I could read and write for completing this work. Let us hope that this volume will get students and researchers thinking and make my task of writing the sequel of this volume, which carries the story from *c.* 1500 to the present day, worthwhile.

Kaushik Roy
Kolkata

MAPS

GLOSSARY

ankush A rod with a curved end; this instrument was used by the *mahout* for controlling the elephant.

atalatl A spear-throwing tool that increased the speed and accuracy of the throw. It was made of a slightly curved piece of wood, ivory or bone. One end of it was hooked, and the hook was fitted into the nock end of the spear shaft. This piece of technology actually used a sort of primitive lever principle and increased the thrust of overhand throw. The precolonial North Americans and the Mesoamericans used this instrument for both hunting and fighting.

bailey The enclosure (courtyard) of a castle.

bakufu Also known as shogunate. This term refers to the military government of pre-modern Japan. At its top was the shogun – the military dictator of medieval Japan, who had reduced the monarch to a mere figurehead.

ballista A pre-modern engine used for throwing stones.

bastion A projecting tower or earthwork structure for defending the curtain wall.

bheda Kautilya's policy of divide and rule.

caliph Spiritual and temporal head of Islam. He is the head of the religion as well as commander-in-chief of the military and also the ruler of the domain. In this set-up, separation of the church and politico-military administration is absent.

caliphate Realm (empire) of the caliph.

catapult A pre-modern engine for shooting missiles with the help of a string drawn by mechanical means. Normally a bolt was placed in the groove of the machine.

chain mail armour This armour comprised small metal rings linked together in a pattern to form a mesh. Either the rings were joined by riveted links or the ring was placed inside another ring. In the latter case, each ring was joined to four other rings. For making such armour stronger, two to three layers of rings were joined together.

chaturanga bala Literal meaning 'four limbed'; the term refers to the ancient Hindu army, comprising infantry, cavalry, elephants and chariots.

composite bow While the self/simple bow emerged in Central Africa, the composite bow emerged in Siberia and in Inner Asia. The composite bow was the favourite weapon of the steppe nomads. A composite bow was manufactured with wood, horn and sinew. In a composite bow, the wooden core functioned as a support to hold the other parts together, and such a bow had horn in the belly and sinew on the outer part. Generally, a composite bow was double curved. When the two sides of the stave were made with horn and the grip was of metal, then a composite bow could be called a reflex bow. When unstrung, the entire length of the bow curved from the belly in a 'c' shape away from the archer. A recurve composite bow was a bow in which the outer parts of the limbs curved away from the archer when unstrung. Overall, a composite bow using the principle of tension (tensile strength of the materials glued in its body) was lighter and had a longer range compared to a self bow and could be used easily from horseback.

crossbow A bow with a mechanism to draw the string. The bow was bound to a wooden stock. The end of the stock was known as the tiller, and the string was drawn back over a nut on the stock. A heavy crossbow was positioned on the ground, and a lighter one was held in hands by the soldier. The string was released by the trigger, which retracted the nut when fired. A large crossbow was called a ballista by the medieval Western Europeans.

curtain It means the outer enclosing wall of the fort or the wall joining the two towers of the fort/castle.

Dar-al Islam Literally meaning 'land of light'. This term refers to the areas under Islam. The opposite of this term is *Dar-al Harb*, meaning land of darkness or the areas under the heathens.

fief The fief was the central element of Western European feudalism. It meant the grant of a piece of land in return for providing military service. The person who granted it became the lord and the person who accepted it became a vassal of the lord. Usually a fief, unlike a *jagir*, was hereditary.

Greek fire A combustible substance (probably naphtha) used by Byzantine soldiers. This material could not be easily removed and on impact burst into flames. It was a sort of flame thrower. The Greeks used it for both naval and ground war from a siphon as well as a catapult.

hastipa Person concerned with the upkeep of war elephant.

hoplite Heavy infantry of Classical Greece wearing armour and equipped with long spears.

hoplon Round wooden shield carried by a hoplite infantryman.

howdah A box like structure made of wood with a protective iron covering which was placed on the back of the elephant. It was tied to the trunk of the elephant with ropes. On the box sat the *mahout* and the warriors.

jagir Also known as *iqta*. It was a grant of a piece of land with the right to extract revenue in order to maintain a cavalry contingent for the Muslim ruler. At certain times, such a land grant became hereditary.

kardakes Indigenous disciplined infantry of the Achaemenid Empire.

Kesselschlacht/Kettelschlacht Literally meaning 'cauldron battle' or 'pocket battle'. In such a battle, the main force is sent to attack both the flanks and rear of the enemy force while holding them frontally with a subsidiary force.

khan This term actually meant 'general of the Mongol army'. A khan probably commanded a *tumen*. Later, khan became an Islamic title.

Kshatradharma Code of conduct of the Rajputs equivalent to the *bushido* of the samurai and chivalry of the knights. The code emphasised conducting just war (*dharmayuddha*).

kumiz Fermented mare's milk consumed by the Central Eurasian steppe nomads (especially the Mongols).

kuriltai Grand assembly of the Mongol chieftains where the Mongol leader (Great Khan) was selected from the heads of the chief families.

legion The basic organisational format of Roman heavy infantry.

legionaries Armoured infantry soldier of the legion.

longbow The typical weapon of the medieval infantry of England. This bow was huge and heavy and could only be used by the infantry. The longbow had a wooden stave of yew and a string. The stave was rounded, with a D-shaped central section. The sapwood of the stave was on the outside and the heartwood towards the archer in order to give the stave a natural spring. Pieces of horn or nocks were added to the ends of the stave to hold the string in placẹ. A longbow required great strength in drawing. The archer drew such a bow to the chest.

macchayantra Literally meaning 'fish machine', this is the ancient Hindu compass. This machine comprised an iron needle shaped like a fish which floated in a transparent container full of oil, and the needle always pointed northward.

Maghreb North-West Africa, which includes Algeria, Tunisia, Morocco and Libya.

mahajanapada Territorial states of ancient India.

mahout Elephant driver.

mamluk Also known as *ghulam*. The term refers to slave soldiers of Islam. Generally, they were paid with *jagir*s (also known as *iqta*).

mangonel Literally meaning 'a throwing engine', it worked by torsion from twisted ropes and had a spoon like arm that revolved on release. The arm hit the crossbar causing the stone on the spoon to be thrown.

mlechchas Those who were outside the *chaturvarna* (four *varna*) caste system (it comprised of four castes: Brahmins, Kshatriyas, Vaishyas and Shudras) and were considered unclean and inferior people – barbarians – by the Hindu intellectuals.

motte Earthwork mound (either artificial or natural) for defending a castle.

nomarch Provincial chief of ancient Egypt in charge of the *nomi*.

nomi Administrative division of ancient Egypt.

osthismos Literal meaning the push or to move forward, it involved pushing and stabbing by the hoplites. Not only did the hoplites push forward against the hostile soldiers, but the rear ranks also pushed the front ranks against the enemy formation. The objective was to break the hostile formation.

peltast Light infantry soldier of ancient Greece who wore no body armour and was armed either with a sling or a javelin. They functioned as skirmishers and generally deployed in front of the heavy infantry and at the flanks. In the Roman army, they were called *velites*.

phalangite Heavy infantry of a Macedonian phalanx.

phalanx Rectangular heavy infantry battle formation of Classical Greece and Macedonia. The soldiers were deployed in several rows.

pharaoh Ruler of ancient Egypt.

pilum Roman javelin which had an iron shank with a conical tip. The iron shank was connected to the wooden handle with a socket or a tang. When the *pilum* was thrown towards the enemy, after hitting the target due to impact, the iron shank bent or broke, so the hostile soldier could not throw it back. If the *pilum* hit an enemy shield, when the hostile soldier tried to withdraw it, the broken or bent shaft ruptured the shield. And when a wounded enemy soldier tried to bring out the broken or bent iron shaft from his body, his wounds became wider and deeper.

plate armour Body armour comprising thin sheets of metal. The main part was the breastplate, which covered the torso. It was inflexible and heavier than scale armour.

polis City state of ancient Greece.

pomeste A land grant to a noble in return for providing cavalry to the grand duke of Russia. The *pomeste* was actually a conditional land grant in medieval Russia (Muscovy) for military purposes and somewhat equal to the *jagir*.

raja Hindu king/ruler.

Rajput Originally known as *thakurs*, who were high caste landowners and became the hereditary warrior community. They acquired Kshatriya status (second highest caste in the fourfold Hindu hierarchical *varna* system).

ratha War chariot of ancient India.

samurai Mounted warrior (who was also a landowner) of medieval Japan somewhat equivalent to the Rajput of medieval India and knight of medieval Western Europe. Most of the samurai were hereditary landowners who had their private mounted forces.

sarissa Iron-tipped long pike made of cornel wood used by a phalangite.

sataghni A big stone-throwing machine used by the ancient Indians for siege warfare. It was an anti-personnel weapon system, and each throw was reputed to have killed a hundred persons.

satrap Achaemenid Persian provincial governor.

satrapy Province in the Achaemenid Persian Empire ruled by a *satrap*.

scale armour This armour comprised of small metal plates or rings interlinked with each other and glued or stitched on a garment made of hide or fabric.

self bow It is also known as simple bow and made from a single piece of wood or bamboo.

senapati General of a Hindu polity.

shaduf A hand-operated device used for lifting water. It comprised a horizontal pole mounted like a seesaw.

shaman Especially found among tribal people, the *shaman* is a person who has religio-magical power. He functions as the intermediary between the natural and supernatural powers and the ordinary mortals. They are quasi-priests of shamanism (spirit centric tribal religions).

trebuchet A throwing machine used for shooting arrows and stones. There were two types of trebuchets: one which worked on the principle of torsion/tension and another which worked on the principle of traction. Torsion trebuchet used the tension inherent in sinew (like hair). Such materials were used for making the spring to generate throwing capacity. In a traction trebuchet, a stone was placed at one end of a pole. When the crew released the pole, it went up and ejected the stone. A complex form of trebuchet was the counterweight trebuchet. It had a container for placing heavy materials at one end of a pole and a sling to hold the stone (missile) at the other end. The pole was on a pivot. The loaded end with the stone was winched down and then released. It made the loaded end rise rapidly and eject its content, and the sling gave added impetus. The range and weight of the missile that could be thrown by a counterweight trebuchet was greater than that of a torsion-based engine or a traction trebuchet.

tulwar A long curved sword with sharpened edges on both sides. It was suitable for use by men on horseback for cutting and slashing the enemy troops.

tumen A Mongol division with a ration strength of 10,000 cavalrymen and more than 30,000 horses.

***varna* system** The fourfold *varna* (caste) system of Hindu society. At the top were the Brahmins, then came the Kshatriyas, then Vaishyas and finally the Shudras. The Brahmins were supposed to engage in religious rituals and teaching and act as advisors to the rulers. The Kshatriyas were supposed to be the traditional fighting class of Hindu India. The Vaishyas were supposed to engage in trade and commerce, and the Shudras were to devote themselves to agriculture. It may be inferred that the Shudras were actually descendants of the defeated Dravidians, the first two *varna*s were descendants of 'pure Aryans' and the third *varna* was the product of intermixture between the Aryans and the Dravidians. This is a theoretical model. In reality there were many subcastes. Liberals and postmodernists claim that this whole model is a fictive imagination of the British Orientalists.

Vijigishu The ambitious Hindu ruler who would dominate the *mandala* (circle of states) and become *ekarat/samrat* (emperor) of India.

volta A sailing technique used to take advantage of the South Atlantic gyre-north-westerly winds which drive eastward-flowing currents.

yam Postal system introduced by the Mongols. This system comprised a relay system of messengers mounted on horses, and along the route at certain intervals there were stations where fresh horses and food and a resting place for the messengers were available.

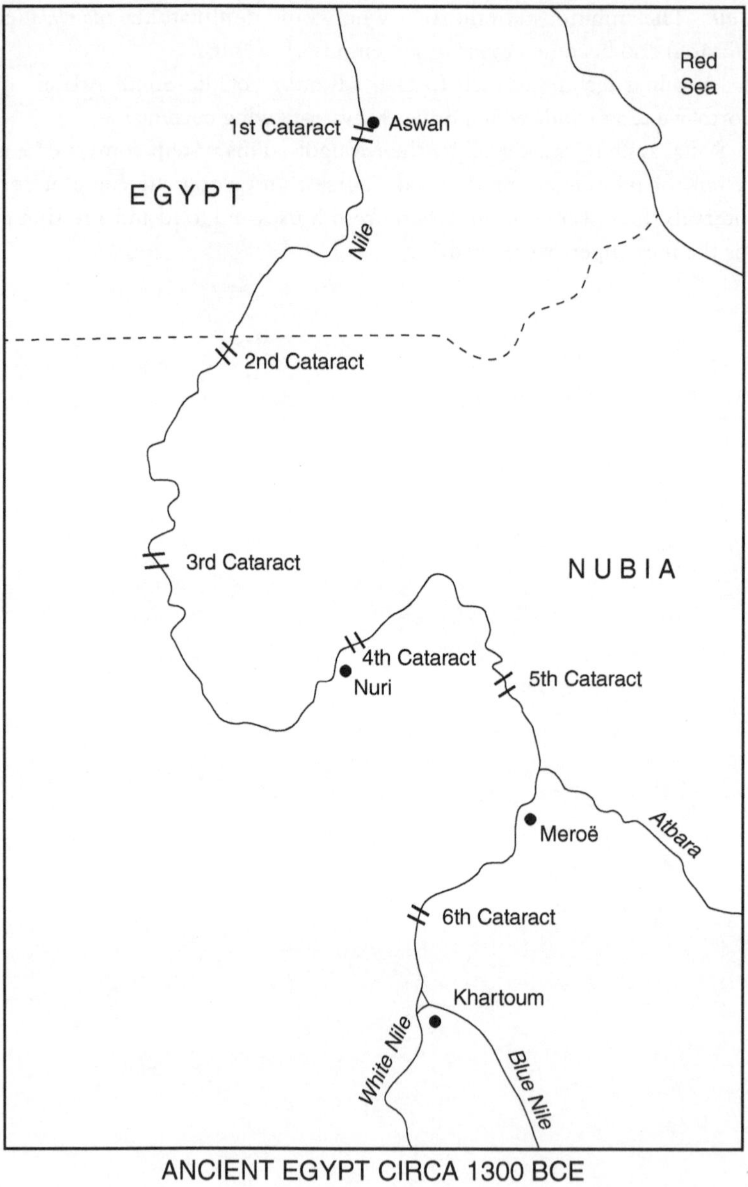

MAP 1 Ancient Egypt, *c.* 1300 BCE

MAP 2 Classical Greece, *c.* 350 BCE

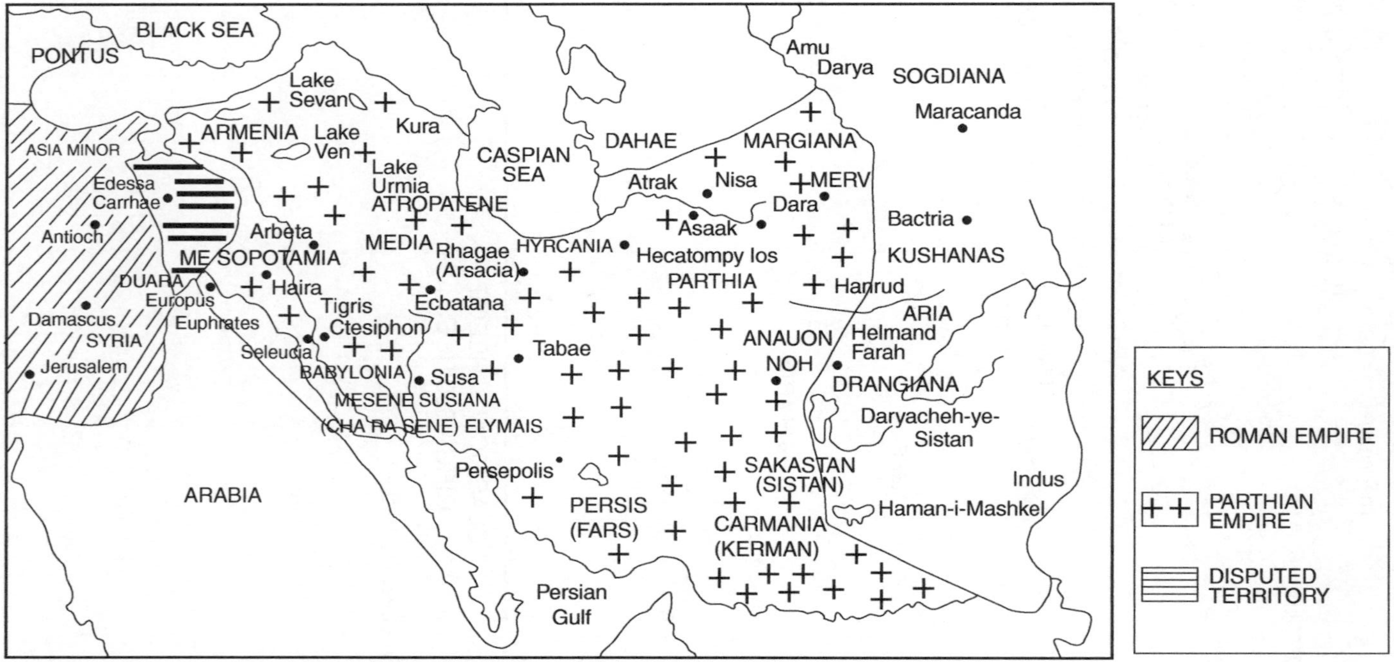

PARTHIAN EMPIRE CIRCA 200 CE

MAP 3 Parthian Empire, *c.* 200 CE

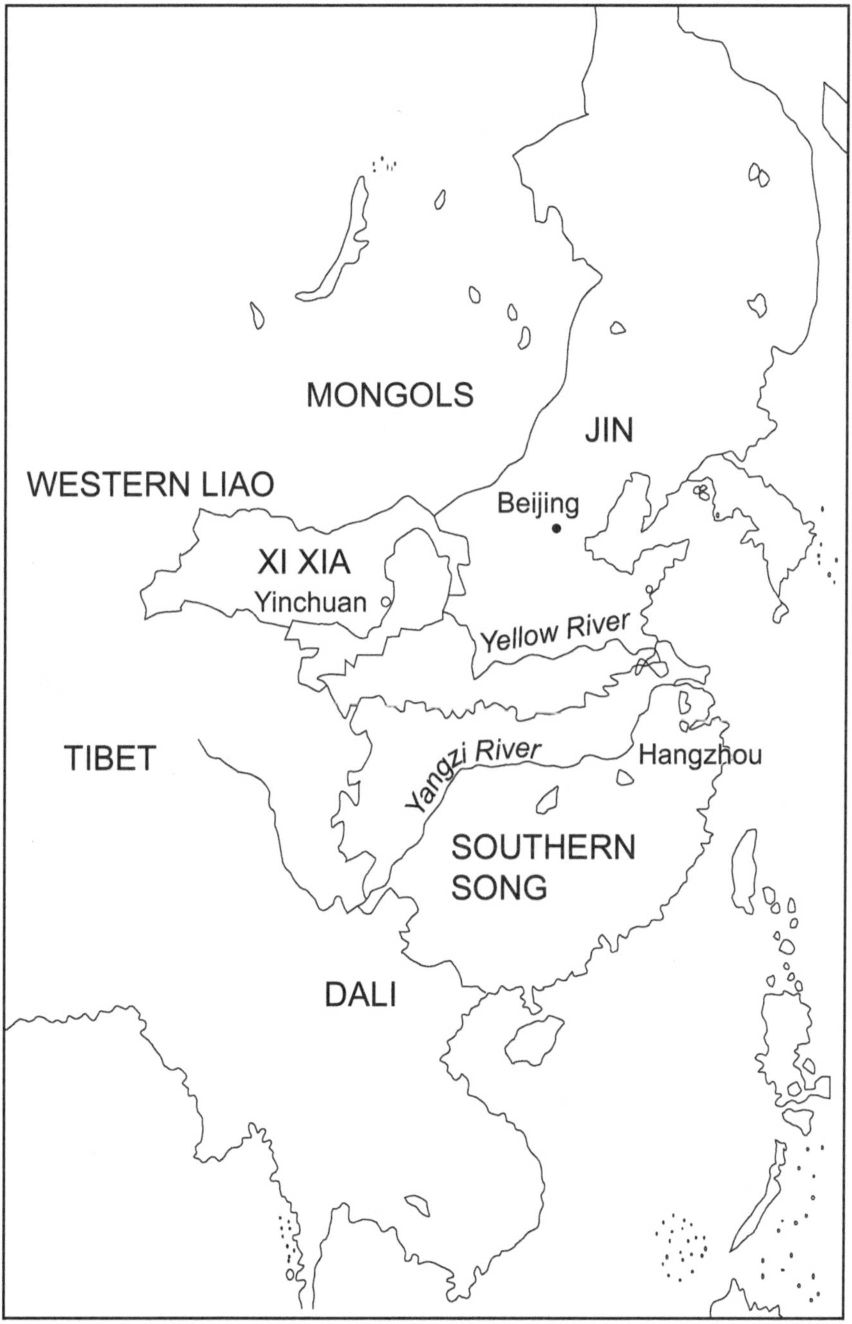

CHINA CIRCA 1200 CE

MAP 4 China, *c*. 1200 CE

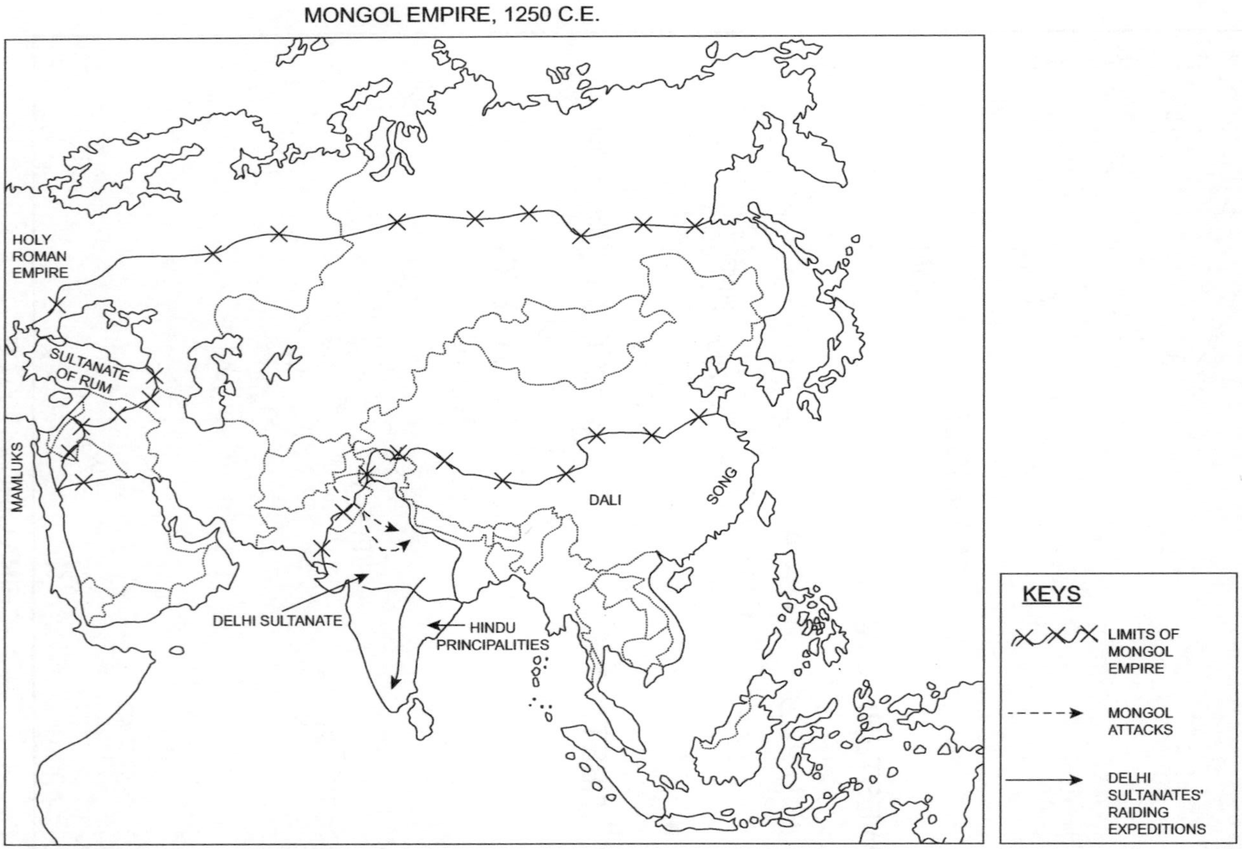

MAP 5 Mongol Empire, 1250 CE

INTRODUCTION

Evolution of warfare before the rise of the West

One may ask whether it is necessary to write about precolonial global military history. To a great extent the answer is yes, because the bulk of writings that purport to offer a global history of warfare focus on the rise of the West from the early modern era. The essential point of this book is to readdress the issue of great military divergence in Eurasia. Western Europe became militarily significant only in the early modern era. And the military divergence that occurred during the early modern era is not unique. This work argues that military divergence and convergence occurred in the Bronze Age, Classical era and the medieval period. The dynamism and innovativeness of the non-Western militaries and interconnectivity which existed among various parts of the Eurasian landmass may be credited with the successive military convergences and divergences in the pre-modern era. Further, those polities that were able to construct a balanced military force by synthesising diverse elements were not only able to survive but also became capable of projecting power across continents.

During the early modern period, Western Europe started to diverge from other parts of the world, and this demographically, economically and militarily marginal area within Eurasia began to dominate the rest of the world. During the medieval era, Western Europe was under simultaneous attacks from the Vikings, the Magyars and Islamic forces. The Western European economy was backward compared to the contemporary Chinese and South Asian economies. The process of Western European expansion in the North and South Americas (New World) had started in the sixteenth century. However, the two Americas and Sub-Saharan Africa were lagging behind Eurasia from the ancient period regarding state building and the generation of military power. For largely ecological reasons, the societies in the two Americas and Australia remained at the level of Stone Age technology till the early modern era. Western Europe's serious competitors were the Islamic, Indic and Chinese civilisations which straddled

DOI: 10.4324/9780429284069-1

the southern and eastern half of Eurasia. The present book charts the varied trajectories of warfare in the regions outside the West before *c.* 1500 CE. Further, it attempts to evaluate whether warfare in the regions beyond Western Europe was primitive/inferior vis-à-vis the overhyped Western warfare.

The pioneering work in explaining the rise of the West vis-a-vis the 'rest' is Geoffrey Parker's 1984 Lees Knowles Lecture, published as *The Military Revolution* in 1988 (Parker 1988). Parker focuses on military-technological innovations (gunpowder, handguns, field artillery, siege artillery, artillery fortress and ocean-going ships) which resulted in massive changes in the military organisation. This in turn forced transformation of the state structure. The gunpowder armies and navies in the hands of nation states characterised by military-fiscalism enabled the West to conquer the whole world between 1500 and 1800. Parker's military revolution model came under sustained criticism from Jeremy Black. Black argues that military changes occurred gradually, and the seventeenth and eighteenth centuries are more important as far as development in military technologies is concerned (Black 1991). Nicholas Hooper and Matthew Bennett claim that several features of Parker's so-called military revolution were present in Western Europe during the Middle Ages (768–1487 CE) (Hooper and Bennett 1996).

In response, Clifford Rogers came up with the punctuated equilibrium theory. He tried to integrate Parker's revolution with Black's evolution. Roger's argument is that Western Europe experienced a series of revolutions interspersed with periods of evolution (Rogers 1995). Several historians working on European military history applied the concept of military revolution further back in time. Even before Parker's model became famous, Arther Ferrill argued for a military revolution which occurred in Greece between the Peloponnesian War and the rise of Alexander (Ferrill 1986). An edited volume published by Cambridge University Press asserts that Western Europe experienced a series of military revolutions from the Middle Ages onwards (Knox and Murray 2003). The concept of military revolution spread like an epidemic not only among the historians working on European history but also among scholars dealing with regions outside Western Europe. Military historians are now finding evidence of military revolutions in Song China (Lorge 2008), Mughal India (Garza 2016) and in eighteenth-century Eastern Europe. The military revolution model is in serious crisis. And nobody understands this better than Parker.

Parker, who had initially broadened Michael Roberts' concept of military revolution (originally used in 1955 to explain the military rise of Sweden under Gustavus Adolphus) to analyse the rise of the Western European maritime empires in the early modern era, came up with a broader concept in 1995 known as the 'Western way of war'. Parker integrated various elements of his military revolution in the wider concept of the Western way of war, which claims that the West had developed a scientific, secular and rational method of conducting warfare from the Classical era onwards (Parker 1995a: 2–9). This argument is a wider version of the approach put forward by a famous Classicist Victor Davis Hanson some years earlier. In 1989, Hanson argued that several key elements of the

Western way of war – the quest for decisive battles, generation of sophisticated military theories, presence of civic spirit among the Western soldiers, emphasis on technology and discipline, etc. – initially emerged in Classical Greece (Hanson 2001). Hanson and Bernard S. Bachrach (an expert of Merovingian-Carolingian armies) assert that the Classical Greek military tradition which embodied the Western way of war continued in the Roman and in the post-Roman Western Europe (Bachrach 1995: 50–91; Hanson 1995: 12–49). Parker's edited volume notes that the Western way of war (a triumphal story of the rise of the West vis-à-vis the rest) continued till the Vietnam War in the 1970s (Parker 1995b). Within this approach, the Asian militaries appear as weak and incompetent. The proponents of the Western way of war portray that from the time of Xerxes, the numerically superior undisciplined Asian hosts had always been beaten by disciplined and technologically superior but numerically inferior Western armies. So Asia was always there for taking by the Western Europeans.

John Keegan, in his global history of warfare, following what could be termed a culturalist approach, asserts that the military cultures of Eurasia could be divided into two categories: Western and Eastern ways of war. Keegan's Western way of war is same as Parker's concept. However, Parker has nothing to say about the Eastern way of war. Keegan argues that the so-called Eastern way of war developed from the steppe nomadic military tradition, which emphasised avoidance of battle and focused on skirmishing plus tactical retreat, among other features (Keegan 1994). The Parker–Hanson–Keegan approach claims that the West from the era of Classical Greece heavily invested in close-quarter combat with disciplined heavy infantry. In contrast, they imply, the Asian military systems centred on light cavalry forces were good for raiding and skirmishing but not for conducting decisive battles and sieges.

This bipolar division of history of warfare into Eastern and Western way of war reminds one of Edward Said's compartmentalisation of the world into Orient and Occident or the far older Herodotean division of the world into Greek (Orient) and Asian (Persian) civilisation (Occident). This book will show that the steppe nomadic cavalry powers frequently waged decisive encounter battles. In fact, this volume asserts that Hanson's claim about the Classical Greeks inventing the concept of decisive battle (an integral part of the Western way of war) (Hanson 2009) is erroneous. Besides the steppe nomads, even the sedentary Asian polities generated sophisticated techniques of warfare and conducted decisive battles with disciplined armies and navies. Again, this volume shows that Parker's assertion about the West generating secular military theories is erroneous. At times, the Western military theories were heavily tinged with religious ethos. And several non-Western polities in history did come up with secular theories of warfare.

The multi-authored *World History of Warfare*, published by Nebraska University Press in 2002, starts with the grand aim of offering a reinterpretation of world military history. Nevertheless, this volume also falls prey to the Parker–Keegan–Hanson bipolar division of Eastern and Western styles of war, which is inherent in the Western way of warfare approach. Japan, China and India are all included within the Eastern style of war. And most of the volume focuses on various stages in the evolution of Western military landscape: Classical West, feudalism and chivalry, gunpowder warfare, trends of war during the Renaissance, absolutism and warfare, industrial war, limited wars in the post-Second World War era, etc. (Archer et al. 2003).

In recent times, two historians, after thoroughly analysing Western military developments, rightly concluded that there was no homogeneous unified Western template of war before the eighteenth century (France 2011; Parrott 2011: 48–63). *A Global History of Pre-Modern Warfare: Before the Rise of the West, 10,000 BCE–1500 CE* studies in detail non-Western societies in comparison with the Western societies and portrays the different types of military systems which existed in different parts of the non-Western regions at various moments in history. Moreover, this volume pushes the argument that the militaries of non-Western societies in several regions and in different time periods were quite combat effective and that the military supremacy of the West is a relatively recent phenomenon and probably temporary. Hence, it is necessary to study the varied history of the non-Western militaries in their own right.

Several historians working in the field of global military history are aware that neither the military revolution model nor the Western way of war approach could properly explain the historical evolution of warfare. Azar Gat, in his voluminous work on world military history, provides lot of space to explain the origins of war with the aid of anthropological theories, harping on the racial motivation behind warfare, but then the bulk of the volume concentrates on capitalism, industrialisation and the nature of warfare conducted by the West (Gat 2006). Gerard Chaliand's *A Global History of War* is a bit confusing, as he merges steppe nomadic warfare with irregular/unconventional/guerrilla war (Chaliand 2014). In tune with the culturalist turn in history writing, the same approach is becoming popular in writing military history too. The latest example of this genre attempts to explain the innovations in military affairs through a cultural lens. Wayne E. Lee, in his book on global military history, argues that capacity and calculation are the two key integers behind military innovations and are shaped by culture (Lee 2016). What 'culture' means is not defined, and by the author's own admission the book mainly concentrates on the West.

The core argument of the culturalist group of scholars is that the concepts of defeat and victory differed between the Western Europeans and the rest of mankind. Further, the Clausewitzian concept of war and peace does not apply in case of the non-Western people. War is the continuation of politics by violent means, and this axiom is applicable only in case of Western European history with strong states. The non-Western societies with weak polities and a different value system (characterised by an overdose of magic and religion), assert the culturalists, cannot be analysed with the aid of the Clausewitzian formula.

A refreshing break from the culturalist approach and the military revolution model is the *longue durée* history of warfare published in 2011 by two political scientists (Levy and Thompson 2011). Instead of military revolution, Jack

S. Levy and William R. Thompson identify three great accelerations which occurred in different parts of Eurasia. The first acceleration occurred in the Middle East/Near East (Mesopotamia-Egypt) between the late fourth and early third millennium BCE. The second acceleration occurred in the last half of the first millennium BCE in Mediterranean Europe (Greece and Rome). And the third acceleration occurred in early modern Western Europe. This thought-provoking book is full of insights. Nevertheless, at another level, this work is conceptually shallow and empirically narrow. The three accelerations of Levy and Thompson are actually three military revolutions packaged in new semantics. Some of the assertions, and especially the tables, hang on weak empirical bases. Siege warfare is non-existent in this work. China and India, which together account for almost half of humanity during recorded history, play a cameo in their account. Central and South Africa too rarely make an appearance. Geography and to a lesser extent political economy are the principal drivers in Levy and Thompson's model. Nevertheless, these two authors fail to account for the importance of the Central Eurasian nomads in shaping the trajectory of Eurasian military history till *c.* 1800. Following somewhat the deep history approach of Jared Diamond and Yuval Noah Harari, Levy and Thompson completely discount the role of individuals in history.

For a historian, the devil lies in the details. So the present work focuses on the small and big changes which amalgamated to generate the push-pull factors that drive the history of warfare from the prehistoric era to the present time. The book argues that Asia, at times, built larger and stronger polities compared to Western Europe. Moreover, the concept of victory and defeat are universal. A bipolar division of mankind by the culturalists is as bad as the compartmentalisation of humanity by the proponents of racial theories. War is a universal phenomenon that involves the use of organised violence by a community to force the other groups to do their will. One does not require a sophisticated cultural lens in order to understand it. Rather than making use of the obtuse concept of culture, the present book emphasises organisational-technological factors along with physical geography to explain the evolution of militaries from the dawn of civilisation till the early modern era. In order to structure the history of evolution of warfare from the dawn of history till the present time, three concepts are used in this volume: convergence, divergence and hybridity. The book stresses that hybridity is the product of amalgamation of selected elements from two or more opposing military systems and the end product is something new/original. And the combat effectiveness of the hybrid military establishment is greater than the ones from which it evolved.

This work claims that the great divergence in matters military which occurred during the early modern era is not exceptional in world history. Divergence and convergence are common if we take a *longue durée* approach of global military history. The model, though somewhat cyclical, takes into account the changes which occurred with time. The origins of warfare can be traced back to the Neolithic era due to convergence of certain politico-economic factors.

The Bronze Age civilisations of Eurasia (including North Africa) witnessed a military convergence, when chariot-centric military systems dominated the military landscape of these societies. After the collapse of the Bronze Age, we notice a divergence in military affairs during the ancient period. While Classical Greece experimented with heavy infantry, the Asian empires generated a cavalry-centric force. However, there were great variations within the different Asian military systems. For instance, while Persia and China came up with the infantry–horse combination, India deployed elephant–infantry force. Hence, it would be wrong to categorise all the military systems of Asia within one broad umbrella. The differences between the ancient Chinese and Indian force structures were as great as between the Classical Greek and Vedic Indian armed forces. Ecological conditions forced the Sub-Saharan African polities to come up with light infantry supported by camel-borne units. This divergence continued during the era of the early Roman Empire. From *c.* 400 CE till *c.* 1500 CE, the horse warriors dominated Eurasia. So we find that during this period, a convergence occurred in military affairs from Korea in the east to France in the west. To a great extent, the horse warriors shaped the convergence and divergence of military systems of Eurasia till the onset of the Industrial era. In this volume, the mounted archers do not merely play a negative role, as in Kenneth Chase's account. Chase shows how the Inner Asian horse warriors prevented the development of gunpowder weapons by the sedentary states of Asia (Chase 2003). In contrast, the book argues how at different moments of history, the various sedentary polities adopted several techniques and tactics from the steppe nomadic horse warriors (who were continuously evolving militarily) in order to upgrade their military organisations. The 'Great Divergence' (Parker's so-called military revolution, or better still the military transformation of Western Europe) started after 1500 CE, an issue that will be discussed in this volume's sequel.

The approach of this book is truly global. The working assumption is that in order to understand the rise of the West, we need to analyse the military developments (or lack of them) in other parts of the world. So instead of a Eurocentric focus, the spotlight needs to be shifted to the non-Western parts of the world. Hence, this work does not privilege Western Europe at the expense of the other regions of the world: the fact that the West dominated the world politically, economically and militarily from the early modern age does not mean it was the dominant region in ancient and medieval eras as well. Hence, besides devoting attention to well-known figures such as Alexander and Julius Caesar, an attempt is made to give space to the relatively lesser-known figures such as Chandragupta Maurya and Wu-ti. Instead of focusing mostly on the Macedonian and Roman empires, considerable space is given to the Mauryan, Han, Aztec and Inca empires. The problem is comparative paucity of sources about these non-European military figures and empires and most of the existing materials are not easily available in English. To an extent, this work attempts to utilise the sources generated during the pre-modern era.

This volume takes into account archaeological perspectives, especially as regards the origins and nature of warfare in the prehistoric era. And geography is the 'mother of military history'. For instance, ecological conditions in Africa obstructed large-scale use of cavalry. The dialectics between geography, force structure and conduct of warfare is emphasised in this work. This volume also touches upon the latest trends in the field, such as the gender issue. This does not mean that the present work neglects the so-called traditional arenas of military history. Due attention is given to strategy, tactics, logistics, command, doctrine and the relationship between warfare and state formation. Politics is the crucial driver in shaping the trajectory of warfare. One cannot understand the changing dynamics of warfare without examining the supporting political setup. The interrelationship between state, warfare and the role of politics is highlighted throughout this monograph. At times, an amalgam of conventional and unconventional wars brought down several empires at different moments of time. Further, conventional and unconventional warfare were interrelated, as the capacity of a state to project military power depended on its ability to control its subjects. So the present work gives some attention to irregular combat. For this reason, we use the term 'warfare' rather than 'war' in this book.

Taking a broad-brush approach, one can discern the several stages in human evolution which in turn affected the trajectory of organised violence. The Neolithic Revolution around 10,000 BCE transformed hunter-gatherer societies (at least some of them) in Eurasia into sedentary agriculturists. The farmers spread out during the Bronze Age from around *c.* 4500 BCE and established cities. And with the onset of the Iron Age from *c.* 1000 BCE, bureaucratic polities supported by intensive farming emerged in Eurasia. The Western European polities rose to prominence when they created a truly global world, establishing communication links between all the continents after 1500 CE.

The principal causative factor in the book's framework explaining the evolution of warfare is geography. It shaped geopolitics and geoeconomics, which in turn to a great extent altered the paradigms of warfare that emerged in different parts of the world at various moments of our timeline. The aim is also to show how disease, climate and economy (all products of geography) shaped the trajectories of warfare in the different continents of the world till the rise of the West. The next level of drivers involved politics and technology. As far as politics is concerned, the role of 'extraordinary individuals' cannot be discounted. We cannot totally do away with human agency. Political and military ambitions of certain individuals at various moments of time generated intense interstate competitions, which in turn did have massive consequences at the supra-regional levels. While analysing such cases, the aim is to link up micro with macro perspectives.

One of the solid academic studies of war and technology from a *longue durée* perspective is the volume published by the noted military historian Martin van Creveld in 1991. He divides the military technological evolution into four phases. The period from the Stone Age till *c.* 1500 CE is categorised by him as the age of animal and muscle power. His categorisation of the history of technology is too mechanical. Moreover, his focus remains on Western Europe (Creveld 1991). The present volume shows that diverse forms of complex technologies evolved in various parts of Eurasia till *c.* 1500 CE. Moreover, the history of technology is not merely a west–east flow. Several times, both in cases of naval warfare and ground combat, cutting-edge technology emerged in the eastern and central parts of Eurasia which were then adopted by Western Eurasia.

As two-thirds of planet earth is covered with water, due attention is given in this volume to maritime combat. Riverine warfare is discussed because landmasses inhabited by the humans are crisscrossed by numerous rivers, estuaries, deltas and lakes. Nevertheless, more space is given to land warfare compared to naval combat due to the fact that humanity inhabits the landmass. Hence, ground war has been more frequent in history and affected larger number of people compared to combat in the sea and the rivers. Overall, this volume gives bulk of its space to analyse the evolution of warfare in Eurasia, since most of world's population lives in this landmass and most of the economic developments and technological progress has occurred in this part of the world.

The present volume has eight chapters which discuss the evolution of warfare before the rise of the West. The first chapter discusses how and when warfare came into existence. Our story begins with the Stone Age. Warfare in the era before the origin of the state was the product of ethnic, religious and tribal tensions. At the present stage of research, we can be sure that warfare existed in the Neolithic period. The weapons used for hunting in the Old Stone Age were used for warfare during the New Stone Age. The causes behind the origins of war and the form of war practised by humans during the New Stone Age throughout Eurasia, Australasia, Africa and the two Americas were more or less similar. And such warfare did not merely comprise bloodless ritualistic duels, but at times proved to be quite lethal as well.

Chapter 2 discusses the evolution of warfare and the linkages between climate change and military innovations in the rise and fall of the Bronze Age societies in Eurasia. This chapter details the beginning of military convergence with the genesis of chariot-centric warfare and its ultimate demise. Some scholars assert that until the advent of the Iron Age, large-scale fighting by regular armies was absent except in Egypt and the Near East. According to them, warfare in the pre-Iron Age era beyond Egypt and Near East comprised mostly small-scale raiding, ritualised combat between the champions/heroes, etc. However, this account shows that pre-Iron Age China and India did witness large-scale warfare.

Chapter 3 shows the military divergence between Greece and the Achaemenid Empire. While the beginning of the Iron Age saw the rise of heavy armoured infantry in Greece; in the Near East, cavalry centred military machines rose. This distinction was mostly due to ecological factors. And this set the path for military divergence in Eurasia. The book is not suggesting a simple watertight bipolar model. Rather, different parts of Asia gave birth to varied military systems. However, they differed from the military systems which came into existence in Europe. While Greece witnessed the genesis of a heavy infantry culture, West Asia saw the rise of cavalry-centric military system. South Asia in contrast gave birth to the chariot–infantry–elephant–cavalry combination. And cavalry plus crossbow-equipped infantry rose to prominence in China. In terms of naval warfare, both Greece and West Asia displayed a convergence. The navies of Greece and Persia were all part of a Mediterranean continuum.

Chapter 4 shows that Alexander was able to overpower Persia because his father Philip II created a hybrid military machine. This chapter also portrays the different trajectories of Chinese and Indian military systems. Philip created a hybrid army by adopting elements from both the Classical Greek military tradition and the Assyrian-Achaemenid military system. This composite army was used by Alexander to overthrow the Achaemenid Empire. Later, Alexander's generals took further the process of synthesis of Asian elements in their military machines. The military machines of both East and West experienced innovations. In all the battles of Alexander, infantry-cavalry combination along with Alexander's daredevil heroic generalship won the day. The scope and intensity of warfare in Eurasia increased in the post-Alexander era. Huge demographic resources and the agrarian bureaucracies of China and India enabled the polities of these two regions to mobilise large armies. The unique Indian military technique of using scythe chariots and war elephants in the battle influenced the Achaemenids and the Seleucids. These polities adopted these military techniques and implemented them in the battles fought in the Near East. The use of war elephants by the Seleucids also forced the Ptolemids to use African war elephants, which in turn influenced Carthage and Rome.

Chapter 5 points to the military divergence in Eastern and Western Eurasia. While in Western Eurasia the Roman military machine was centred on heavy infantry, in Eastern Eurasia, cavalry played the dominant role. There were certain variations within this broad categorisation. For instance, though the Romans continued the Classical Greek tradition of heavy infantry centric warfare, the former came up with a more nimble and manageable form of infantry organisation. This was exemplified in the transition from hoplite organisation to the legionary formation. And in the eastern half of Eurasia, certain variations in the military practices of different regions were present. While elephants and cavalry held sway in South Asia, cavalry supported by auxiliary infantry was the dominant military paradigm in China. The two contrasting military cultures (theory and practice of warfare), one based on heavy infantry and another based on cavalry, confronted each other in the Middle East, as exemplified in the Roman–Parthian Wars. The limits of Roman infantry culture became evident in Mesopotamia, where the cavalry centric forces (which were also continuously evolving in terms of arms, armour and techniques) of Parthia checked the Roman legionaries.

Chapter 6 portrays military convergence in Eurasia. The period between *c.* 400 CE and 1500 CE could roughly be categorised as the era of horses. When the steppe nomads ran roughshod over the agrarian bureaucratic civilisations of Eurasia, the latter responded by raising their own cavalry armies. But the sedentary agriculturists in terms of stamina and horsemanship came far short of the standards exhibited by the mounted nomads. From Korea to the coastline of the Atlantic in Western Europe, men on horseback held sway. When the steppe nomads (for instance, Huns and Mongols) were able to conquer sedentary polities, they adapted their political and military infrastructures by integrating selected military elements from the defeated sedentary polities. The Arabs from the desert of Arabia rose to military prominence by creating a hybrid military machine based on camels, horses and light infantry. Several sedentary polities of Eurasia, like the Guptas and the Song, in order to check the supremacy of the horse riders also raised cavalry forces. But sedentary cavalrymen could not become effective horse archers, as they were not born and brought up in pastoral nomadic settings of the steppes. The sedentary polities in turn came up with heavily armoured lancers and swordsmen as the second-best solution. However, they were no match against superb horse archery and lance-sword equipped heavy cavalry of the steppe nomads.

Chapter 7 compares and contrasts naval warfare conducted by the Chinese, Indian and Islamic polities with the maritime activities of the Western European polities during the medieval era. It concludes that, before *c.* 1500 CE, Western navies did not enjoy any superiority (managerial or technical) over their Asian opponents. And as regards riverine and amphibious warfare, the Asian powers like the Cholas and the Mongols were superior compared to the Norsemen. This chapter compares and contrasts Mongol invasions of Japan and South-East Asia with the Norman invasion of Britain. In fact, the carrying capacity of the Mongol navy was greater compared to the Western European navies. However, from *c.* 1400 CE onwards, the Western European navies started experimenting with high sea voyages. Physical geographical factors and economic necessity explains the experimentation with ocean-going navies by certain Western European polities. Here we find seeds of the divergence which flowered after *c.* 1500 CE.

The last chapter turns the spotlight on the genesis of military systems in the two Americas, Africa and Australasia. And these military organisations were much different from those that had emerged in Eurasia. In precolonial America (except in Mesoamerica and Andes), Australasia, and Africa (except Maghreb, Egypt and the mid bend of the Niger), high intensity wars waged by bureaucratic standing armies supported by a hierarchical political set up were absent. From the perspective of deep history, we have two structuralist arguments: a cultural one and a geographical one. To me, the geographical one seems to carry more traction. Agriculture was more developed in Mesoamerica and in parts of South America compared to North America. Moreover, in South America, llamas are available for carrying baggage. Agricultural surplus generated trade which could be carried on the back of the llamas. Hence, the emergence of big empires such as the Aztec, Maya and Inca in Latin America. Availability of camels, gold and salt resulted in long-distance overland trade, which in turn accelerated the formation of big polities in parts of North and West Africa. And they were able to mobilise big armies. But in precolonial North America and in South Africa, the political unit remained in the form of tribes and bands. Again, warfare was not static. The introduction of recurve bows among the North American tribes and horses in certain regions of Africa raised the intensity of conflict. Warfare was not always a ritualised cosmetic affair among the tribes and chiefdoms, and in the case of Latin American empires. However, absence of iron weapons and baggage carrying animals severely limited the casualties (at least in absolute terms) in case of the precolonial societies of North America, Latin America and Australasia. Instead of merely summing up the arguments advanced in the book, the Conclusion also provides some pointers about how warfare developed in the post-1500 CE world.

References

Archer, Christon I., Ferris, John R., Herwig, Holger H. and Travers, Timothy, *World History of Warfare* (2002, reprint, London: Cassell, 2003).

Bachrach, Bernard S., 'The Roman Way of War 250 BC-AD 300', 'On Roman Ramparts 300–1300', in Geoffrey Parker (ed.), *The Cambridge Illustrated History of Warfare: The Triumph of the West* (Cambridge: Cambridge University Press, 1995), pp. 50–63, 64–91.

Black, Jeremy, *A Military Revolution? Military Change and European Society, 1550–1800* (London: Macmillan, 1991).

Chaliand, Gerard, *A Global History of War: From Assyria to the Twenty-First Century*, translated by Michele Mangin-Woods and David Woods (Berkeley, CA: University of California Press, 2014).

Chase, Kenneth, *Firearms: A Global History to 1700* (Cambridge: Cambridge University Press, 2003).

Ferrill, Arther, *The Origins of War: From the Stone Age to Alexander the Great* (1985, reprint, London: Thames and Hudson, 1986).

France, John, *Perilous Glory: The Rise of Western Military Power* (New Haven/London: Yale University Press, 2011).

Garza, Andrew de la, *The Mughal Empire at War: Babur, Akbar and the Indian Military Revolution, 1500–1605* (London/New York: Routledge, 2016).

Gat, Azar, *War in Human Civilization* (Oxford/New York: Oxford University Press, 2006).

Hanson, Victor Davis, 'Genesis of the Infantry 600–350 BC', 'From Phalanx to Legion 350–250 BC', in Geoffrey Parker (ed.), *The Cambridge Illustrated History of Warfare: The Triumph of the West* (Cambridge: Cambridge University Press, 1995), pp. 12–31, 32–49.

Hanson, Victor Davis, *Carnage and Culture: Landmark Battles in the Rise of Western Power* (New York: Doubleday, 2001).

Hanson, Victor Davis, *The Western Way of War: Infantry Battles in Classical Greece* (1994, reprint, Berkeley/Los Angeles: University of California Press, 2009).

Hooper, Nicholas and Bennett, Matthew, *Cambridge Illustrated Atlas Warfare: The Middle Ages, 768–1487* (Cambridge: Cambridge University Press, 1996).

Keegan, John, *A History of Warfare* (1993, reprint, New York: Vintage, 1994).

Knox, MacGregor and Murray, Williamson, *The Dynamics of Military Revolution, 1300–2050* (2001, reprint, Cambridge/New York: Cambridge University Press, 2003).

Lee, Wayne E., *Waging War: Conflict, Culture, and Innovation in World History* (Oxford/ New York: Oxford University Press, 2016).

Levy, Jack S. and Thompson, William R., *The Arc of War: Origins, Escalation and Transformation* (Chicago/London: University of Chicago Press, 2011).

Lorge, Peter, *The Asian Military Revolution: From Gunpowder to the Bomb* (Cambridge: Cambridge University Press, 2008).

Parker, Geoffrey, *The Military Revolution: Military Innovation and the Rise of the West, 1500–1800* (Cambridge: Cambridge University Press, 1988).

Parker, Geoffrey, 'Introduction: The Western Way of War', in Geoffrey Parker (ed.), *The Cambridge Illustrated History of Warfare: The Triumph of the West* (Cambridge: Cambridge University Press, 1995a), pp. 2–9.

Parker, Geoffrey (ed.), *The Cambridge Illustrated History of Warfare: The Triumph of the West* (Cambridge: Cambridge University Press, 1995b).

Parrott, David, 'Had a Distinct Template for a "Western Way of War" Been Established Before 1800?', in Hew Strachan and Sibylle Scheipers (eds.), *The Changing Character of War* (Oxford: Oxford University Press, 2011), pp. 48–63.

Rogers, Clifford J. (ed.), *The Military Revolution Debate: Readings of the Military Transformations of Early Modern Europe* (Boulder: Westview, 1995).

van Creveld, Martin, *Technology and War: From 2000 BC to the Present* (London: Brassey's 1991).

1

THE ORIGINS OF WARFARE

> I looked, and there before me was a pale horse! Its rider was named Death, and Hades was following close behind him. They were given power over a fourth of the earth to kill by sword, famine and plague, and by the wild beasts of the earth.
>
> (*The Holy Bible*, Revelation 6:7–8)

Introduction

In this chapter, how, why and when organised violence emerged on this planet is discussed. We begin with the premise that warfare (organised violence) is the product of both biology and culture. Doyne Dawson argues in an erudite way that the cultural model of the social scientists needs to take into account the fact that human groups evolved by natural selection. In the early twentieth century, Mendelian genetics challenged the assumption that acquired traits can be inherited. This in turn generated the trend of cultural determinism in the humanities discipline. Objective scientific knowledge was challenged, and culture is regarded as the product of social learning. In the 1970s, the British biologist William Hamilton showed that natural selection promotes the survival and reproduction not only of individuals but also of their close relatives who share much of the same genetic heritage. And so culture is partly the product of natural selection. The same decade also witnessed development in other fields that proved that chimpanzees organised small game hunting and engaged in a sort of raiding warfare. Both humans and chimpanzees have descended from a common ancestor that inhabited East Africa in the Late Miocene epoch, about 8 million years ago. A territorial instinct was present in the common ancestor of the chimpanzees and humans. So probably the genetic seeds of warfare were present in our common ancestor. Dawson writes that certain cultural features such as a communal ritual, a powerful myth, charismatic leadership etc., result in group formation. And the success

DOI: 10.4324/9780429284069-2

of such groups in the long run predisposes people to more cooperative behaviour towards their own ethnicity and deeper hostility to outsiders (Dawson 1999: 79–100). In recent times, practitioners of deep history have given due importance to the role of genetics in shaping human behaviour (Harari 2014). Hence, culture and biology together, like a DNA double helix, shape human behaviour.

This chapter attempts to see whether warfare emerged in different parts of the globe for similar reasons. Probably, we *Homo sapiens* (in this book by *Homo sapiens* we mean *Homo sapiens sapiens* but for the sake of brevity the former term is adhered to) inherited homicidal tendencies, like the apes and chimpanzees, from our common ancestor. But we could fight more efficiently and effectively than the apes and chimps because of the development of segmented groups and language, which enabled the *Homo sapiens* to effectively cooperate, especially during hunting. Hunting skills and tools were used with decisive effect for intraspecies combat. The shift in the politico-economic fabric, that is, the transition from hunting-gathering to agriculture, not only resulted in greater density of population but also exacerbated clashes for resources. This in turn requires a re-evaluation of the origins and nature of prehistoric warfare.

Most archaeologists, in reaction to the carnage wrought by the Second World War, have painted a picture of 'peaceful' societies before the rise of the state. Prehistoric fortifications are categorised by them as ritualistic markers. These archaeologists, in their attempt to portray a warless prehistoric past, argue that weapons and armour found in the graves were mere status symbols (Keeley 1997: 19; Thorpe 2003: 145–65). Quincy Wright in his *A Study of War* writes: 'Primitive peoples only rarely conduct formal hostilities with the object of achieving a tangible economic or political result. Their hostilities are seldom conducted by a highly professional military class using distinctive instruments and techniques' (quoted in Wright 1942: 58). Anthropologist Lawrence Keeley is of the opinion that prehistoric burials, when excavated, show evidence of a high homicide rate. The skeletons show that death had occurred due to insertion of projectile points. Keeley's Hobbesian approach appears to be much more fruitful than Jean Jacques Rousseau's 'noble savage' perspective. Keeley rightly challenges Wright and Harry Turney-High's assertion that prehistoric warfare was primitive, non-serious and a low-key business with extremely low casualty rate (Keeley 1997: 5–16). Keeley says: 'Primitive warfare is simply total war conducted with very limited means' (quoted from Keeley 1997: 175). One can say that prehistoric warfare involved the participation of all the adult males of the tribes and forager bands, and in a way it was total. On the other hand, the wars were fought mainly with hunting tools. The population of the different segmented groups engaged in conflicts was small. Further, due to logistical deficiencies, such conflicts did not last long. Hence, prehistoric conflicts were indeed limited. So the characteristics of both total and limited wars were present in prehistoric conflicts. Our story properly begins with the Stone Age.

Old Stone Age

The Stone Age is divided into Paleolithic (Old Stone Age), Epipaleolithic (Late Old Stone Age)/Mesolithic (Middle Stone Age) and Neolithic (New Stone Age).

The Paleolithic Age (Pleistocene for the geologists) started between 250,000 and 500,000 years ago, when the *Homo erectus* became food gatherers (Childe 1946: 22). *Homo erectus* appeared in Java and Africa between 1.7 million and 1.6 million years ago, and *Homo sapiens* emerged some 100,000 years ago. Around 60,000 BCE, *Homo sapiens* started dispersing from Africa.

The environment of South-East Asia was different from the environment of East Africa and Europe. In the latter two areas, large grasslands provided ideal habitat for big game animals that lived in herds. Humans developed bones and specialised stone tools to hunt these animals and process their meat, hide, sinews and bones. However, there were no herds of big game in South-East Asia. For acquiring food and raw materials in this region, the humans mostly used a bamboo toolkit rather than stone (Dawson 2001: 39; Miksic and Goh 2017: 77–9; Scott 2017: 4). Robert O'Connell asserts that the use of primitive tools for hunting initiated a cultural break for the *Homo sapiens* and paved the way for the emergence of warfare (O'Connell 1989: 19–21).

Mesolithic period

The Middle Paleolithic Age started about 300,000 years ago when *Homo erectus* started using fire, and the Upper Paleolithic Period started 40,000 years ago. During the latter period, in Europe and Africa, the emergence of arrowheads and spear points reflected sophisticated hunting techniques (Miksic and Goh 2017: 80). The Middle Stone Age started in North Europe, around 11,500 BP (before present). The Epipaleolithic/Mesolithic Period in the Near East can be stretched from 18,000 to 8500 BCE. It must be noted that all these chronological frameworks are highly speculative.

About 35,000 years ago, the Neanderthals disappeared from Europe and the Levant, leaving *Homo sapiens* as the dominant species. The Neanderthals lived in small groups and engaged in foraging and hunting. They knew how to make fire, tools and weapons from stone and bone. They also practised funeral rites, scattered flowers on their graves and placed special objects with the dead. However, they did not make permanent dwellings and did not know pottery. The Neanderthals and *Homo sapiens* interbred with each other, and some traces of their genes remain with us even today. It is estimated that between 1 to 4 per cent of the unique human DNA of modern populations in the Middle East and Europe is Neanderthal DNA (Harari 2014: 17; Wileman 2014: 4, 52).

The Neanderthals were bulkier and more muscular compared to *Homo sapiens* (read: humans). They, like humans, were able to hunt and kill animals larger than themselves. The Neanderthals were in Europe and in the Levant long before *Homo sapiens* entered these two geographical zones. Still, the Neanderthals became extinct, while the humans survived. The Neanderthals did not use the bows and arrows that humans started using 71,000 years ago. The use of projectile weapons gave humans an upper hand in hunting vis-à-vis the Neanderthals. Later, bows and arrows proved to be an excellent weapon of war for the humans. Some 40,000 years ago, Europe was covered with dense vegetation, and the Neanderthals used close contact hunting techniques – that is, they crept close to the game and then stabbed them with spear. In contrast, in the wide and open grasslands of Africa, the humans used projectile technology to bring down fast-moving small animals. About 39,000 years ago, due to climatic changes, the dense vegetation of Europe was reduced by about 75 per cent. Then the Neanderthals' close contact hunting became ineffective. But humans had an edge with their bows and arrows. Further, the Neanderthals were less adaptive than the humans. They mainly depended on big animals for protein and, unlike humans, did not consume aquatic animals. Some 45,000 years ago, humans used bone needles to prepare clothing from hides and skins of the hunted animals. Then, about 27,000 years ago, humans were able to manufacture snares and nets to catch small animals. These innovative techniques were absent among the Neanderthals. Further, humans equipped with projectile technology worked in packs with dogs and thus were able to dominate the Eurasian ecosystem, as the domesticated dogs aided the humans in hunting. Anthropologist Pat Shipman asserts that the symbiotic relationship between humans and dogs, among many other factors, enabled the former to overcome the Neanderthals. And most of the modern dogs' descent can be traced back to the European/Asian Arctic wolf. The dogs located the prey and held the animal in place till the human hunters arrived at the scene. The dogs also warned the humans about the arrival of other competing predators. So the humans could hunt effectively with less expenditure of energy and thus spend most of their time and energy reproducing offspring (Shipman 2017).

Did warfare exist before the advent of *Homo sapiens*? Don Brothwell suggests that the Neanderthals fought against each other. Such conflicts were linked to hunting rights over game, etc. (Brothwell 1999: 31). Similarly, Julian Maxwell Heath also suggests that in Belgium, some 45,000 years ago, there were no humans, but there is evidence of Neanderthals fighting against each other in this region (Heath 2017: 26). In contrast, archaeologist Slavomil Vencl asserts that while there is no evidence of warfare in the Old Stone Age era, armed violence (intergroup conflict) started in the Mesolithic period (Vencl 1999: 58–9). Similarly, the Marxist archaeologist Paul M. Dolukhanov claims that there is no evidence of intergroup conflicts in Paleolithic Eastern Europe. Low population density combined with the abundance and diversity of available resources enabled the humans to avoid warfare. However, warfare emerged during the Mesolithic era in Eastern Europe (Dolukhanov 1999: 77–9).

Warfare among the Neanderthals is an ongoing debate. However, the Mesolithic era probably witnessed some sort of crude warfare among *Homo sapiens*. Besides material, non-material factors might also have played a role in the origin of armed conflicts. Humans might have fought for prestige, for the choicest game or for women even before the Mesolithic period. Further, scarcity of resources is a relative concept. In Mesolithic Southern Scandinavia, conflict occurred not because of competition for economic resources but due to a sense of a strong degree of internal territoriality among the different human groups.

Slights on personal honour resulted in feuds, which in turn escalated into inter-community conflicts. The people fought with clubs, bows and arrows (Thorpe 2003: 157–60).

Large herd animals like the mammoth, musk ox and woolly rhinoceros evolved in Eurasia during the second half of the Upper Pleistocene period, which is about 35,000 to 30,000 years ago. The transition from Pleistocene to the Holocene or from the later glacial to the post-glacial epoch occurred roughly 10,500 years ago. This time period witnessed the extinction of several large Pleistocene mammals, including mammoths, and the emergence of animals such as elks, deer, wild pigs and beavers, etc. In the Upper Paleolithic era, in Eastern Europe, hunting provided 30 per cent of the total food intake of the humans. The rest was provided by food gathering and fishing. The hunters of Eastern Europe restricted the size of their population through primitive birth control and infanticide. Nevertheless, population increased with time in certain regions due to the spread of farming, which supplemented and increased the volume of food production. According to one calculation, during the Paleolithic era, the density of population in the Near East was about 3 persons per 100 sq km. It rose to 12.5 persons per 100 sq km in the same region during the Mesolithic era. The population density of the agricultural Neolithic communities of the Near East came to be about 25 persons per sq mi. (Dolukhanov 1979: 19, 23–5). Population explosion might have functioned as a trigger for the emergence of warfare.

The Zagros Mountains were comparatively densely populated in Upper Paleolithic times. A wild ancestor of cultivable barley was native to this region. The Upper Paleolithic people mainly depended on hunting migratory herd animals like wild goats and wild ox. The tools of the inhabitants comprised long, thin stone blades. The earliest villages appeared in the lower belt of the Zagros Mountains around 10,000 years ago. Each village covered about 3.2 acres of land, and the houses were built of clay (Dolukhanov 1979: 39–41).

Around 10,800 BCE, a millennium-long cold period occurred due to glacial melt in North America. The cold water poured through the St. Lawrence River into the Western Atlantic. The cold spell somewhat checked the trend towards sedentary agriculture in the Fertile Crescent. Then, around 9600 BCE, warmer and wetter conditions prevailed. The average temperature increased by 7 degrees Celsius within a decade. Trees, mammals and birds spread throughout Eurasia, and human population rose again (Scott 2017: 43). In Britain and Denmark, the Mesolithic people started making boats. The Mesolithic hunters roamed from Denmark through Doggerland into Britain. Around 6500 BCE, due to melting of ice, the sea levels rose and Britain was cut off from North-Western Europe. Doggerland was flooded and disappeared (Wileman 2014: 5–6). Britain was then separated from continental Europe.

The *Homo sapiens*, during their march from West Asia in the eastern direction, entered north-west India roughly around 50,000 years ago. Then they spread to Central India and finally moved into Sri Lanka about 29,000 years ago. At that time, there was a land bridge between South India and Sri Lanka. Later it vanished with the rising sea level. *Homo sapiens* must have fought and exterminated related species like *Homo erectus* and *Homo habilis* in South Asia (Habib 2016: 25–37).

About 30,000 years ago, flaked tools including projectile points appeared in Vietnam. Some scholars categorise this development as the beginning of the Upper Paleolithic Age in South-East Asia. Arrowheads were also coming into use during this period in Sulawesi. Roughly 15,000 years ago, some groups in East Asia adopted a sedentary lifestyle based not on agriculture but on sophisticated exploitation of aquatic resources including both animals such as fish and birds and wild plants that grew in swamps. This adaptation enabled them to find adequate food to make it unnecessary to move every few months (Miksic and Goh 2017: 81–2, 94).

According to one interpretation, Australia was populated by humans around 40,000 BCE. The sea level was much lower than today. At that time, Borneo, Sumatra, Java etc., were all parts of mainland South-East Asia. New Guinea and the adjacent islands were connected to Australia. New Guinea, Australia, Tasmania and many other smaller islands constituted one great landmass that extended from east of Timor in the Indian Ocean to north-eastwards through the Banda Sea, west of New Guinea, east of the Moluccas into the Pacific. Between the south-eastern extension of Asia and Greater Australia, there was an archipelago of islands known as Wallacea. The Wallacean Archipelago lay between the Sunda Shelf and the Sahul Shelf, which are now under the sea. The islands between Borneo and Australia could be seen at that time and this enabled humans to undertake short voyages from Asia in rudimentary boats and cross over to Australia. This migration occurred over several generations. The migrants came either from Taiwan or the coastal areas of South-East China. There were two possible routes through the Wallacean Archipelago: a northern route from Borneo via Sulawesi and the Moluccas and a southern route from Java through Flores and Timor to North-West Australia. The sea level was lowest between *c*. 53,000 and 18,000 BCE. During this period, the longest marine voyage would have been within a range of 100 km or 60 nautical miles. Using rudimentary paddle boats, such a journey would have lasted two days. The initial population of Australia was about 3,000 persons. About 200 nautical miles separate the easternmost Solomon Islands and the Santa Cruz Islands. This long stretch of out-of-sight land could not be crossed by migrants with the nautical technology at their disposal in 40,000 BCE. Hence, Micronesia and Polynesia were not peopled at that time (McGrail 2015: 78–81, 129; Shipman 2017: 15).

Tasmania and New Guinea became separated from Australia around 7000 BCE. The Tasmanians used scrapers and bone tools. They failed to develop hafted tools, which came into existence in Australia around 3000 BCE. The hafted tools enabled the humans in Australia to build advanced forms of log rafts and bark boats. Why did no further development in stone technology occur in Tasmania? We could hazard a guess. Isolation from the Australian mainland resulted in lack of competition from other human groups. This probably resulted in social and cultural stagnation, which in turn stopped the wheels of technology from moving further. We have some idea about the boats built by the Tasmanians: they were able to construct bundle rafts. Individual pieces of bark were bundled together and were bound by a rope made of grass. In the shallow water, these rafts were steered and propelled by poles. The indigenous inhabitants of Australia constructed river rafts by lashing together several mangrove saplings. They were also able to construct sea-going rafts. Each was about 6–7 feet long, made of mangrove logs fastened together by hardwood treenails. They also made bark boats. The bark came from stringy bark tree or river red gum. Three types of bark boats were made: simple, lashed and stitched. A stitched bark boat had the seams caulked with gum and measured between 16 and 18 feet and carried a crew of eight (McGrail 2015: 82–7).

Humans entered North America about 15,000 years ago. Archaeologists are still divided on the opinion whether the humans walked across the land bridge (3,218 km-long ice sheets known as Berengia) between Siberia and Alaska or used rafts following the ice-free corridor. The hunter-gatherer population inhabited the region which later became the Maya Lowlands as early as 9000 BCE (Haas 1999: 13; Webster 1999: 335; Zeitlin and Zeitlin 2000: 50–1). However, some scholars believe that the first humans had crossed over into North America from North-East Asia around 50,000 BCE and that these people reached Chile and Brazil roughly around 20,000 BCE (Stannard 1993: 261–4).

Neolithic period

Agriculture began in this period and functioned as a stimulant for the rise of large-scale organised violence among humans. The time span of the Neolithic Period differed from region to region. In Macedonia, the Early Neolithic could be dated to the sixth millennium, Middle Neolithic during the first half of fifth millennium, Late Neolithic around the second half of the fifth millennium (4600–4000 BCE) and Final Neolithic around the fourth millennium BCE (Kokkinidou and Nikolaidou 1999: 89, 92). The Neolithic era in the Near East lasted from 8500 to 4500 BCE. The development of pottery around the seventh millennium allowed farmers to store food surplus, a technique which the Neanderthals lacked. Grain properly sealed and stored in pottery could be preserved for years. And this allowed the population to expand. During the Late Neolithic Age in the Near East, all the weapons were made from flint and obsidian (Hamblin 2010: 5, 25).

Farming communities probably originated in the Levant around *c.* 9000 BCE. Around 7000 BCE, farmers from the Levant and Anatolia started arriving in Greece and Cyprus (Heath 2017: xi). A group of archaeologists assert that between *c.* 7000–3400 BCE, Neolithic Greece experienced warfare. Competition for grazing rights, arable land and water were the contributory causes behind the origins of warfare in Neolithic Thessaly. The size of the villages in Thessaly varied from 2–9 acres each. They were surrounded by walls and ditches. And it would be wrong to claim that these walls and ditches had no defensive utility but were constructed merely for ritual purposes and as markers for space and animal enclosures. The perimeter walls were constructed with stone or mud bricks. In addition, the village fortifications had baffle gates, flanking buttresses, wall projections and towers. Again, many of the villages were surrounded by two concentric ditches, and some of the ditches had v- or y-shaped sections. V- or y-shaped ditches were impractical for non-military purposes, as they eroded more quickly than any other form and were more difficult to dig. The migration of peasants from Anatolia due to the beginning of colder and more arid conditions further intensified the struggle for land, water and grazing rights (Runnels et al. 2009: 165–94). Some 5,000 to 4,500 years ago, there was a general cooling of climate and reduction of the broadleaved forests throughout the temperate zone of the world (Dolukhanov 1979: 19).

The reduction of forest and game animals reduced the scope for hunting, and *Homo sapiens* were forced to go for farming. Organised violence between different groups of farmers – and also between farmers and the hunter-gatherers who were pushed to the wall – rose in intensity and frequency. And this accelerated the emergence of village fortifications. Neolithic villages in Central Europe during 4000 BCE were protected with double moats and ramparts (Gronenborn 2006: 21). Early Neolithic mass graves in Germany show that even women and children had been killed with axes (Keeley 1997: ix–x).

We have a rough estimate of population density of Neolithic Greece. In Messenia, it was 1.77 persons per sq km, 1.53 persons per sq km in Crete, 1.35 persons per sq km in Euboea, 1.2 persons per sq km in Cyclades and 0.84 persons per sq km in Central Macedonia (Dolukhanov 1979: 25). Double ditches for protecting human settlements were evident in the Early Neolithic sites of Macedonia. In Central and Eastern Europe, in the Neolithic era, the farmers surrounded their settlements with ditches, banks, palisades and stone walls (Chapman 1999: 117). In Early Neolithic North-Western Europe, bows and arrows became the premier weapon for conducting warfare. Archers were at a premium for defending the settlements against raiders from hunting-gathering and pastoral communities (Mercer 1999: 156).

The indigenous Mesolithic communities of Europe were not willing to peacefully accommodate the Neolithic farming groups who established themselves in territories that had long been held by the hunter-gatherer communities. The Neolithic farmers required large amount of land for planting their crops and grazing their livestock, which put them in conflict with the hunter-gatherers. Wright's assertion that primitive (read: prehistoric) conflicts were mainly social duels rather than wars fought for economic objectives is just not tenable.

Minor variations in military practices existed among the different regions. In Neolithic Near East and Anatolia, slingshots (fired clay and stone balls) were common for hunting and warfare, although this weapon was rare in the Balkans (Chapman 1999: 110). A club (a heavy stick) was probably the earliest human weapon. The Paleolithic axe consisted of a rock bound to the club. A mace, unlike the axe, had no cutting edge. Rather, the Neolithic mace was a rounded weight fastened to a wooden shaft. The mace could be used for dispatching a wounded prey. But this job could have been done with a knife or an axe. Further, the mace was useless in killing or wounding a fast-footed uninjured animal. The real job of the mace was skull crashing and was probably the first real weapon for close-quarter combat introduced by humans (Hamblin 2010: 20–1).

Between 4300 and 4000 BCE, the Neolithic farmers reached Britain and Ireland (Heath 2017: xii). In the Neolithic era, the bow in Britain was made from yew. The staves made from this wood generated good tension under torsion. Yew grows well in the temperate climate of Britain. Even in the Middle Ages, bows in Britain were made from yew, and in its absence from elm and ash (Mercer 1999: 145). Around 3500 BCE, Neolithic communities in Britain had to defend themselves from raiders. Herds of cattle were the principal objective of loot (Heath 2017: 92). Brian Fagan writes that the ditches and palisaded enclosures of the prehistoric sites of Britain were places for communal burials. Some archaeologists even claim that copper and bronze axes found in the Neolithic settlements in Europe were not weapons of war but actually a form of currency for exchange. Keeley rightly challenges such a view. For him, the so-called common burial grounds surrounded with palisades and ditches were actually fortified enclosures which at times experienced intense combat (Keeley 1997: 18–19).

It would be wrong to assert that the human bones found in ditches of Neolithic enclosures and tombs are evidences of ancestral rituals rather than signs of warfare. Most of the combat fatalities which occurred in Neolithic Europe were caused by bows and arrows. Flint arrowheads were made for fighting and not for hunting because the latter occupation was marginal for the Neolithic communities. Rather, the Neolithic farmers depended on domesticated animals. Around 2500 BCE, the Neolithic farmers of Europe were using barbed and tanged arrowheads for killing other competing groups of humans (Heath 2017: xvii–xxi, 41).

Were all the prehistoric warriors male? To a large extent, the answer is yes. Female warriors were few. Maiden warriors were present in Anglo-Saxon society (Shepherd 1999: 219–43). But why do bulk of the warriors remain male till this date? There was no 'male conspiracy' to keep the females out of combat. Warfare became mostly a male affair due to genetic factors and reasons of pragmatic politics. One of the factors which enabled *Homo sapiens* to outwit the other species and related hominids was our capacity to replicate ourselves to a greater degree. This in turn resulted in a division of labour. While the males took care of acquiring food and the job of protection, the females, thanks to the reproductive organs present in their body, remained at home producing little *Homo sapiens* in greater numbers and caring for them in the homes. And this trend accelerated with the onset of agriculture. The domestication and sedentarisation of females and monopolisation of the security apparatus by the males resulted in a skewed balance of power in favour of the males. Secretion of testosterone hormones in males which made them bulkier and aggressive in comparison to female *Homo sapiens* further strengthened the trend towards male dominance.

Rudimentary agriculture and permanent settlements in the Middle East started about 12,000 years ago. Around 10,000 BCE, *Homo sapiens* numbered between 2–4 million. Around 6500 BCE, Southern Mesopotamia was not arid. The tip of this region constituted the wetlands of the Persian Gulf, and the latter extended up to Ur. Between 6500 and 3800 BCE, the diet of the humans in this region mainly comprised fish, birds and turtles caught from the wetlands. The first states probably started in the rich alluvium of Mesopotamia around 6,000 years ago. In around 3000 BCE, the population of Uruk was between 40,000 and 45,000 (Scott 2017: 3, 47–9, 159).

Some scholars both from the right and left make a critique of state formation. After all, they assert, states made war. Further, they write that the state-agriculture equation made people unhappy. Previously, hunter-gatherers roamed around and were happy. They enjoyed a healthy diet of meat and fruits. In contrast, with the onset of agriculture, people were restricted to mostly a carbohydrate diet, resulting in malnutrition. Agriculture resulted in an increase in the population of humans, who congregated together in unhealthy cities where epidemic disease had an easy playing ground. And the unhealthy populace fell easy prey to such marauding diseases (O'Connell 1995: 53–83; Scott 2017).

This romantic picture is unsustainable. It is as if the hunter-gatherers had a choice. The only way for survival was to shift to the agricultural mode of production which enabled most of them to survive and expand. Hunting and gathering could only sustain a small number of populations at the rudimentary level. Whatever may be the limitations of agriculture and state, this combination enabled humanity to survive and progress till the eighteenth century, when the Industrial Revolution took off. Hunting and gathering was, after all, a dead end. Several anthropologists have noted that life within a hunting-gathering band was no less dangerous and brutal compared to the experiences of the common mass in agriculture-based cities such as Uruk and Harappa.

Roughly 5,500 years ago, large-scale forest clearance for agriculture started in Java. In Sarawak and Malaysia, stone tools were used for harvesting rice and chopping wood from around 4,300 years ago. Rice was cultivated in Sumatra around 2,600 years ago (Miksic and Goh 2017: 82–3). Hardwood and bamboos used for building rafts and boats were easily available in South-East Asia. From *c.* 8000 BCE, mats, ropes and baskets were constructed from rattan palms and leaves. The boats were built using basketry techniques. Resin from the tropical trees was used for making the boats watertight. By 3000 BCE, the coastline of South-East Asia took the shape which it retains more or less even today. Each of the seagoing rafts of Northern Vietnam and Taiwan was made of several large bamboos lashed together in the shape of boat (McGrail 2015: 89–91). Increasing population pressure, which encouraged migration, and the requirements of fishing were the imperatives behind the construction of boats.

Between 4500 and 4000 BCE, Austronesian speaking Neolithic people migrated from South China to Taiwan. In 3000 BCE, they had expanded to Luzon in the Northern Philippines. By 2500 BCE, the whole of the Philippines was settled (McGrail 2015: 91–2). The migration was partly due to the rising sea level, and the submerged land became South China Sea. Around 1,000 BCE, the Indonesians invented outriggers in order to stabilise their boats. The technique of building outriggers reached South Asia from South-East Asia around 500 BCE (Miksic and Goh 2017: 88, 97).

Large-scale settlement started in China with the end of the last Ice Age, which occurred between 18,000 and 12,000 years ago. The Han Chinese people probably came from their original homeland around the Caspian Sea. It is probable that some intermixing with the 'Peking Man' (*Homo erectus*) occurred (Major and Cook 2017: 30). From an ecological point of view, roughly, we could differentiate China into two zones. Northern China centred on the Yellow River (Huang He) and the lands south of the Yangtze River. North China was influenced by the steppe nomads of Eurasia. And Southern China, without any pastureland for horses, was largely a rice-growing region, and most of the communications were dependent on water (Peers 2006: 8).

Neolithic settlements arose in China around 10,000 BCE near the rivers and marshes. For the next 3,000 years, Neolithic settlers gradually expanded agricultural practices. Low stone walls came up around the settlements of Inner Mongolia in the Early Neolithic Period. In North China, along the sparsely populated arid border with Mongolia, walls were constructed with rocks and boulders. Around 7000 BCE, ditches appeared around the settlements which grew out around the soft alluvial soil of the Yellow and Yangtze rivers of China. These ditches were probably constructed to prevent domesticated livestock from venturing out and also to keep out the raiders bent on stealing the livestock. The period from 7000 to 5000 BCE constituted the Middle Neolithic Period in China. Around 6500 BCE, some of the settlements in the Yangtze River Basin developed circular ditches protected by earthen walls. The villagers grew millet, cabbage and hemp and domesticated pigs, dogs and chicken. Rice cultivation started around 6000 BCE. The period from 5000 to 3500 BCE is considered as the Late Neolithic Period of China. During the Lungshan Period (3000–2000 BCE), immediately prior to the Hsia dynasty, the Chinese settlements had massive earthen walls, some exceeding 80 feet in width. These walls were constructed from layers of pounded soil that had ash and fibrous plants intermixed in it for strength (Yates 1999: 9–10; Sawyer 2011: 19–33). Construction of such walls required use of an extensive labour force. And these massive walls had a defensive military function. The Liangzhu Culture (3200–2300 BCE) in the lower reaches of the Yangtze River was ruled by hereditary chieftains supported by religious and military leaders. There were many towns and cities, and one city had 30,000 inhabitants. It is evident that the rulers of this culture, thanks to the spread of rice cultivation, could mobilise large number of labourers for constructing defensive works (Major and Cook 2017: 43–5).

Conclusion

Whether hunter-gatherers during the Mesolithic era fought among themselves for the regions which contained the heaviest concentration of game is open to debate. We do not have firm archaeological evidence to prove the existence of warfare among the hunter-gatherers. At this stage, we can be sure that warfare existed in the Neolithic Period. The weapons used for hunting in the Old Stone Age were used for warfare during the New Stone Age. The only exception was the use of the mace, a weapon designed for close-quarter combat. Nutritional insecurity among the hunter-gatherers caused by climate change and the production of substantial surplus by the agriculturists somewhat accelerated the rise of organised violence among humans. Thus, we can say that the causes behind the origins of war and the form of war practiced by humans during the New Stone Age throughout Eurasia were more or less similar. Hunter-gatherer bands and pastoralists conducted raids, and the emergence of agriculture resulted in population expansion and the production of surplus. Then, intergroup combat increased in scope and intensity. Hence, we can speak about a convergence of genetic and political-economic factors that gave rise to warfare in the New Stone Age.

Keeley is right in asserting that primitive/prehistoric warfare at times could be extremely lethal. Moreover, in such warfare even children, women and old men were executed. Considering the small demographic bases of the tribes and settled communities, even a small number of absolute casualties resulted in a high percentage of fatalities for such societies. We do not have any evidence of naval and riverine war during the Stone Age. Still, this chapter has devoted some space to the study of rudimentary boats. This is because with the passage of time, the humans would build upon these primitive crafts and would manufacture naval crafts for combat. And combat became more complex with the onset of the Bronze Age in Eurasia.

References

Brothwell, Don, 'Biosocial and Bio-Archaeological Aspects of Conflict and Warfare', in John Carman and Anthony Harding (eds.), *Ancient Warfare: Archaeological Perspectives* (Gloucestershire: Stroud, 1999), pp. 25–38.

Chapman, John, 'The Origins of Warfare in the Prehistory of Central and Eastern Europe', in John Carman and Anthony Harding (eds.), *Ancient Warfare: Archaeological Perspectives* (Gloucestershire: Stroud, 1999), pp. 101–42.

Childe, Gordon, *What Happened in History* (1942, reprint, New York: Penguin, 1946).

Dawson, Doyne, 'Evolutionary Theory and Group Selection: The Question of Warfare', *History and Theory*, vol. 38, no. 4 (1999), pp. 79–100.

Dawson, Doyne, *The First Armies* (London: Cassell, 2001).

Dolukhanov, Paul M., *Ecology and Economy in Neolithic Eastern Europe* (London: Duckworth, 1979).

Dolukhanov, Paul M., 'War and Peace in Prehistoric Eastern Europe', in John Carman and Anthony Harding (eds.), *Ancient Warfare: Archaeological Perspectives* (Gloucestershire: Stroud, 1999), pp. 73–87.

Gronenborn, Detlef, 'Climate Change and Socio-Political Crisis: Some Cases from Neolithic Central Europe', *Journal of Conflict Archaeology*, vol. 2, no. 1 (2006), pp. 13–32.

Haas, Jonathan, 'The Origins of War and Ethnic Violence', in John Carman and Anthony Harding (eds.), *Ancient Warfare: Archaeological Perspectives* (Gloucestershire: Stroud, 1999), pp. 11–24.

Habib, Irfan, *A People's History of India*, vol. 1, *Prehistory* (2001, reprint, New Delhi: Tulika, 2016).

Hamblin, William J., *Warfare in the Ancient Near East to 1600 BC: Holy Warriors at the Dawn of History* (2006, reprint, New York/London: Routledge, 2010).

Harari, Yuval Noah, *Sapiens: A Brief History of Humankind* (2011, reprint, London: Vintage, 2014).

Heath, Julian Maxwell, *Warfare in Neolithic Europe: An Archaeological and Anthropological Analysis* (Barnsley: Pen & Sword, 2017).

The Holy Bible, New International Version Containing The Old Testament and The New Testament (Colorado Springs: International Bible Society, 1984).

Keeley, Lawrence, *War Before Civilization: The Myth of the Peaceful Savage* (1996, reprint, New York/Oxford: Oxford University Press, 1997).

Kokkinidou, Dimitra and Nikolaidou, Marianna, 'Neolithic Enclosures in Greek Macedonia: Violent and Non-Violent Aspects of Territorial Demarcation', in John Carman and Anthony Harding (eds.), *Ancient Warfare: Archaeological Perspectives* (Gloucestershire: Stroud, 1999), pp. 89–99.

Major, John S. and Cook, Constance A., *Ancient China: A History* (New York/London: Routledge, 2017).

McGrail, Sean, *Early Ships and Seafaring: Water Transport Beyond Europe* (Barnsley: Pen & Sword, 2015).

Mercer, R.J., 'The Origins of Warfare in the British Isles', in John Carman and Anthony Harding (eds.), *Ancient Warfare: Archaeological Perspectives* (Gloucestershire: Stroud, 1999), pp. 143–56.

Miksic, John N. and Goh, Geok Yian, *Ancient Southeast Asia* (London/New York: Routledge, 2017).

O'Connell, Robert, *Of Arms and Men: A History of War, Weapons, and Aggression* (New York/Oxford: Oxford University Press, 1989).

O'Connell, Robert, *Ride of the Second Horseman: The Birth and Death of War* (New York/Oxford: Oxford University Press, 1995).

Peers, C.J., *Soldiers of the Dragon: Chinese Armies 1500 BC-AD 1840* (Oxford/New York: Osprey, 2006).

Runnels, Curtis N., Payne, Claire, Rifkind, Noam V., White, Chanter, Wolff, Nicholas P. and LeBlanc, Steven A., 'Warfare in Neolithic Thessaly: A Cade Study', *Hesperia*, vol. 78 (2009), pp. 165–94.

Sawyer, Ralph D., *Ancient Chinese Warfare* (New York: Basic Books, 2011).

Scott, James C., *Against the Grain: A Deep History of the Earliest States* (New Haven/London: Yale University Press, 2017).

Shepherd, Deborah J., 'The Elusive Warrior Maiden Tradition: Bearing Weapons in Anglo-Saxon Society', in John Carman and Anthony Harding (eds.), *Ancient Warfare: Archaeological Perspectives* (Gloucestershire: Stroud, 1999), pp. 219–43.

Shipman, Pat, *The Invaders: How Humans and Their Dogs Drove Neanderthals to Extinction* (Cambridge, MA/London: Harvard University Press, 2017).

Stannard, David E., *American Holocaust: The Conquest of the New World* (1992, reprint, New York: Oxford University Press, 1993).

Thorpe, I.J.N., 'Anthropology, Archaeology, and the Origins of Warfare', *World Archaeology*, vol. 35, no. 1 (2003), pp. 145–65.

Vencl, Slavomil, 'Stone Age Warfare', in John Carman and Anthony Harding (eds.), *Ancient Warfare: Archaeological Perspectives* (Gloucestershire: Stroud, 1999), pp. 57–72.

Webster, David, 'Ancient Maya Warfare', in Kurt Raaflaub and Nathan Rosenstein (eds.), *War and Society in the Ancient and Medieval Worlds: Asia, the Mediterranean, Europe, and Mesoamerica* (Washington, DC: Center for Hellenic Studies, distributed by Harvard University Press, 1999), pp. 333–60.

Wileman, Julie, *Warfare in Northern Europe Before the Romans: Evidence from Archaeology* (Barnsley: Pen & Sword, 2014).

Wright, Quincy, *A Study of War*, vol. 1 (Chicago, IL: University of Chicago Press, 1942).

Yates, Robin D.S., 'Early China', in Kurt Raaflaub and Nathan Rosenstein (eds.), *War and Society in the Ancient and Medieval Worlds: Asia, the Mediterranean, Europe, and Mesoamerica* (Washington, DC: Center for Hellenic Studies, distributed by Harvard University Press, 1999), pp. 7–45.

Zeitlin, Robert N. and Zeitlin, Judith Francis, 'The Paleoindian and Archaic Cultures of Mesoamerica', in Richard E.W. Adams and Murdo J. MacLeod (eds.), *The Cambridge History of the Native Peoples of the Americas*, vol. 2, *Mesomaerica*, Part 1 (Cambridge: Cambridge University Press, 2000), pp. 45–121.

2

MILITARY CONVERGENCE AND THE BRONZE AGE CIVILISATIONS OF EURASIA

Introduction

With the end of the Neolithic Age, copper and bronze came to be used extensively. This culture of use of copper led this period to be called the Chalcolithic Age, followed by the Bronze Age. Copper is superior to stone because the former can be sharpened to use in cutting. It can also be bent and shaped by hammering and beaten into sheets, which then can be cut. Further, molten copper assumes the shape of the container into which it is poured. On cooling, it retains this shape, becomes hard and can be given a cutting edge. The characteristic feature of the Bronze Age was that people learnt to alloy copper with either tin or arsenic to make a stronger metal. This alloy is known as bronze; alloying 10 per cent tin with 90 per cent copper results in bronze. Bronze Age Eurasia witnessed a gradual convergence of the military systems of various polities. Chariots emerged as the core of the Bronze Age military system. From the north of Caucasus, chariot technology spread into Anatolia. From Anatolia, use of chariots spread into Mesopotamia and the Indus Valley region. Further, from the borderlands of Caucasus through Central Asia, the chariots spread into West China. Another line of diffusion occurred from North Caucasia into Anatolia and then southward towards Egypt. The trajectory of chariot warfare also spread westward from Anatolia into Greece, finally reaching Britain. Minor variations of course were present in the different regions. The principal medium for the spread of war chariots was the expansion of the Indo-European (Proto-Indo-European/PIE) speakers across Eurasia. And in the Eastern Aegean region, a uniform set of naval techniques emerged due to continuous interactions between the coastal communities of Asia Minor, Egypt and Greece.

Archaeologists John Carman and Anthony Harding write that until the advent of the Iron Age, large-scale fighting by regular armies was absent except in Egypt

DOI: 10.4324/9780429284069-3

and the Near East. According to them, warfare in the pre-Iron Age era beyond Egypt and the Near East comprised mostly small-scale raiding, ambushes and combat between the champions/heroes, etc. (Carman and Harding 1999: 247). However, this account will show that pre-Iron Age China and India did witness large-scale warfare and massive fortifications. Victor Davis Hanson asserts that decisive infantry battles, which constituted an integral part of the Western way of warfare, first emerged in Classical Greece (Hanson 2009). We will see that decisive chariot battles took place in the Near East and momentous infantry battles occurred in India long before the advent of hoplite battles in Classical Greece. Further, disciplined heavy infantry, another characteristic element of Western warfare, also originated in Mesopotamia before the emergence of the 'men in bronze' in Greece. Let us see the trajectory of emergence and finally the decline of chariot warfare in the different parts of Bronze Age Eurasia.

North, West and Central Europe

In 2000 BCE, the population of the world was about 25 million (Scott 2017: 6). The chronology of the Bronze Age varied with regions. Around 2000 BCE, while Central Europe was undergoing the Early Bronze Age, Denmark was in the Late Neolithic Age. Around 1500 BCE, the Middle Bronze Age dawned in Central Europe, and then Denmark moved into the Early Bronze Age. Around 1000 BCE, Central Europe was experiencing the Late Bronze Age and by 500 BCE was in the Early Iron Age. At the latter date, Denmark was still in the Late Bronze Age (Randsborg 1999: 191). Between 2500–1200 BCE, in the Rhone Valley of Switzerland, the warriors used bows and arrows for long-distance combat and daggers for close-quarter combat. In Southern France, the battle axe was the weapon of choice for close-quarter fighting (Harding 1999: 160–1).

Fortified sites in Northern and Western Europe are often found in locations that dominated land and riverine passes, which were actually important trade routes. Some of the fortified sites besides protecting the trade routes also produced trade goods. Fortifications in Bronze Age Britain were mainly constructed on the Neolithic sites. The fortification at Gloucestershire in England had huge timber structures. Hilltop sites with defensive fortifications that gradually evolved into hillforts became common in England during the Middle Bronze Age. As a point of comparison, fortified hilltops were also common in the precolonial Andes. Some of the hilltop sites in Britain during the Late Bronze Age were protected with steep-sided and flat-bottomed ditches. In Wales, hillforts came up between the eleventh and ninth centuries BCE. The ramparts were constructed with layers of rubble revetted with palisades (Osgood et al. 2010: 9, 11–14).

Greece

The Bronze Age lasted in Greece from *c.* 3000 to 1000 BCE. The Early Bronze Age lasted from 3000 to 2100 BCE, the Middle Bronze Age from 2100 to 1600

BCE and the Late Bronze Age from 1600 to 1150 BCE. The date of the coming of the Greeks is a matter of debate among historians and archaeologists, and this debate is linked with a larger debate about the origins of the Indo-Europeans. Historian Robert Drews argues that the notion of the PIE being a very large community that was spread along the South Russian steppes is problematic. The argument that between *c*. 4500 and *c*. 2000 BCE, the PIE migrated to Anatolia in search of land is also erroneous. He suggests that at the end of the third millennium, the PIE-speaking community was no larger than the Hurrian, Sumerian, Hattic and the Proto-Anatolian speakers. The PIE speakers were smaller in number than the Semitic people. During the late seventeenth and early sixteenth century BCE, the PIE-speaking community started leaving their original homeland in North Caucasia. Their reason for migration was the knowledge that their mastery of chariot warfare would give them dominance over 'native' societies, and they would be able to gain power and wealth by subjugating the pre-existing communities (Drews 1989: 197–8).

In his latest book, Drews persuasively argues that the PIE people were actually Neolithic settlers (Indo-Hittites) who came from Anatolia and settled around the Don estuary of the Sea of Azov during the seventh millennium BCE. During the fifth millennium BCE, these Chalcolithic settlers spread along the Kuban River in Northern Caucasia. Towards the end of the fourth millennium BCE, they started working with bronze. And the PIE people who had spread around the middle Volga and Don rivers started constructing wagons between 3500–3000 BCE. The war chariots gradually evolved from the wagons. The wheeled vehicles enabled the PIE people to dominate the wide grassland between Southern Ural in the east up to the Dnieper River in the west (Drews 2017: 13–20).

After the mid-fifth millennium BCE, the climate of the Eurasian steppes became colder and drier. Climate change was probably another reason that pushed the PIE speakers from their original homeland towards the surplus-producing agrarian enclaves located along the Eurasian rimland. The two-wheeled war chariots developed from the four-wheeled horse- or donkey-drawn carts used for transporting goods (Renfrew 2005: 205–29). Probably the chariots first emerged in the borderland between the Eurasian steppes and the agrarian civilisations situated along the Eurasian rimland. J.P. Mallory, using linguistic evidence, asserts that PIE society routinely engaged in organised violence and celebrated warlike deeds. Their social institutions were geared for continuous battles and raiding (Mallory 2006: 77–98). Along with chariots, the Central Eurasian steppe nomads also discovered the deadly composite bows. The PIE people eagerly adopted the horse-drawn chariot–composite bow combination and invaded the neighbouring sedentary societies. In fact, evidence from South-Eastern Europe shows that the fortified villages faced extensive external attacks during this time (Ivanova 2006: 33–48).

Drews argues that the invaders, equipped with composite bows and riding horse-drawn chariots, entered Greece around 1600 BCE. These invaders were the PIE people who had established sway from Indus Valley in the east to the Danube in the west. They were the Aryans who conquered North-West India between 1500 and 1200 BCE. The Kassites who took over Southern Mesopotamia and the Hyksos who conquered Egypt were actually Aryans. The PIE people established the Mycenaean civilisation in Greece. Drews continues that these invaders who established Mycenae were taller and broader as compared to the inhabitants of Middle Helladic Period (2000–1650 BCE) Greece. Drews writes that the coming of the Indo-Europeans involved tens of thousands of people but not hundreds of thousands. In India, Egypt and Greece the invaders did not number more than one-tenth of the original inhabitants. In most of the cases, the subjugated people accepted the language of the invaders. However, in the Levant, the Aryan language was submerged within the pre-existing indigenous languages. At the beginning of the sixteenth century BCE, the population of the Thessalian Plain, Argolid, Eurotas Valley, Messenia, Boeotia and Attica was roughly three-quarters of a million people, and the PIE-speaking conquerors numbered probably 75,000 men, women and children. Superior weapon systems (horse drawn chariots and composite bows) enabled the PIE to win against their numerically superior opponents. In Messenia, the original inhabitants were reduced to the status of helots (serfs), and the Spartiates (invaders) became the ruling community (Drews 1989: 158–96).

Drews proposes that the PIE speakers probably came from Anatolia to Greece by ship. They established their strongpoints at Argos, Eleusis, Pylos and across the Corinthian Gulf. The PIE speakers' initial base in Greece was Thessaly, and from there they went on to subjugate central and southern parts of this country. At times, the Greek chariots charged in mass formation, and the charioteers fought with thrusting spears and bows. During the Mycenaean Period (Late Helladic Period), chariots and bows were important. Horses and chariots, continues Drews, were introduced in Cyprus and Crete around 1450 BCE. Between 2000 and 1490 BCE, the Minoan civilisation flourished in Crete. The Mycenaeans had knowledge and skill of sailing, and they were able to send an invasion fleet to Minoan Crete around 1490 BCE. Transport ships for carrying horses were an innovation of the Bronze Age (Drews 1989: 165–92). Marine archaeology throws light on the technology of shipbuilding during the Bronze Age. Around 1350 BCE, the coastal polities of Asia Minor were constructing cargo ships with planks which were fastened to each other by mortises and tenons (Casson 1994: 35). Therefore, the hypothesis of the PIE-speaking people arriving by ships is quite tenable.

We can conclude that the first speakers of Greek language arrived in Greece from Anatolia and North Caucasus roughly around 2000 BCE. During the Late Bronze Age, they established a number of kingdoms in the mainland of Greece. The most notable among them were Mycenae, Thebes, Pylos, etc. The Mycenaeans' principal rival was the Hittite Kingdom in Anatolia (Strauss 2006: xviii–ix). The Hittite Empire, which extended from the Aegean to the Euphrates, lasted from 1380 to 1180 BCE. The armies of these polities were centred on war chariots.

The Trojan War

Heinrich Schliemann's excavations along with blind poet Homer's *Iliad* provide the materials for studying the Trojan War. The Trojan War is a classic case of conflict between two Bronze Age civilisations. Homer uses the terms Achaeans, Dorians, Argives and Hellenes but never 'Greeks'. And modern scholars refer to the Greeks of the Late Bronze Age as Mycenaeans. The Trojan War probably occurred between 1210 and 1180 BCE and was the last great event before the decline and fall of the Mycenaean civilisation. The Greek palaces were destroyed around 1180 BCE. The Greek 'Dark Age' started from 1150 BCE and lasted until 750 BCE. Scholars date the composition of *Iliad* and *Odyssey* somewhere between 1000 and 500 BCE (Wood 1996: 128; Strauss 2006: xiii–ix). We could safely assume that the *Iliad* portrays the Mycenaean world as it was in around 1250 BCE.

Troy (also known as Wilusa and Ilion) was established around 3000 BCE in North-Western Anatolia. Troy was actually Hisarlik, and this city was part of the Hittite state of Wilusa. The city of Troy covered 75 acres and was surrounded by wheat fields. Its population varied from 5,000 to 7,500 people, drawing sustenance from Troy's rich hinterland. Troy was a city with lofty gates, big towers and wide streets and had a protected harbour. Troy exported woollen textiles and horses. Its wealth also depended on the mackerel available in the waters around the Dardanelles. Moreover, the city was a transit point for goods from Syria, Egypt and the Caucasus region (Wood 1996: 20–187; Strauss 2006: 1–9).

According to Homer, the Greeks attacked Troy because the Queen of Sparta, Helen, eloped with Paris and also took away the treasures of the kingdom. Paris was one of the sons of Priam, the ruler of Troy. However, there were other reasons for attacking Troy. Paris's seizure of Helen was actually a Trojan raid on Lakonia to capture treasures and women. The political economy of the Greek polities was geared towards raiding, as loot and plunder were required to keep the rickety economy afloat. Slaves were needed to work in the flax fields. These slaves were acquired by launching raids on Asia Minor. In fact, the Greek chieftains were regularly launching plundering raids on Asia Minor. Therefore, the Trojan raid was probably a retaliation to prior Greek raids on the coast of Asia Minor. *Iliad* is actually a compressed picture of many Greek forays in Asia Minor (Homer 1986: xiv; Wood 1996: 2, 167, 208).

Dislocation and destruction of the polities in the Near East due to climate change and invasion of Sea Peoples had a negative effect on the Mycenaean civilisation. The Sea Peoples operated in the Aegean between 1200 and 1100 BCE. To a great extent, the Mycenaeans depended on trade with the Near Eastern polities. This overseas commerce declined in an era of violence and insecurity in the Eastern Aegean. Further, communication with Cyprus was cut, and the latter was the source of copper for Mycenae. The Dorian invasion (probably PIE speakers) further accelerated the collapse of Mycenae. The final destruction of Mycenae occurred around *c.* 1150 BCE (Bury and Meiggs 1989: 44–58).

The Greek poleis (city-states) lasted roughly from 750 to 323 BCE. The period between 750 and 480 BCE is known as the Archaic Age, and the period between 480 and 323 BCE is called the Classical Period. A lot of ink has been spilled over the issue of whether Homer in his epic *Iliad* is describing warfare as it was fought during the Dark Age or in the Archaic Age. Probably, the *Iliad* is referring to the period of pre-poleis Dark Age Greece. According to Homer, the Trojan War lasted for 10 years. The *Iliad* covers the last two months of the ninth year of the Trojan War. In reality, the Trojan War may not have lasted for 10 years. Nevertheless, it was a protracted war. The Greek force that attacked Troy was a collection of the war bands of various warrior chiefs and not an impersonal institution of any Greek superstate. Agamemnon (ruler of Mycenae and elder brother of Menelaus, King of Sparta) was the overlord of Greece. Each of Agamemnon's generals was the leader of a band of warriors. Strong personal ties and loyalty to their chieftain bound the warriors of a band (Strauss 2006: 3, 35–6). The Greek soldiers and sailors did not receive wages but volunteered for service and expected a combination of food, drinks, gifts and a share of booty (Wood 1996: 22, 158).

The Greeks mainly used the penteconter, which was a 50-oared galley. It was 65 feet long, the beam was about 3.5 feet and the vessel had a draught of about 2.5 feet. The vessel had a single sail. The prow and the stern were decked, and the intervening space amidships was occupied by the rowers' benches. The crew of such a ship varied from 80 to 120 men (Wees 2010: 211–15). There was nothing uniquely 'Greek' in the penteconter. In fact, galleys emerged from a common pool of knowledge generated by the seafaring people who inhabited the coastal regions of Asia Minor, Egypt and Greece. Around 2000 BCE, in the Eastern Aegean, long and slender galleys emerged for raiding. Initially, they had paddles. Then they had sails. By 1000 BCE, rowers became common in the galleys. Rams were attached to the galleys from 800 BCE onwards (Casson 1994: 10–51).

Historian Barry Strauss asserts that the Trojan War consisted mainly of raids, that is, it was a low-intensity conflict. The Greek force lacked siege techniques. Only a trick allowed them to enter the city of Troy (Strauss 2006: 5–8). Both sides used chariots and infantry. The men fought with bronze swords, shields, spears and javelins and wore bronze armour (Wood 1996: 23). The shafts of the spears were made from the wood of the ash tree. Homer tells us that the favourite weapon of the Trojans was the spear. Although bows were used too, fighting with bows in Troy, unlike in Vedic India (1500–500 BCE), was considered unmanly. The *Iliad* tells us: 'Pyraechmus commanded the Paeonians with their curving bows. They had come from far, from Amydon and the banks of the broad River Axius' (quoted from Homer 1986: 62). Probably, the Paeonians were equipped with composite bows, a military technology acquired from the Scythian steppe nomads.

Archaeological evidence points to the fact that chariots in the Mycenaean culture came from the Eurasian steppe through the medium of the PIE speakers.

There is a debate about whether the Greeks used the chariots during the Trojan War as battlefield taxis for carrying the champions to the battlefield or as actual instruments of war in the front line. Homer provides some details about the chariots used in warfare. Each chariot had two wheels, and each wheel had eight spokes (Homer 1986: 111; Drews 2017: 53).

Monomachia, that is, a single duel between the heroes/champions, was probably practised during combat in the Bronze Age. Homer tells us that just as the opposing forces were to clash, Paris armed with some bronze-tipped spears and a curved bow stepped out from the Trojan ranks and challenged any Argive champion to meet him in a one-on-one duel (Homer 1986: 64). In fact, the Indian epic *Mahabharata* (the longest epic poem in the world), composed around 300 BCE, describing the mythical Mahabharata War between the Kaurava and the Pandava clans also contains examples of duels between the heroes riding on chariots (Roy 2004: 36).

What about the 'face of battle' in the Bronze Age? Homer provides us with some examples. We are told that Menelaus, using a javelin, struck Pylaemenes, commander of the Paphlaonian infantry, and the latter's collarbone broke. Another example tells us that Telamonian Aias of the Achaeans struck Acamas the Thracian with a spear. The bronze tip of the spear hit Acamas's forehead, pierced the bone and entered his brain (Homer 1986: 107, 117). The battlefield was bloody, grisly and nasty. Now let us see the evolving trajectory of warfare further east of Greece.

Near East

In the Near East, the Early Bronze Age lasted from 3300 to 2000 BCE, the Middle Bronze Age from 2000 to 1600 BCE and the Late Bronze Age from 1600 to 1200 BCE. Copper smelting first occurred at Catal Huyuk in Anatolia in the early sixth millennium BCE and in Iran/Persia in around 4500 BCE. Copper came mainly from Oman in the Persian Gulf. During the Early Bronze Age, most of the metal weapons were made from arsenic–copper rather than a tin–bronze combination. Arsenic–copper alloying creates a melted liquid that is less viscous and hence easier to cast. The principal source for tin during the Early and Middle Bronze Ages was in Afghanistan. Badakshan in Afghanistan also supplied the Near East with lapis lazuli, a semi-precious stone. The 'Tin Road' from Afghanistan to Elam and then to Sumer became a key strategic corridor. From Mesopotamia, tin was shipped to Anatolia, Syria and Egypt. Tin-bronze became predominant only during the Middle Bronze period. By the end of the Middle Bronze Age, bronze body armour came into existence for the elite chariot warriors (Hamblin 2010: 5, 21–3).

The sixteenth century BCE witnessed the final emergence of fully developed chariot warfare, thus ushering in the 'Chariot Age' in the Near East. The Hittite charioteers fought with bows, and they carried spears as the last-ditch weapon (Dawson 2001: 148). Between 2500 and 1800 BCE, Sumer and Akkad were divided among many small city-states. The Akkadians during the Sargonid Period (*c.* 2340–*c.* 2100 BCE) used composite bows (Carey et al. 2005: 12). The composite bow appeared in the Near East around 2500 BCE. The composite bow comprised a central wood core with thin strips of horn and leather laminated around the belly. A string of twisted gut powered the bow. Due to the long drying time required when gluing together the multiple layers of wood, horn and sinew, the composite bow took years to be manufactured. A trained archer could shoot one arrow every six seconds. The bow was 4.2 feet long and, when drawn to the ear, could send a reed shaft fletched arrow with a bronze arrowhead through an ingot of copper three fingers thick (Archer 2010: 61).

The Sumerian city-states of Southern Mesopotamia during 2500 BCE engaged in pitched battles with trained heavy infantry organised in linear formations. These formations were the precursors of phalanx formations of Classical Greece. The Stele of the Vultures of Lagash provides a representation of the Sumerian army under Eannatum in action sometime around 2400 BCE. The army comprised light and heavy infantry plus chariot corps. We see that the unarmoured light infantry was equipped with spears and battle axes. What is interesting is that armoured heavy infantry equipped with spears was deployed in several ranks. A wall of interlocking shields covered them fully, and the spear points of the first rank of infantry projected in front of the wall of shields. Sumerian armies also used chariots drawn by onagers. Horse-drawn chariots came later (Watkins 1989: 19).

Around 1900 BCE, horses attached to the wagons and chariots were controlled with nose rings. The bronze bits introduced by the Hyksos allowed wagons and chariots to be drawn efficiently by the horses. The PIE speakers then experimented with the spoked wheel, which replaced the disc wheel and made the chariots lighter and more mobile. Use of horses as draught animals for chariots became more effective with the introduction of the jointed bronze snaffle bit in the Fertile Crescent during the fourteenth century BCE (Drews 2017: 43–8).

Egypt

Around 4000 BCE, small provincial entities like the *nomi* in Egypt were ruled by the *nomarchs* (chiefs). They raised their private military forces and controlled the trade routes. They were assembled in a loose federation around what became the Upper and Lower Kingdoms. Around 3200 BCE, the King of Upper Egypt known as Narmer or Menes united the two kingdoms and established a single Egyptian state. His successor, Hor-Aha, focused on irrigation and founded the capital at Memphis. Thus, the Early Dynastic period started, which lasted for 500 years. By 3000 BCE, Egypt's population was about 1 million and its agriculture was able to support on average 450 people per sq mi. The Old Kingdom lasted from *c.* 2650 to 2160 BCE. Raids against the Nubian (Sudanese) tribes were undertaken in order to acquire cheap manpower. The Old Kingdom army comprised the standing army of the ruler and forces raised by the *nomarchs*. The standing army comprised auxiliaries (Nubians and Libyans) and conscripted Egyptians. One man in 10 had to serve for a year (Gnirs 1999: 77; Gabriel 2009: 27–9). Long before the Greeks, the Egyptian army during the 12th dynasty (twentieth century BCE) used rams to batter down the walls of the cities and strongholds held by rebels and enemies (Herzog and Gichon 1997: 222).

The Middle Kingdom (not to be confused with the pre-modern Chinese concept of Middle Kingdom) period lasted from 1975 to 1640 BCE. In 1650 BCE, the Hittites/Hyksos were able to conquer Egypt mainly because they had war chariots. The Hyksos developed heavy chariots. Each chariot had three crew: a driver, a fighter equipped with a bow and a shield bearer. The Indians would develop such heavy chariots in around 350 BCE. The Hyksos came from Canaan with an invasion force numbering about 30,000 men (Goetze 1963: 125–7). They had acquired the latest military technology, such as helmets, body armour, composite bows and war chariots from Anatolia and Mesopotamia.

The Egyptian army lacked all these advanced paraphernalia. The archers of Egypt used the simple or self bow, whose range was between 164 and 196 feet (Drews 1995: 105). The simple or self bow was made from a single strip of flexible wood and had a short lifespan. The self bow tended to warp if used intensively over a long period. This reduced tensile strength as well as range and hitting power. The Hyksos conquered the fertile Nile Delta region and pushed the Egyptians far south to Thebes (not to be confused with Thebes of Greece). Simultaneously, the Nubians attacked the weakened Egyptian power and established themselves around the First Cataract of the Nile. The Hyksos also introduced the bronze sickle sword in Egypt (Gabriel 2009: 4–31).

The population of Egypt during *c.* 1500 BCE was 7 million, and feeding them was possible because of the expansion of agriculture. Introduction of a device known as *shaduf* by the Hyksos raised agricultural production. Thus, we see that both advanced military technology and civilian technology were introduced in Egypt from Anatolia, where the latest military technologies were in vogue due to constant warfare. Further, the invading army was the medium for the transfer of technology. The Egyptian archers adopted the composite bows from the Hyksos. *Pharaoh* (King) Amosis I (r. 1539–1514 BCE) waged war successfully against both the Nubians and the Hyksos. This was possible because Amosis I had raised the combat effectiveness of his army by adapting his army through adopting the latest foreign military technology. He founded the 18th dynasty after expelling the Hyksos. Nubia's eastern desert was the source of gold. Moreover, Nubia was famous for highly prized export items like ivory, ebony and panther skins. Additionally, the alluvial soil of Nubia also supported an extensive farming and animal production. The rich soil around the Third Cataract produced cereals. Therefore, the *pharaohs* were desirous of extending control over Nubia. Thutmose became the sole ruler of Egypt in 1481 BCE (Herzog and Gichon 1997: 36; Gabriel 2009: 9, 16–17, 31–4, 44, 203).

The Battle of Megiddo was fought in 1458 BCE. When the ruler of the Kingdom of Qadesh moved south from Syria and occupied the city of Megiddo in

Palestine, Thutmose decided to strike. Credit is due to Thutmose for following the 'daredevil' approach. Disregarding the advice of his generals, the *pharaoh* went for a direct approach to Megiddo through the narrow Musmus Pass. The Syrians never anticipated that Thutmose would advance through this narrow pass directly towards Megiddo and hence left it unguarded. The Egyptian advance column from the Musmus Pass came into the plains of Megiddo and deployed from column to line slowly, taking seven hours. This was the opportunity for the Syrians to strike, when the Egyptians were vulnerable. However, the Syrians missed their chance. Next, the Egyptian centre where the main infantry force was deployed pushed the Syrians back against their camp. The Egyptian chariots then penetrated through the Syrian defence between their right and centre, and the Syrians retreated to the city. After plundering the Syrian camp, the Egyptians began their seven-month-long siege of the city. If the victorious Egyptian force had not paused to plunder the Syrian camp, the defeated Syrian army would not have been able to retreat within the city (Carey et al. 2005: 15–17).

The Egyptian infantry of Ramses II (b. 1279–d. 1213 BCE) carried pikes, sickle swords and daggers. The pikes were probably used as thrusting spears. The infantry had non-commissioned officers who carried swagger sticks. However, infantry played a secondary role compared to chariots in the Egyptian order of battle. Infantry was important for conducting siege operations. Ramses II's army used ramps and scaling ladders for conducting sieges (Herzog and Gichon 1997: 36, 147).

In 1285 BCE, Ramses II decided to contain the Hittites by advancing towards the city of Qadesh on the Orontes River. The result was the Battle of Qadesh. The Egyptian field army was organised in four divisions of 5,000 personnel each. Each of these divisions comprised chariots, archers, spearmen and axe-wielding infantry. King Muwatallis II's Hittite army included 19,000 men supported by 3,000 chariots. Qadesh was probably the world's largest chariot battle, just as Kursk (1943) was the world's biggest tank battle. The heavy Hittite chariots attacked and scattered the light chariots of the Egyptians. Then the Hittite chariot force moved towards the Egyptian camp and engaged in looting. This development gave an opportunity to Ramses to rally his reserve chariot force and attack the flanks of the Hittites. As the Hittites retreated, the Egyptian infantry joined in the pursuit. In response, Muwatallis sent his reserve chariot force to check the advancing Egyptians. As the Hittite reserve chariot force moved towards the Egyptian camp, they were threatened on their north by the Egyptian Sutekh Division, which was just entering the battlefield. The Hittites, after losing 1,000 chariots, decided to retreat. The Hittites disengaged and garrisoned the city of Qadesh. Ramses had gained a victory in the battlefield but lacked siege engines to take Qadesh (Carey et al. 2005: 16–21). The necessity of making a quick long march through the desert of Sinai probably prevented Ramses from concentrating the large number of baggage animals and porters required to carry his siege engines to the battlefield in Palestine. At times, the *pharaohs* used the navy to transport heavy war materials to the coast of Canaan.

While the invasion of the Sea Peoples seriously weakened Egypt, the Hittite Empire collapsed due to internal rebellions, attack by the Sea Peoples, plague, decline in agricultural productivity, demographic collapse and changes in climatic conditions (Bryce 2019: 254–67).

South Asia

Around 3500 BCE, the Indus Valley civilisation came into existence. There was no superstate in the Indus Valley. Rather, there were several small city-states/principalities. This civilisation covered some 700,000 sq km, and the population varied from 1 to 5 million. Vast demographic resources of South Asia and China have been a crucial driver of history in these two regions. The rulers of the Indus Valley civilisation lived in the cities. It is estimated that the total urban population came to about 250,000 people, and of them about 150,000 lived in the two biggest cities: Mohenjo Daro and Harappa. These cities and others, like Kot Diji, Kalibangan etc., were protected with defensive walls against the threat posed both by the external invaders and the exploited peasantry of the countryside. Mohenjo Daro had a 50 foot-high citadel fortified by a baked brick wall and solid towers. The fortified citadel mound of Harappa had a rampart of mud brick wall whose base was 40 feet wide (Verma 2003: 1).

Agriculture was practised in the Helmand Basin by 4000 BCE. The Helmand Basin was influenced by the Bronze Age culture of Mesopotamia. We can speculate that agricultural practices spread from Helmand Basin to the Indus Valley. The Indus people raised winter and summer crops. Cotton was cultivated in North-East Baluchistan before 4000 BCE. From there, cotton diffused eastwards into Mohenjo Daro. Grapes and melons came from the Helmand Valley. Pigs, goats, sheep and humped cattle were domesticated in the Indus Valley (Habib 2002: 2–28).

The inhabitants of the Indus Valley civilisation spoke a Dravidian language, and Tamil is an offshoot of it. Whether the Dravidians were of a single racial stock or a mixture of several ethnic stocks like Proto-Elamites, Austronesians etc. remains conjecture at best. The traditional view is that the Aryan speakers were of a Caucasoid race, and they defeated and destroyed the Dravidic Indus Valley civilisation. The Dravidians were pushed eastwards and later southwards into the interior and southern part of the Indian subcontinent (Basham 2011: 20–37). Modern politically correct scholars are uncomfortable with the idea of linking race with language. They argue that the PIE speakers migrated in several waves, and there was no invasion. Rather, a peaceful intermingling of the different language groups occurred (Thapar 2005: 106–28).

Political correctness must not allow us to write out the role of organised violence from the equation altogether. In fact, the *Rig Veda* (composed around 1200 BCE by a motley group of sages) repeatedly speaks of the dark and flat-nosed *dasyus* (Dravidians?) who lived in *puras* (cities) and were defeated by the horse-drawn-chariot-riding light-skinned *Aryas* (Aryans?) (*Rig Veda* 1994: 19, 112,

160–2). The Indus people lacked horses and chariots. They fought as infantry and had both missile weapons (bows and slings) and weapons for close-quarter fighting. Stone arrowheads had been discovered at Harappa. Clay pellets, each weighing about 6 ounces, had been found in the Indus sites. The infantry also used double-edged swords. The blade was attached to the rectangular-shaped tang with rivet holes. Knives of copper with wooden handles and maces with wooden handles and cooper or bronze heads were used for close-quarter combat. In addition, spears were found in the Indus sites. The spearheads made of copper were leaf-shaped or pointed (Pant 1978: 33–40).

One could surmise that after the fall of the Indus Valley civilisation in 1200 BCE, some of the people from the Indus Valley migrated into West, Central and North India. And these people brought the ideas of fortification from the Indus Valley to these regions of India. The Chalcolithic communities of Central and West India were capable of constructing massive fortifications. Some examples are given. The Jorwe culture (*c.* 1400–700 BCE) near the bank of the River Banas in Rajasthan constructed walls made of mud and strengthened with burnt brick pieces. Further, each town surrounded with a wall had a citadel. The construction techniques and ideas were influenced by the Harappans. Some construction techniques also evolved autonomously. For instance, Balathal (2500–1600 BCE) had walls made of mud but rivetted with stones on both the inner and outer sides. Rectangular square bastions were also constructed. The Chalcolithic culture in Malwa on the left bank of the Bina River region of Central India constructed moats and behind them they raised two walls separated by 50 feet. The width of each wall came to about 55 feet (Singh 2011: 100–10). Innovations occurred in the fields of both military architecture and hand weapons. The Dravidic Chalcolithic people of Central India came up with a curved battle axe, a weapon which was absent in the Indus Valley (Pant 1978: 46). We can speculate that the Dravidians, after the collapse of the Indus Valley civilisation, migrated towards West India and Central India and for defensive purpose constructed fortresses. However, the invading Aryans attacked these fortresses. For instance, one hymn of *Rig Veda* mentions the Aryan war god Indra attacking a mountain fortress (*Rig Veda* 1994: 161–2).

Drews challenges the traditional theory that the Aryans came to India from Iran through the overland passes of North-West Afghanistan. The northern overland route across Afghanistan and Hindu Kush Mountains are not conducive to large-scale migration even today. The southern land route across the arid Baluchistan (characterised by unavailability of drinking water) through the Kirthar Range is worse. He proposes that the Aryans came into India from Mesopotamia. The possible route of the PIE/Aryans was from Armenia to the Persian Gulf, and then they sailed and reached the lower reaches of the Indus River (Drews 1989: 182–3). In fact, it is probable that the early Aryans had some knowledge of sea travel. One hymn of the *Rig Veda* says: 'make a ship whose oars will carry us across; make the weapons ready and set them in place' (quoted from *Rig Veda* 1994: 66). In several hymns, there is mention of ships sailing in the ocean far away from the shores (*Rig Veda* 1994: 182). The Harappan ports in the Makran Coast were fortified (Ratnagar 1981: 47). Probably the fortifications were in response to the threat posed by the pirates/Sea Peoples or by the seaborne invading Aryans.

The Aryan society (Vedic society) which came into existence after the collapse of the Indus Valley civilisation was highly militaristic. A hymn of the *Rig Veda* notes that those who fought as heroes and sacrificed their lives in combat went straight to heaven. This reflects the high esteem in which the warriors were held in Vedic society. Warrior status became important in Northern and Western European Bronze Age societies also. Large amounts of weaponry were buried with the dead. Therefore, we can say that the intrusion of the PIE speakers resulted in further militarisation of Bronze Age societies. The *Rig Vedic* hymns mention the emergence of a fourfold caste system in the Indian society. At the top were the Brahmins who had knowledge of the *Vedas*. Then came the warrior class or Kshatriyas, who were the ruling class (the Brahmins functioned as their advisors). Next came the Vaishyas, who engaged in trade and commerce. The lowest rung of the society comprised the Shudras, who engaged in agricultural pursuits (*Rig Veda* 1994: 31, 54; Osgood et al. 2010: 2–23). Probably, the Shudras were the defeated Dravidians. This fourfold caste system was an ideal picture. In reality, there were many overlaps and sub-castes.

The *Rig Veda* repeats the importance of horse-drawn chariots, archers and body armour for waging battles. We are also told of wagons that were used for transporting weapons and armour of the soldiers during campaigns (*Rig Veda* 1994: 67, 157, 237). A Vedic army comprised of the Kshatriya military leaders equipped with bows and covered with leather body armour, riding horse-drawn chariots. However, the mass of the ground force comprised infantry recruited from the villages and various tribes. The Vedic military leaders took care to recruit men from one particular village within a single unit, and men from one particular tribe were also put together in one unit. The objective was that the men (including their relatives, friends, kinsmen etc.) who knew each other should fight together within a single formation. This was an attempt to generate what latter-day sociologists term as primary group solidarity. The infantry played a secondary role in battle and plundering raids. However, infantry was important for siege warfare. The infantry was equipped with bows and arrows, spears, swords and axes. Against enemy strongholds, the infantry at times launched fire arrows (Singh 1989: 9–11).

The Vedic Aryans (1200–500 BCE) waged big and bloody battles. The Vedic literature describes several battles between different confederations of tribal leaders and monarchs. For example, in the Battle of Ravi, Sudas, the leader of the Bharata tribe, inflicted a defeat on a confederation of 10 tribes. Hence, this battle is also known as the Battle of 10 Kings. The forces of the confederacy of 10 kings deployed on the right bank of the Ravi River had planned to attack Sudas by crossing the deep river. Therefore, they started digging channels in order to divert the water. It seems that a lot of non-combatants, who in this case were used for digging channels, accompanied the armies. However, this piece of military engineering went wrong, and the water flowed back to its natural bed and then flooded the right bank. This put the forces of the confederacy in disarray, and Sudas took advantage of the situation. He crossed the river and suddenly attacked them. Three of the kings of the confederacy of 10 were killed in the course of the battle. Next, Sudas defeated the confederation of Aja, Sigru and Yaksu tribes led by King Bheda on the banks of Yamuna River. The biggest battle was probably fought by King Susrava against a confederation of 20 kings. This confederacy was supposed to have fielded, says the *Rig Veda*, some 60,000 men (Thapliyal 2010: 27, 157). Even if the Vedic hymns are overestimating the number of soldiers deployed by the confederacy of 20 kings, we still can safely assume that huge numbers of combatants were involved in this decisive conflagration. Besides battles, raids for capturing cows and women were an integral feature of Vedic warfare. For protecting the cows, walled enclosures known as cow pens emerged (*Rig Veda* 1994: 67–8). Cow pens gradually evolved into forts and towns. We could surmise that the Aryans, after settling down, gradually adopted the defensive military architecture of the Indus people.

China

The Bronze Age in China started around 2600 BCE. Around 2500 BCE, walls were built with earth intermixed with stones in order to provide extra hardness (Sawyer 2011: 9, 35). The Shang dynasty lasted roughly from 1500 to 1027 BCE. Increasing cultivation of wheat and large-scale raising of domesticated sheep, goats and bovines characterised the Shang era. A hierarchy of officials and lineage chiefs constituted the government (Major and Cook 2017: 75). In the Shang era, the populace was divided into four groups: ruling families, royal clan members, nobility and common people who were slaves and serfs. The nobility lived in the cities with massive palaces and temples, as in case of Mycenaean Greece. The common people were conscripted for labour duties as well as for providing military service.

Before the rise of the Shang dynasty, warfare consisted of raids and engagements between Neolithic villages. The war bands were led by the clan chiefs. However, with the rise of the Shang, a central authority came into existence which maintained a standing army whose number could be expanded during crisis. The royal army, which was 1,000 strong, was supplemented by the chiefs' levies. A rudimentary bureaucracy also emerged to administer this force. A campaign lasted for a few days to a maximum period of three months. The army was organised into three divisions; right, left and centre, and each division was composed of infantry and chariots. In general, Shang expeditionary armies comprised 3,000 to 5,000 men, but during emergencies, forces comprising 30,000 men were also raised. The objectives of Shang warfare were imposition of royal suzerainty, mass capture of prisoners, pillage and plunder. Control over the areas outside the core was imposed through tributary networks. Many prisoners were sacrificed as part of the Shang religious ceremonies (Sun Tzu 1994: 33–5; Peers 2006: 7, 18). This practice was absent in India, Persia and Greece. War as usual was mostly a male 'game'. One exception in ancient China was Fu Hao, one of the consorts of Shang ruler Wu Ding. She had commanded the Shang army during campaigns against many enemies. Since she was a military leader, her tomb contained battle axes and other weapons (Major and Cook 2017: 79–80).

The weapons of the Shang soldiers included halberds, bows, spears and *ko* (dagger-axes). The spears' length came to about 7 feet, and each of them had a bronze blade. The *ko* consisted of a blade mounted at right angle to a shaft which was between 3 and 6 feet in length. The *ko* was generally used by an infantry soldier with a shield. Only the elites carried bronze weapons and helmets, while the common soldiers carried weapons made of stones and animal bones and leather shields. The soldiers used compound bows which shot bronze-tipped arrows (Sun Tzu 1994: 34–5; Peers 2006: 19).

Chariots emerged as the key weapon in around thirteenth century BCE under the Shang. The chariot was introduced from the region north-west of China. Around 1500 BCE, pastoral nomadic cultures east of the Ural Mountains were using chariots. Horse-drawn chariots first emerged among the nomads between Caucasus and Urals. And from this region, this technology, as we have discussed earlier, spread into the Middle East, India and China. Both in Shang China and in Vedic India, each chariot was drawn by two horses. The crew of a Shang chariot comprised an archer and a driver, and occasionally a third person equipped with a spear or *ko*. Each wheel was about 5 feet in diameter and had 18 or more spokes. The vehicle, which was a box-like structure, was enclosed with rails. Probably, horses were acquired from Mongolia. And the horses were controlled with a bit and bridle made of rope. Chariots constituted cutting-edge military technology in the Bronze Age. Chariot warfare included a lot of paraphernalia: the technique of constructing chariots, the art of horse breeding and taming, the skill required for driving the chariots and training the horse teams, use of bridles etc. Chariot warfare in Vedic India gave rise to the professional class of *rathakaras* (specialist artisans associated with chariot construction) and *rathins* (chariot drivers). The Shang rulers acquired all the necessary tools and the associated skills connected with the war chariots either through trade with the steppe nomadic neighbours or from the prisoners of war captured during the clashes with the pastoral nomads (Shaugnessy 2005: 1–49; Peers 2006: 20; Major and Cook 2017: 97–8).

The Zhou dynasty existed from roughly *c.* 1027 to 256 BCE. The Zhou from the Wei River Valley overthrew the Shang. Initially, what started as an insurgency against the Shang monarch escalated into a full-blooded conventional campaign. In a great battle at Muye (1046 BCE?) in Henan province, the Zhou destroyed the Shang force. In this battle, the Zhou deployed 3,000 chariots (Major and Cook 2017: 90, 98). One might suspect the number of chariots given by the chroniclers for this battle. By that logic, the number of soldiers and various weapon systems given by the chroniclers who documented Greek and Roman military activities could also be challenged. By all accounts, Muye was a great chariot battle, like Megiddo and Qadesh.

In Shang times, elephants were present in the Yellow River Valley. The wild elephants were captured and tamed by the Shang people. From around 800 BCE, the Vedic Indians also started domesticating elephants, and the use of elephants for domestic and military purposes spread to South-East Asia. Elephant trainers came to China from Annam, Champa and Cambodia. However, elephants vanished from North China when the Bronze Age ended (Schafer 1957: 289–90). Nevertheless, the pachyderms survived in Central and South China. In ancient China, unlike in Vedic India, however, elephants never became an integral component of the military system. The chariots became more important under the Zhou regime. The Western Zhou around 800 BCE used chariots, each of which were drawn by a team of four horses. These chariots constituted the principal shock weapon of the Western Zhou army. The Zhou army, like the Shang army, was divided into three sections: left, right and centre. The Zhou general took position either in the centre or in the left wing of the army. The chariots were deployed in composite units, and each chariot was protected by 25 men. Therefore, we see a rough infantry-chariot combined arm tactics was emerging. The Eastern Zhou, based in their capital Lo-yi, lasted from 770 to 256 BCE. This polity was able to deploy 700 chariots in 632 BCE, and by fifth century BCE could muster some 4,900 vehicles. Between the eighth and second century BCE, the nature of warfare was gradually transformed. Mass infantry armies gradually replaced chivalrous aristocratic chariot warfare (Peers 2006: 7–27). This transformation in warfare was completed in the era of Warring States (403–221 BCE), when the chariot was replaced with an infantry and cavalry combination. The Zhou army comprised contingents raised by the vassals of the Zhou ruler. And as usual, many of these vassals occasionally became disloyal. In the long run, this resulted in the decline of Zhou and the beginning of the Spring and Autumn period. It was the prelude to the Warring States era.

Collapse

Between 2000 and 1000 BCE, the world's population doubled to 50 million (Scott 2017: 6). In addition, the Bronze Age civilisations collapsed around *c*. 1200 BCE. As regards the collapse of the Bronze Age societies in the Eastern Mediterranean region, we have the catastrophe theory by Drews and the complexity theory as propounded by Eric H. Cline. According to Drews, the 'barbarians' (from the fringes of civilisation; they were poor people who came from mountainous regions) attacked the Bronze Age polities. They were successful because they introduced a new type of warfare. The barbarians were equipped with long slashing iron swords and javelins. Their skirmishing infantry proved superior to the Bronze Age states' armies centred on bow-equipped warriors on chariots. Drews writes that this military revolution ended the Bronze Age and ushered in the Iron Age. The Bronze Age armies' tactics were based on bowmen from chariots shooting arrows for disorganising the enemy force. Between 1600 and 1200 BCE, the bow was the main weapon of the Mycenaean chariot warriors, who were protected with knee-length corselets for protection against enemy missiles. Drews continues that the armies of the Bronze Age kingdoms did not coordinate chariot attacks (core element) with the supporting infantry, which played a marginal role in the battlefield. The 'barbarian' infantry, equipped with long iron swords, javelins and spears and equipped with round shields proved superior to the chariot-borne archers in close-quarter combat. Drews notes that the round shields were introduced into the Aegean by the 'barbarian' invaders from temperate North Europe. Overall, the barbarian infantry would throw their javelins (which were used previously for hunting wild animals) for disabling the chariot drivers, and then they would close in and cut down the bowmen on the chariots with slashing iron swords (Drews 1995). Here, we may refer to Kristian Kristiansen's argument, discussed next, that warriors equipped with long swords had emerged in North Europe during the Late Bronze Age. When this paradigm of war was introduced in the Bronze Age Aegean, the latter's chariot-based military system collapsed.

Archaeologist Kristiansen writes that from the eighteenth to seventeenth century BCE in Europe, a new panoply of weapon system (long sword, lance and chariot) emerged. Between the Middle Bronze and Late Bronze ages, the evolution from dagger to rapier to sword occurred in Northern and Western Europe. By the end of Late Bronze Age, in the Atlantic Zone of Europe, a carp's tongue sword that was capable of stabbing and slashing emerged as the core weapon for close-quarter combat. The ratio of infantry equipped with lances to infantry armed with long swords was about 10:1. These new weapons and innovative military tactics resulted in the use of heavy body armour. Defensive armour was already in use from the second millennium but was made of organic materials like leather, wood and bone. However, from the thirteenth century BCE, bronze armour came into vogue. And all these required specialised skill on the part of the warriors, which in turn could be inculcated through thorough training. Thus, professional soldiers and their supporting social set up came into existence. In temperate Europe, the new weapon system gave birth to warrior aristocracies (Kristiansen 1999: 177–8; Osgood et al. 2010: 23–4).

Kristiansen is actually implying a 'military revolution' in the Bronze Age without actually using the term. Drews however uses the term military revolution. Like Geoffrey Parker's military revolution of early modern Europe, it was technology driven. Technological changes (hardware development) gave birth to new tactics and new social groups, which in turn generated political changes. To an extent, Kristiansen's advocacy of the introduction of new weapon systems during the Bronze Age is similar to Drews' theory, which links military change with the rise and fall of the Bronze Age civilisations in West Asia. Drews is implying two military revolutions. The first military revolution, based on chariot system during the Middle Bronze Age, gave birth to the palace-centred Bronze Age polities. The second military revolution (about which Kristiansen speaks), based on long sword-equipped skirmishing infantry around 1200 BCE, destroyed the Bronze Age polities and ushered in the Iron Age. We could attempt a fusion of Kristiansen's and Drews' arguments. We could speculate that infantry equipped with long slashing bronze swords, round shields and javelins emerged in North Europe during the seventeenth and eighteenth centuries BCE. By 1200 BCE, bronze long swords were replaced with iron swords. And this paradigm of infantry warfare, when introduced in the Aegean region, proved disastrous for the Mediterranean Bronze Age polities.

Cline rightly points out that a combination of factors (both military and non-military) was responsible for the collapse of the Bronze Age societies in the Eastern Aegean. Cline writes:

> but it is much more likely that a concatenation of events, both human and natural – including climate change and drought, seismic disasters known as earthquake storms, internal rebellions and 'system collapse' – coalesced to create a 'perfect storm' that brought this age to an end.
>
> *(quoted from Cline 2015: 11)*

Cyprus provided most of the copper ore to the Bronze Age civilisations of the Eastern Aegean. Egypt imported cedar wood from Phoenicia (Lebanon) for constructing chariots and bows and exported gold extracted from the mines in Nubia. Ugarit, located at the coast of North Syria, was the transit point for the export of horses from the Hittite Empire to Egypt. Trade in horses was a state-protected business venture in Ugarit (Cline 2015: 3, 27, 57, 102, 108). We have discussed the importance of the 'Tin Road' from Afghanistan to the Middle East earlier. The Bronze Age polities were dependent on long-distance trade. Decline of this commerce due to disruption of the trade routes, the collapse of agriculture and external invasions plus internal uprisings generated a chain reaction which resulted in a system collapse. Let us review the various factors one by one.

The attacks by the Sea Peoples partly contributed to the collapse of the Aegean Bronze Age polities. In 1179 BCE, Ramses III had to fight a land and sea battle in the Eastern Nile Delta against them. The Sea Peoples comprised of diverse groups from different regions. They were armed with sharp swords, wooden spears with metal tips and bows and arrows and came on boats, wagons, ox carts and chariots. Sometimes, their families also accompanied them. Desire for spoils, slaves and also population pressure in their original homelands probably forced the Sea Peoples to attack the Bronze Age polities in the Eastern Aegean. The Sea Peoples attacked Egypt twice: 1207 and 1177 BCE. Both their invasions were turned back. However, it proved to be a Pyrrhic victory for Egypt. The New Kingdom in Egypt declined to the level of a second-rate power (Drews 1995: 51; Cline 2015: 1–2, 9).

A drought in North Europe pushed the Sea Peoples from their original homelands into the Eastern Mediterranean. The thirteenth and twelfth centuries BCE witnessed a drier climate in the Mediterranean region. The drier climate also adversely affected the agriculture of the polities of the Bronze Age civilisations. For instance, the level of groundwater in Syria declined. Declining rainfall also reduced the productivity of agriculture in Mycenaean Greece (Cline 2015: 142–7). High solar activity resulted in a warmer and drier climate, resulting in drought. Reduced rainfall dislocated agriculture. For the pastoral nomads, even a slight reduction in precipitation caused disaster. The collapse of agriculture due to crop failures resulted in the spread of disease due to malnutrition and internal strife (Gorny 1989: 91). To cap it all, external invasions occurred. Internal disturbances as well as external attacks resulted in the burning of several of the cities, a fact which is noted by the archaeologists.

In fact, the driest period in the Bronze Age occurred between 1250 and 1100 BCE. The North Atlantic sea surface temperature oscillation was responsible for the drastic climate change in the Eastern Mediterranean lands. The highest temperature in Greenland during the last five millennia occurred in 1200 BCE. This warming resulted in an increase in glacial melt that caused an influx of cold freshwater in the North Atlantic and Mediterranean Basin. This reduced the sea surface temperature. The decline in Mediterranean sea surface temperature reduced freshwater flux into the atmosphere, which in turn reduced precipitation over land. Rainfall decreased, which resulted in shrinkage of forest in the Mediterranean region. Strong westerlies blowing towards Northern and Western Europe carried moisture and heat from the Atlantic, leaving the Mediterranean with higher atmospheric pressure, reduced precipitation and colder winters. And when weaker westerlies blew towards the Mediterranean, atmospheric pressure was lower and the winters were warmer. In the northern steppe lands, extreme cold winters destroyed the agricultural output. Droughts occurred in the region stretching from Khorasan in the east to Danube in the west and in the south up to Egypt. This probably motivated the Indo-Europeans (the second wave) to move south at a time when the agriculture-based Bronze Age polities were suffering from a crisis. Urbanism and intensive agriculture resulted in steady deforestation of the upstream watershed of riverine areas, subsequent siltation and flood. The latter resulted in salinisation of the soil, lower yields and abandonment of the arable land. And the adverse effects of this process were exaggerated by drier climate and reduced rainfall (Scott 2017: 31). In fact, famines raged across the Hittite lands. Grain prices rose in Egypt. The second group of PIE people might be the Sea Peoples. This second wave of PIE speakers probably equipped with long iron swords fought as infantry and overcame the already weakened chariot-based Bronze Age palatial cultures around Eastern Aegean.

However, the higher level of moisture from 1100 BCE resulted in the expansion of olive cultivation and horticulture in the Mediterranean area (Langgut et al. 2013: 149–75). We can speculate that this favourable change in climate aided the emergence of city-states in Greece and the Assyrian Empire in the Near East, an issue which is dealt with in the next chapter.

In addition to adverse climatic changes, internal rebellions and external invasions, the Eastern Mediterranean region was struck by a series of earthquakes which started around 1225 BCE and lasted till 1175 BCE. This phenomenon is known as a 'seismic storm'. Drought and earthquakes resulted in the reduction of food supplies, which in turn activated internal rebellions. This in turn further debilitated the governments of the Bronze Age polities (Cline 2015: 141–8). The Bronze Age polities were already seriously weakened by earthquakes, drought, internal rebellions and dislocation of the trade routes when the Sea Peoples/'barbarian' infantry attacked.

Both Drews and Cline focus on the Aegean and Near East. They have nothing to say about the collapse of the Bronze Age Indus Valley civilisation. But we can attempt to bring the downfall of the Indus Valley civilisation within the ambit of catastrophe and complexity theories. One group of scholars argues that the Dravidian language, which is not found anywhere else outside India, may be Proto Indo-European in nature. We can assume that the Indus Valley civilisation was the product of Dravidians who were PIE speakers or belonged to the first group of Aryan invaders. They established this culture between 3500 and 2500 BCE. By 1500 BCE, salination of the soil, increasing aridity, deforestation of the Indus Valley and earthquakes weakened the agrarian basis of this society (Trautmann 2005: xiii–liii; Lal 2011: 11–19). As a result, gradual migration of the city dwellers into North and Central India had already started. Climate change and probably demographic pressure pushed the second group of PIE speakers from Central Eurasia towards the urban Bronze Age civilisations along the rimland of Eurasia. And the second group of PIE speakers (Indo-Aryans, an offshoot of the Indo-Iranians, who in turn branched from the Indo-European stock) gave the final death blow to the already weakened Indus Valley civilisation around 1200 BCE. The Dorians who invaded Mycenaean Greece were probably the second wave of PIE-speaking people.

Conclusion

Chariots constituted the core of the Bronze Age military systems, from China in the east to Greece in the west. The chariot-centred military system was introduced by the first wave of PIE speakers. The Bronze Age Greek forces were incapable of conducting effective siege warfare. If the conquest of a city like Troy took several years for an alliance of the Greek city-states, then it could be said that these forces were good for skirmishing, raiding and ambushes rather than conquest. The Bronze Age civilisations collapsed due to combination of several factors. Climatic changes resulting in increasing aridity, intensive agriculture causing salinity of soil, and the law of diminishing returns caused social strife in the cities that weakened their defensive potential. Further, increased competition for arable land and water resulted in rising struggles among the various Bronze Age polities. And then the Indo-Europeans (second wave) struck from Central Eurasia. The migration of the Sea Peoples (whether they were refugees or invaders or Drews' barbarians or the second wave of PIE speakers is still debated) further exacerbated the scenario. As regards military techniques, the 'barbarian' infantry with javelins and long swords were able to overwhelm charioteers. Iron was available in greater quantities and was cheaper compared to bronze, so a larger number of iron swords could be manufactured. Further, iron swords being sharper and stronger were better for cutting and slashing compared to bronze swords. Probably the chariot-centric armies of the Bronze Age cultures would have been able to offer more effective resistance if their polities had remained stable. But the collapse of agriculture, internal strife resulting in widespread rebellions, fire, breakdown of trade, earthquakes, etc., had seriously weakened the fragile agrarian base of the Bronze Age societies when the iron-equipped invaders came.

To sum up, speaking about a 'military revolution' during the Middle Bronze Age and another during the Late Bronze Age is too much. We can conclude that the Bronze Age saw the rise and fall of the chariot system. The military landscape of Asia, unlike that of Europe, witnessed gigantic chariot battles like Megiddo and Qadesh and big infantry battles like the Battle of Ten Kings. While the beginning of the Iron Age saw the rise of heavy armoured infantry in Greece, in the Near East, cavalry-centred military machines rose. Cavalry dominated the military system of Mesopotamia despite the fact the origins of disciplined heavy infantry could be traced back to this region. This distinction was mostly due to ecological factors. And this set the path for military divergence in Eurasia from *c.* 1000 BCE onwards.

References

Archer, Robin, 'Chariotry to Cavalry: Developments in the Early First Millennium', in Garrett G. Fagan and Matthew Trundle (eds.), *New Perspectives on Ancient Warfare* (Leiden/Boston: Brill, 2010), pp. 57–79.

Basham, A.L. (ed.), *A Cultural History of India* (1975, reprint, New Delhi: Oxford University Press, 2011).

Bryce, Trevor, *Warriors of Anatolia: A Concise History of the Hittites* (London/New York: I.B. Tauris, 2019).

Bury, J.B. and Meiggs, Russell, *A History of Greece to the Death of Alexander the Great* (1900, reprint, Houndmills: Macmillan, 1989).

Carey, Brian Todd, Allfree, Joshua B. and Cairns, John, *Warfare in the Ancient World* (Barnsley: Pen & Sword, 2005).

Carman, John and Harding, Anthony, 'Epilogue: The Future Study of Ancient Warfare', in John Carman and Anthony Harding (eds.), *Ancient Warfare: Archaeological Perspectives* (Gloucestershire: Sutton, 1999), pp. 245–50.

Casson, Lionel, *Ships and Seafaring in Ancient Times* (London: British Museum Press, 1994).

Cline, Eric H., *1177 BC: The Year Civilization Collapsed* (2014, reprint, Princeton/Oxford: Princeton University Press, 2015).

Dawson, Doyne, *The First Armies* (London: Cassell, 2001).

Drews, Robert, *The Coming of the Greeks: Indo-European Conquests in the Aegean and the Near East* (1988, reprint, Princeton, NJ: Princeton University Press, 1989).

Drews, Robert, *The End of the Bronze Age: Changes in Warfare and the Catastrophe CA 1200 BC* (1993, reprint, Princeton, NJ: Princeton University Press, 1995).

Drews, Robert, *Militarism and the Indo-Europeanizing of Europe* (London/New York: 2017).

Gabriel, Richard A., *Thutmose III: The Military Biography of Egypt's Greatest Warrior King* (Washington, DC: Potomac Books, 2009).

Gnirs, Andrea M., 'Ancient Egypt', in Kurt Raaflaub and Nathan Rosenstein (eds.), *War and Society in the Ancient and Medieval Worlds: Asia, the Mediterranean, Europe, and Mesoamerica* (Washington, DC: Center for Hellenic Studies, distributed by Harvard University Press, 1999), pp. 71–104.

Goetze, Albrecht, 'Warfare in Asia Minor', *Iraq*, vol. 25, no. 2 (1963), pp. 124–30.

Gorny, Ronald L., 'Environment, Archaeology, and History in Hittite Anatolia', *Biblical Archaeologist*, vol. 52, no. 2/3 (June–Sept. 1989), pp. 78–96.

Habib, Irfan, *A People's History of India*, 2, *The Indus Civilization* (New Delhi: Tulika, 2002).

Hamblin, William J., *Warfare in the Ancient Near East to 1600 BC: Holy Warriors at the Dawn of History* (2006, reprint, London/New York: Routledge, 2010).

Hanson, Victor Davis, *The Western Way of War: Infantry Battles in Classical Greece* (1994, reprint, Berkeley/Los Angeles: University of California Press, 2009).

Harding, Anthony, 'Warfare: A Defining Characteristic of Bronze Age Europe?', in John Carman and Anthony Harding (eds.), *Ancient Warfare: Archaeological Perspectives* (Gloucestershire: Sutton, 1999), pp. 157–73.

Herzog, Chaim and Gichon, Mordechai, *Battles of the Bible: A Military History of Ancient Israel* (1978, reprint, London: Greenhill, 1997).

Homer, *The Iliad*, translated by E.V. Rieu (1950, reprint, Middlesex: Penguin, 1986).

Ivanova, Mariya, 'Tells, Invasion Theories and Warfare in Fifth Millennium BC North-Eastern Bulgaria', *Journal of Conflict Archaeology*, vol. 2, no. 1 (2006), pp. 33–48.

Kristiansen, Kristian, 'The Emergence of Warrior Aristocracies in Later European Prehistory and Their Long-Term History', in John Carman and Anthony Harding (eds.), *Ancient Warfare: Archaeological Perspectives* (Gloucestershire: Stroud, 1999), pp. 175–89.

Lal, B.B., 'The Indus Civilization', in A.L. Basham (ed.), *A Cultural History of India* (1975, reprint, New Delhi: Oxford University Press, 2011), pp. 11–19.

Langgut, Dafna, Finkelstein, Israel and Litt, Thomas, 'Climate and the Late Bronze Collapse: New Evidence from Southern Levant', *Tel Aviv*, vol. 40 (2013), pp. 149–75.

Major, John S. and Cook, Constance A., *Ancient China: A History* (New York/London: Routledge, 2017).

Mallory, J.P., 'Indo-European Warfare', *Journal of Conflict Archaeology*, vol. 2, no. 1 (2006), pp. 77–98.

Osgood, Richard, Monks, Sarah and Toms, Judith, *Bronze Age Warfare* (2000, reprint, Gloucestershire: History Press, 2010).

Pant, G.N., *Indian Arms and Armour*, vol. 1 (New Delhi: Army Educational Stores, 1978).

Peers, C.J., *Soldiers of the Dragon: Chinese Armies 1500 BC-AD 1840* (Oxford/New York: Osprey, 2006).

Randsborg, Klavs, 'Into the Iron Age: A Discourse on War and Society', in John Carman and Anthony Harding (eds.), *Ancient Warfare: Archaeological Perspectives* (Gloucestershire: Sutton, 1999), pp. 191–202.

Ratnagar, Shereen, *Encounters: The Westerly Trade of the Harappa Civilization* (Delhi: Oxford University Press, 1981).

Renfrew, Colin, 'Archaeology and Language', in Thomas R. Trautmann (ed.), *The Aryan Debate* (New Delhi: Oxford University Press, 2005), pp. 205–29.

The Rig Veda, An Anthology: One Hundred and Eight Hymns, Selected, Translated and Annotated by Wendy Doniger O'Flaherty (1981, reprint, New Delhi: Penguin, 1994).

Roy, Kaushik, *From Hydaspes to Kargil: A History of Warfare in India from 326 BC to AD 1999* (New Delhi: Manohar, 2004).

Sawyer, Ralph D., *Ancient Chinese Warfare* (New York: Basic Books, 2011).

Schafer, Edward H., 'War Elephants in Ancient and Medieval China', *Oriens*, vol. 10, no. 2 (1957), pp. 289–91.

Scott, James C., *Against the Grain: A Deep History of the Earliest States* (New Haven/London: Yale University Press, 2017).

Shaugnessy, Edward L., 'Historical Perspectives on the Introduction of the Chariot into China', in Peter Lorge (ed.), *Warfare in China to 1600* (Aldershot: Ashgate, 2005), pp. 1–49.

Singh, Sarva, Daman, *Ancient Indian Warfare* (1965, reprint, Delhi: Motilal Banarasidass, 1989).

Singh, Umesh Kumar, *Fortifications in Ancient India: A Study of Protohistoric Cultures* (Delhi: Kaveri Books, 2011).

Strauss, Barry, *The Trojan War: A New History* (New York: Simon and Schuster, 2006).

Sun Tzu, *The Art of War*, translated with Introductions and Commentary by Ralph D. Sawyer with the collaboration of Mei-chun Lee Sawyer (Boulder: Westview, 1994).

Thapar, Romila, 'Some Appropriations of the Theory of Aryan Race Relating to the Beginning of Indian History', in Thomas R. Trautmann (ed.), *The Aryan Debate* (New Delhi: Oxford University Press, 2005), pp. 106–28.

Thapliyal, Uma Prasad, *Warfare in Ancient India: Organizational and Operational Dimensions* (New Delhi: Manohar, 2010).

Trautmann, Thomas R., 'Introduction', in Thomas R. Trautmann (ed.), *The Aryan Debate* (New Delhi: Oxford University Press, 2005), pp. xiii–xliii.

Verma, Amrit, *Forts of India* (1985, reprint, New Delhi: Publications Division, 2003).

Watkins, Trevor, 'The Beginnings of Warfare', in John Hackett (ed.), *Warfare in the Ancient World* (London: Sidgwick & Jackson, 1989), pp. 15–35.

Wees, Hans van, '"Those Who Sail Are to Receive a Wage": Naval Warfare and Finance in Archaic Eretria', in Garrett G. Fagan and Matthew Trundle (eds.), *New Perspectives on Ancient Warfare* (Leiden/Boston: Brill, 2010), pp. 205–26.

Wood, Michael, *In Search of the Trojan War* (1985, reprint, Berkeley/Los Angeles: University of California Press, 1996).

3

MILITARY DIVERGENCE IN EURASIA

Classical Greece against Persia

Introduction

The Iron Age in Eurasia witnessed the origins of a military divergence between Asia and Europe. This is not to suggest a simple compartmentalised bipolar model but rather that different parts of Asia gave birth to varied military systems. However, they differed from the military systems that came into existence in Europe. While Greece witnessed the genesis of a heavy infantry culture, West Asia saw the rise of a cavalry-centric military system. South Asia in contrast gave birth to composite armies comprising chariots, infantry and elephants. In terms of naval warfare, both Greece and West Asia displayed a convergence. As we will see, the navies of Greece and Persia were all part of a Mediterranean continuum. The varied military divergence collapsed only with the end of Classical Age in Greece *c.* 340 BCE. Let us see how the process of military divergence started unfolding after the end of Bronze Age in Eurasia. This chapter focuses especially on the diverging Greek and Persian ways of warfare.

Greece

The Aegean Bronze Age civilisations collapsed around 1200 BCE. The Iron Age in Greece started *c.* 1000 BCE. In North Europe, bronze swords and lances with bronze heads disappeared. They were replaced by cheaper lances and spears with iron heads. This transition was completed between 700 and 500 BCE. It paved the way for the rise of the Greek phalanx (Randsborg 1999: 199–202). Poor pasturage in Greece made it unsuitable for large-scale horse breeding. Hence, cavalry did not play a premier role in Greek warfare. Further, the uneven terrain of Greece, with mountain ranges and lack of wide fertile plains, was not conducive to chariot or cavalry warfare. According to Victor Davis Hanson, between

DOI: 10.4324/9780429284069-4

the seventh and sixth centuries BCE most of the fighting among the Greek states was done by heavy infantry composed of farmers wearing bronze armour and equipped with thrusting spears (Hanson 1999b: 203), typically known as hoplite warfare.

Hoplite warfare reached its zenith during the Classical Period in Greece (480–323 BCE). However, historians remain divided over its origins and impact on Greek society. The term Dark Age, with regard to Greece and post-Roman medieval Western Europe, is now under attack. Traditionally, the Dark Age in Greece referred to the period from the eleventh to eighth century BCE. Now historians prefer the term Early Iron Age. This period witnessed the introduction of iron weapons. During the eighth century BCE, Ionian Greeks functioned as piratical raiders and pillaged and plundered the coastal cities in the Levant (Hale 2013: 181). This was a continuation of the Mycenaean attacks on Troy and other cities of the Hittite Empire during the Bronze Age.

As regards the origins of hoplite warfare, historians can be divided into two camps: 'gradualists' and 'radicals'. The gradualists claim that the hoplites emerged slowly. Fluid, open warfare slowly gave rise to close-order hoplite combat around 650 BCE. H. van Wees writes that infantry combat remained fluid and individualistic throughout most of the seventh and sixth centuries BCE. The infantry, armed with large round shields, hoplite arms and armour, did not fight as a densely packed mass of uniformly equipped soldiers. Rather, combat was organised around the ties of kinship and dependence. Relatives and retainers who may or may not be armed as hoplites surrounded leading men. The critical change occurred with the rise of poleis (city-states), especially when the latter acquired power to override communal ethos and forced the citizens to take their place in uniformly armed infantry formations regardless of who they were stationed next to (Rosenstein 2010: 292).

In contrast, the radicals (whose most ardent representative is Hanson) claim that hoplite warfare emerged suddenly and it had a revolutionary impact on the Greek poleis (Kagan and Viggiano 2013a: xi–xxi). Gregory F. Viggiano asserts that the introduction of the hoplon (double grip shield) was crucial in the emergence of the hoplite phalanx (Viggiano 2013: 115). The shield was concave and circular and on average 3 feet across. The core of the shield was made of either willow or poplar wood. Poplar was lighter in weight compared to oak or ash. And willow wood had the tendency to contract somewhat in the event of penetration, helping to minimise damage from an edged weapon during combat. Often the shield was covered with a thin bronze sheathing. The bronze sheathing offered extra protection against the wood splintering during contact with weapons and helped to dissipate the force of the enemy blows over a large area. The inner surface of the shield was padded with a glued layer of leather, which reduced the wear and tear on the wooden core and also prevented the wood from snagging or pinching the bearer's arm (Schwartz 2013: 157–9).

Hanson points out the linkage between the rise of the polis and the emergence of hoplite warfare. Farming and fighting constituted the basis of Classical Greek civilisation. The broadening of the property-owning class positioned between the landless poor and the wealthy landowners who rode on horses aided the emergence of the hoplite warriors (Hanson 2007: 12; Hanson 2013: 258). Greece experienced a steady population growth of about 0.25 per cent per annum from the tenth to the fourth centuries BCE. The growth of population between the eighth and fifth centuries BCE was about 1 per cent per annum. Thus, in most of the poleis, there were more farmers than aristocrats who were able to arm themselves. To an extent, economic growth aided the emergence of the hoplites (Viggiano 2013: 123). The hoplites were small farmers who were citizens of the polis, with each farmer owning about 10 acres of land (Wees 2013: 224). Therefore, the hoplites (except in the case of Sparta) were a citizen militia and not a standing regular army. The abolition of slavery also played a part in the rise of the hoplites. In the seventh century BCE, Attica had an aristocratic government. Solon, appointed as archon in 594 BCE, granted freedom to all men enslaved for debt. He declared it illegal to accept the person of a debtor as security for a loan (Adcock 1960: 37). Coinage came into the Greek cities in the mid-sixth century BCE, and the Greeks adopted the use of coins from Lydia (Asia Minor) (Trundle 2010: 227). There were roughly 700 poleis in the Greek world, and the size of their citizen populations (those males who enjoyed full political rights; women, slaves and metics lacked the right to vote) varied from several hundred in the smaller ones to 30,000 in the biggest polis, Athens (Thommen 2012: 21).

What about the actual dynamics of a hoplite battle? The first three ranks of the eight rows of the classical phalanx (linear formation) reached the enemy in the first assault with their spears. When one side's phalanx broke due to the thrusting by spears, hand to hand combat occurred with swords and butt spikes (Hanson 1999a: 204). Historian Peter Krentz believes that opposing groups of hoplites did not actually crash into each other. Some men ran into each other and some literally shoved the enemy hoplites. However, the bulk of the hoplites broke ranks just before collision with the enemy force. And then combat became the sum of numerous individual combats (Krentz 2013: 147–8). Krentz, unlike Hanson, asserts that the *osthismos* (push of the two opposing hoplite formations against each other) was not literal (Krentz 2010a: xiii).

Hanson challenges this assertion and argues that hoplite battles involved collision of opposite bodies of massed infantry deployed in a phalanx. The hoplites were grouped in a close formation, with 3 feet of space per man, and they advanced to contact and push forward against the enemy files (Krentz 2010a: xiii; Hanson 2013: 263). Since historians do not possess a time machine and the true nature of hoplite combat can never be discerned from the existing archaeological and literary sources, we will probably never know what really happened during combat among the hoplites. Nevertheless, we can safely say that in general hoplites formed in close order advanced at a snail's pace towards the enemy. The view put forward by some scholars that the hoplites ran against the enemy seems untenable to me. The case of the hoplites during the Battle of Marathon may be an exception. Moreover, after the collision of opposing bodies of hoplite infantry, individual close-quarter combat broke out. Before starting a battle, the Greek soldiers followed certain rituals that strengthened their group solidarity, reinforced their morale and prepared them to encounter the inhuman 'face of battle' during a hoplite encounter (Parker 2004: 142–3). Xenophon the Athenian general (b. 431–d. 354 BCE) tells us that the Spartan army encouraged homosexual relationships in order to strengthen the male bonding within the units (Fox 2004a: 193).

Can one argue that the development of disciplined heavy infantry was purely a Western/Classical Greek construct? As we have seen in the previous chapter, the emergence of disciplined and drilled heavy infantry could be traced back to ancient Mesopotamia. The Royal Standard of Ur and the Stele of the Vultures from Girsu (both dating between 2500 and 2460 BCE) depict disciplined infantry armed with spears. The infantry soldiers were deployed in ranks/files, and they were uniformly equipped with helmets, cloaks, stabbing spears and battle axes. On the Stele of the Vultures, the infantry was organised in proto-phalanx formation with spears projecting forward from a wall of shields (Fagan 2010: 86). The question is why the succeeding Assyrian and Achaemenid polities did not pursue this line of development. The answer is ecology, availability of military assets and architecture of the opposing ground forces. In the wide-open ground of Asia Minor and Mesopotamia, and also due to the vast distances that had to be traversed by the armies, cavalry was more suitable than slow-marching infantry. Besides physical geography, easy availability of horses in the Middle East also functioned as a crucial integer behind the military divergence. Lastly, the force structure of the polities in the Middle East was also influenced by the nature of armies threatening them. Unlike the Greek city-states, the Middle Eastern polities were frequently threatened by the mounted steppe nomadic archers. And for opposing them, cavalry was more suited than heavy infantry geared for 'ritualistic' hoplite battles. A completely different polity and military system in contrast to the city-states of Greece emerged in the Middle East under Persian Achaemenid Empire. However, the Persians built upon the political and military systems of their predecessor, the Assyrians.

Neo-Assyrian Empire: 934–612 BCE

The Assyrians were a Semitic-speaking people of Northern Mesopotamia (Iraq). The Neo-Assyrian Empire succeeded the Old Assyrian Empire (2025–1378 BCE). The Neo-Assyrian Empire lasted roughly from 930 to 612 BCE. During the ninth century BCE, the Assyrian army comprised chariots with a crew of three and cavalry armed with lances, spears and bows. Besides the driver, one charioteer was equipped with a bow and other two with spears. The fighting men in the chariots wore mail coats. In battle, the Assyrians used mixed units of cavalry and infantry. The spear-equipped infantry's duty was to protect the chariot from infiltrating enemy foot soldiers. The siege train included four to six wheeled rams covered with metals or hides and scaling ladders. For river crossing, the soldiers used boats, rafts and inflated animal skins (Herzog and Gichon 1997: 159, 195; Melville 2016: 22). Further, the empire made provisions for supplying the pack animals and the soldiers with daily rations. Grain and fodder were stored in the fortified places past which the army marched. The soldiers were provided with barley and bread. The logistical apparatus was capable of supplying a force of 35,000 men and the animals accompanying them (Fales 1990: 23–34). This raised the combat effectiveness of the Assyrian army, as the troops and the war animals marched to the battlefield with their bellies full and did not have to spend time and energy on foraging duties. Soldiers dispersed for gathering food and wood for themselves and the animals accompanying the army were vulnerable to sudden hostile attacks. Further, foraging was bad for military discipline. In addition, the elaborate logistical infrastructure also increased the strategic reach of the Assyrian army. The Greeks could match it only much later under Alexander.

The replacement of the chariot with cavalry was not completed during the ninth century BCE. This was because the Assyrian cavalrymen had problems controlling the horses without stirrups in the battlefield. So the cavalrymen rode in pairs: one holding the reins of the horse and the other using the composite bow from horseback. The composite bow was adopted from the nomads north of Caucasus. It was definitely an unwieldy combination. And the paradigm of cavalry war was similar to that of chariot war to an extent: the driver held the reins of the chariot while the other charioteer drew the bow using the chariot as a mobile platform. Cavalry was cheaper than the chariot and could be used more effectively in rough terrain. So cavalry was the weapon of the future. Historian Robin Archer writes that the emergence of new hostile social groups forced the Assyrian Empire to restructure its army around cavalry, resulting in marginalisation of the chariots. Around the start of the first millennium BCE, several communities like the Medes and Manneans rose in the Zagros and Taurus Mountains. Chariots were useless for combat in these regions. Brittle construction of the chariot and lack of suspension meant that at high speed on uneven ground, a chariot would either bounce uncontrollably, which would damage the wheels, or the machine would overturn. So the Assyrians were forced to develop cavalry from the reign of Shalmaneser III (r. 858–824 BCE) (Archer 2010: 71; Fagan 2010: 90).

Why the Assyrians did not develop light infantry equipped with composite bows and javelins that would have been more effective than cavalry for fighting in the rough terrain? It can be argued that, for fighting battles and conducting raids on the level grounds of Mesopotamia and Syria (core strategic areas for the Assyrian Empire), cavalry was more effective than chariots or slow-moving infantry. By 700 BCE, the Assyrians had been able to master the horse, and the single horseman was armed with either a spear or bow or both (Dawson 2001: 176, 189). The chariot by then was playing a secondary role. Horses were initially imported in Assyria from the Hittites in Anatolia between 1500 and 1200 BCE. By the seventh century BCE, good quality horses from the Taurus mountains were available in Assyria (Gladitz 1997: 13–35). The availability of good breeds and also the ecological conditions favouring horse breeding were crucial elements in the emergence of cavalry as the shock arm in the Neo-Assyrian army and later the Achaemenid Persian army.

The Neo-Assyrian soldiers used light round shields that were made of wicker and covered with leather. Each shield had a single grip in the middle and could be carried by a strap round the neck. This shield emerged in the Late Bronze Age in Anatolia and Assyria. The Sea Peoples in Late Bronze Age Greece also used this type of shield. The archers comprised the core of the Assyrian infantry, with spearmen playing a second fiddle (Raaflaub 2013: 98–9). This was probably because in the wide plains of West Asia, archers with a longer reach than the spearmen were militarily more effective. Since the Assyrian archers had to be mobile to change the angle of their shots in relation to the advancing hostile cavalry and the archers were using a long-distance weapon (bow and arrow) and not gearing up for close-quarter combat with the enemy, they did not accept heavy bronze armour and heavy shields like the hoplites. A few centuries later, the Persian Achaemenid light infantry, following the same logic, adopted archery and wicker shield. Under Sargon II (r. 721–705 BCE), Assyrian heavy infantry wore mail shirts and conical helmets (Melville 2016: 17). Ionian Greek soldiers served as mercenaries in the Assyrian army during the first half of the eighth century BCE (Hale 2013: 180). The Ionian military mercenaries probably functioned as a medium for transfer of infantry techniques between the Middle East and Greece.

The principal objectives of Assyrian imperialism were to acquire metal, wood and stone which the core/heartland lacked and to prevent the neighbouring polities from gaining power, which would threaten the Assyrian security paradigm. Before Tiglath-pileser III (r. 745–727 BCE), the objectives of Assyrian campaigns were to launch annual raids for collecting plunder and to force the targeted polities to pay tribute. Under Tiglath-pileser III, the Assyrian king's personal guard was transformed into a standing army. It was then that the nature of Assyrian imperialism changed. It was the standing army under Sargon II which made possible annexation of neighbouring polities and the emergence of a territorial empire with a provincial administrative apparatus in contrast to the hegemonic empire, in which the Assyrian king held sway over semi-independent tribute-paying clients (Melville 2016: 23–7). The attack by the semi-nomadic Medes in 614 BCE ended the Neo-Assyrian Empire. The Medes won due to their superior horses and better horsemanship (Melville 2011: 13–33). However, the Medes were soon overthrown by the Persians, who founded an empire bigger than the Assyrians.

Persian Achaemenid Empire: 550–330 BCE

> I am Darius the Great King, King of Kings, King of countries containing all kinds of men, King in this great earth far and wide.
>
> – Inscription at the tombs of Naqs-i-Rustam (quoted in Cawkwell 2006: 61)

The Achaemenid Persian Empire lasted from *c.* 550 to 330 BCE. This empire was much larger than the Assyrian Empire and under Darius I (r. 522–486 BCE) extended from the Indus River in the east to the Danube River in the west. This empire encompassed about 2 million sq mi, and its core was South-Western Iran (ancient Parsa) (Waters 2014: 5–6, 20). At its peak, the Achaemenid Empire ruled over 70 million people (Krentz 2010a: xii). The rise of Persia started with Cyrus (r. 557–530 BCE), who established the Achaemenid dynasty. He, along with Cambyses (r. 530/529–522 BCE) and Darius I, established the Persian 'World Empire', which, like the later Ottoman Empire, included parts of the three continents: Asia, Europe and Africa.

Coinage was introduced in the Persian Empire when Darius started the circulation of darics. The Greek world used talents, minae and drachmas. One talent was equivalent to 60 minae, and one mina was equal to 100 drachms (Hill 1960: 129, 131). Under Darius, the empire was divided into 20 *satrapies* (provinces). The twentieth *satrapy* was India, which actually included the Indus Basin. Each *satrapy* was under a *satrap* (governor). The Achaemenid Empire had two capitals: Susa and Persepolis. The Royal Road connected Sardis with Susa (Cawkwell 2006: 1, 6). The Achaemenids built upon the road system constructed by the Neo-Assyrians (Thommen 2012: 71). Greece had nothing comparable to the road system of the large agrarian bureaucratic empires of Asia.

The Achaemenid Persian army was formed in divisions, each of which comprised 10,000 men. Each division included 10 battalions of 1,000 men each. Each battalion was divided into units of hundred men that were further subdivided into subunits comprising 10 men each. Even the Mongols continued this organisational framework based on the decimal system. During war, the army was drawn from all the communities in the empire. This had its advantages as well as disadvantages. The army was numerically strong but somewhat heterogeneous, which created problems in command and cohesion. The standing army comprised Medes and Persians. The kernel of the Achaemenid army was made up of cavalry and elite infantry. The 10,000 strong Immortals (elite royal infantry) comprised the best infantry unit in the Achaemenid military establishment. Most of the generals were Persians. They commanded the non-Persian contingents like Ethiopians, Assyrians and Indians, etc. (Gray and Cary 1960: 190). The Achaemenid army, like the Greek polis, followed a battle-centric strategy: aiming to wipe out the hostile force in a single or series of great battles.

Historian Pierre Briant asserts that the Achaemenid Empire possessed an indigenous hoplite phalanx. The Achaemenid hoplites were known as *kardakes*. They were recruited from the different subject communities of the empire and were officered by the Persians. They constituted the regular infantry and were armed uniformly and trained intensively (Briant 1999: 120–1). However, during confrontation with the Greeks, the *kardakes* did not prove to be as effective as the Greek hoplites. Even the Immortals did not wear bronze armour like the Greek hoplites. Achaemenid infantry was actually light infantry, noted for its swiftness and mobility. The Persian and Median infantry wore soft felt caps, multi-coloured tunics with sleeves and fish scale armour. They wore trousers and carried rectangular wicker or leather shields. They were equipped with short spears, single edged curved swords, quivers, bows and arrows made of reeds and daggers (Krentz 2010a: 23). The Achaemenid Empire hired mercenary Greek hoplites in order to make up for the lack of heavy infantry. In most of the battles fought by the Achaemenid army, infantry played second fiddle. Christopher Tuplin attempts to make a case that cavalry was not the predominant arm in the Achaemenid army. He bases his argument on the fact that in the public iconography of the Achaemenid Empire, depiction of cavalry did not occupy a prominent place. Tuplin then reinterprets the battles and campaigns fought by the Persians and claims that in all these encounters, cavalry did not play a premier role. This work tends to disagree with Tuplin's reinterpretation of the Persian battles and campaigns. However, Tuplin has a case when he writes that the Persian cavalry fought mainly with throwing and thrusting spears, and not as bowmen. Further, he continues that Persians did not develop cuirassiers, that is, heavily armoured horsemen (Tuplin 2010: 101–82).

The Achaemenids had access to Saka/Scythian and Bactrian cavalry who fought as mounted bowmen. Since the steppe nomads were born horsemen and adept in fighting with bows from horseback, there was no need to transform the Iranians into mounted bowmen. And sedentary people could never be as effective as steppe nomads in operating as mounted bowmen. The Achaemenids did develop quasi-cataphracts or cuirassiers later. So despite Tuplin and Briant's assertions, the Achaemenid Persian army in terms of force structure and doctrine remained a different organisation compared to the heavy infantry-centric Classical Greek military system.

Greco-Persian wars: 491–479 BCE

Most of the written sources for the Persian wars with Greece were generated by the Greeks. Our main source for the Persian invasion of Greece is the Ionian historian Herodotus's (b. 484–d. 425 BCE) *History*. About one-third of Herodotus's account is taken up by Xerxes's invasion of Greece during 480–479 BCE. Herodotus probably had access to the eyewitness accounts. Diodorus Siculus's (b. 90 BCE–d. 31 BCE) work was written much later.

The Persian conquest of Lydia, which was partly due to the gold of the Pactolus River which ran through this region, brought the Greeks and the Achaemenids face to face. Greece and Persia clashed over control of the Ionian cities (which were rich sources of revenue) along the coast of Anatolia, just as Mycenaeans and the Hittite Empire had clashed over the control of the Western Anatolian Coast, which resulted in the Trojan War. The Persians wanted to control the narrow Dardanelles Strait, which posed a serious threat to Athens's lifeline. The Dardanelles Strait was known as the Hellespont during antiquity. This narrow waterway separates North-Western Turkey in Asia from European

Turkey. Actually, the Strait of Dardanelles connects the Aegean Sea with the Sea of Marmara, and the latter is connected with the Black Sea through the Bosphorus. So the Dardanelles Strait, like the Malacca Strait, is an important geographical chokepoint with enormous strategic significance. The soil of Athens was not fertile enough to produce grain in large quantities. Athens grew wine and olives and imported grain. Athens imported wheat from the Black Sea region. Athenian ships carrying wine and oil during their onward voyage to the Black Sea region and loaded with wheat during their return voyage passed through the Strait of Dardanelles. Achaemenid Persia, by controlling the Dardanelles Strait, could actually throttle the Athenian supply line. By taxing the Athenian ships carrying grain, the Persians could influence fluctuations in the Athenian grain market and, at worse, by closing the Dardanelles Strait, could initiate a famine in Athens. When Cambyses annexed Egypt to the Achaemenid Empire, the lucrative Greek trade with the Nile Delta came under Persian control (Waters 2014: 30). Economic rivalry was the principal reason behind the Greco–Persian conflict. As early as 525 BCE, when the Persians defeated the Egyptians at the Battle of Pelusium, Greek mercenaries fought on the side of the Egyptians against the Achaemenids. Around 513 BCE, in order to prevent any further Greek intervention in North Africa, Persia and Carthage fixed a common boundary along the Libyan Coast. Greece was only one of the serious security threats faced by the Persians. Egypt was vital because it was the principal granary of the Achaemenid Empire. The prospect of rebellion in Egypt and nomadic incursions along the *satrapies* of Bactria and Sogdiana (the Achaemenid Empire's north-east frontier) also forced the Great Kings to give attention to these theatres (Ure 1960: 100, 112; Cawkwell 2006: 1, 171).

Since Athens had supported the Ionian rebellion (499–93 BCE) against the Achaemenids, Persia, after crushing the rebellion, wanted to punish her. At this stage, Darius I had no plan to conquer Greece. That would come later, with his successor Xerxes. It could be argued that the Achaemenid navy came into existence under Darius I. He took great interest in the sea. He rebuilt the canal of *Pharaoh* Necho II (r. 610–595 BCE) which connected the Nile and the Gulf of Suez and sent the mariner Scylax to explore the Indus River (Hirsch 1986: 224). Darius appointed Datis as commander of the expeditionary force designed to punish Athens. A fleet comprising 300 triremes was put at his disposal. Athens lacked the naval power to fight the approaching Achaemenid fleet. Further, Datis was accompanied by the ex-tyrant of Athens, Hippias, who assured him rather optimistically that once the Achaemenid force reached Athens, the enemies of Athenian 'democracy' would rally to him. Athens was one of the largest city-states of contemporary Greece (Krentz 2010a: 2–3). The total population of Attica was about 250,000 persons (Thommen 2012: 21). Still, economically, demographically and militarily, the Achaemenid Empire completely outclassed 'puny' Athens.

In 490 BCE, the Achaemenid expeditionary force landed at Marathon, and within 24 hours, the Athenian force came up to challenge them. The Athenian force, including their Plataean allies, comprised 15,000 soldiers, while the Achaemenid army numbered 30,000 men (Munro 1899: 188–9). At Marathon, two opposing military systems clashed. The Athenian army had 10 generals, each leading a particular tribe. Miltiades (b. 550–d. 489 BCE) was one of the generals who played a principal role during the Persian invasion. Each Athenian tribal contingent comprised fathers and sons, brothers and cousins, neighbours and drinking mates. Therefore, primary group solidarity was very strong in the Athenian force. And this was vital for maintaining cohesion during the intense close-quarter hoplite combat. Miltiades had served with the Achaemenids previously and knew about their tactical system. The Achaemenid army was deployed in several rows made up of archers, with cavalry as backup. While the archers would shoot arrows to disrupt the enemy formation, the cavalry would harass the hostile formation. And if the enemy broke, then the Persian cavalry would ride them down. Miltiades ordered the Athenians to rush the Persians. The Athenians jogged towards the Persian line and broke into them. It took about 10–12 minutes for the Athenians to reach the Persian line. And within that short time, Persian archers were not able to inflict significant casualties on the heavily armoured Greek infantry. In close-quarter combat, heavily armoured hoplites had an edge over the lightly equipped Persian infantry. Nevertheless, the Persian centre, comprised of crack troops, counterattacked, but the Greeks enveloped their two flanks. About 6,400 Persians died in this battle. In addition, the Athenians were able to capture seven Achaemenid ships before they could escape to the sea. The Athenian casualties came to only about 192 (Krentz 2010a: xiv, 1, 3). However, the Persian cavalry were able to reembark to their ships and sailed away.

Was it possible for a heavily equipped armoured hoplite to rush at the static Persian battle line? Krentz writes that a hoplite equipped with a helmet, cuirass, shin guards, shield, spear and sword carried a total weight of 18–22 kg in the seventh century BCE. During the Persian wars, the helmets and the shin guards became thinner and leather and linen corselets had mostly replaced the bronze plate cuirasses. So the total weight carried by a hoplite dropped to 14 kg. As a point of comparison in the late nineteenth century, a European soldier carried 28–31 kg. of load while going to battle (Krentz 2013: 135). However, it must be noted that on average the Classical Greeks were smaller in stature and bodyweight and less nourished as compared to nineteenth-century European males. The average height of a Greek male during the Classical Period was 5.5 feet, with a bodyweight varying from 65–67 kg. The average weight of a Western male at present is about 80 kg. This was because the Greeks were brought up mainly on a diet of cereals, pulses and vegetables (Schwartz 2013: 166). Krentz continues that at Marathon, the hoplites did not wear greaves and chest armour. They only carried shields and spears. So the total weight carried by a hoplite was only 13 kg. Hence, the hoplites were able to run or at least jog towards the Persian line, which was a short distance away. All the Athenians, including the hoplites, light armed infantry and dismounted horsemen (including citizens and slaves freed due to the emergency of the Persian invasion), charged with spears and swords towards the Persians (Krentz 2010b: 196–201). The Persian cavalry was away on a separate mission. It was trying to secure the road that ran from the plains of Marathon towards Athens. It returned to the battlefield too late, and it was too few in number to attack the flanks and rear of the Greek phalanx (Evans 1986–7: 105–6). If larger number of Persian cavalry had attacked earlier, then the battle might have turned in favour of the Achaemenids.

Darius died in the autumn of 486 BCE. Just after ascending the throne, his son and successor, Xerxes (r. 486–65 BCE), was preoccupied with the revolt in Egypt and Babylon. Only in 481 BCE was Xerxes ready to march against Greece. The size of Xerxes's invasion force continues to baffle modern historians. The lowest estimate is 65,000 and the highest estimate is 180,000 soldiers. F. Maurice concludes that the Great King marched with 150,500 combatants (including 10,500 cavalry) and 60,000 non-combatants supported by 15,000 camels for logistical purpose. In total, more than 35,000 animals marched with the invasion force. Xerxes spent the winter of 481–80 BCE at Sardis. On 25 March 480 BCE, the army started marching from Sardis. On 24 June, the lead elements reached Therma. A road was constructed in Thrace for Xerxes's army. On 1 August, the army stopped at Pieria to construct a road through Mount Olympus. On 13 August, the army halted opposite Thermopylae (Maurice 1930: 210–28, 233; Keaveney 2011: 41).

The threat of Xerxes's invasion resulted in most of the Greek city-states uniting to form the Hellenic League. The Hellenic League reacted to the Persian advance by sending a contingent of 10,000 hoplites (which included both Athenian and Spartan contingents) into Thessaly to occupy the passes at Tempe. Later, the Greeks decided to make a stand at Thermopylae, because this pass was narrower than that of Tempe. The Spartan commander Leonidas (b. 540–d. 480 BCE), with 1,000 Spartans and 3,000 soldiers of the Greek allies, decided to hold the Thermopylae Pass. Due to the narrowness of the pass, the Persians could not take advantage of their numerical superiority. Xerxes sent the Median contingent in a frontal attack against the Greeks at Thermopylae. Diodorus writes that the Greeks fought valiantly due to Leonidas's heroic leadership and because they were fighting for their freedom (Diodorus 2010: 12–16; Keaveney 2011: 54). Diodorus gives an account of the combat:

> Since the battle line was shoulder to shoulder, the fighting hand-to-hand, and the combatants in dense array. . . . But, the Greeks had the advantage . . . in the great size of their shields, and so the Medes were gradually forced back: large numbers of them were killed, and not a few wounded.
>
> *(quoted from Diodorus 2010: 16–17)*

Each of the Spartan hoplites equipped with an 8–9-foot-long spear was more than a match for the lightly equipped Persian infantry soldiers, who carried shorter and lighter spears (Lazenby 2007: 9).

However, Xerxes fed one contingent after another while tiring out the Greeks, who were continuously in combat. Western historians have made much of Leonidas. The truth is that he was a mediocre commander who had failed to conduct a thorough reconnaissance of the surrounding countryside and provide protection to his flanks and rear. In fact, Leonidas fought as a non-commissioned officer rather than a general. Leonidas was all brawn and no brain. Leonidas was outthought, out-manoeuvred, outfought and outgeneralled by Xerxes. Leonidas had failed to guard the route over Mount Oeta. The failure could be attributed to the Phocians, but Leonidas should be blamed for putting too much faith in them. That the Persian light infantry were able to traverse the narrow mountainous path in the night reflects their skill as mountain warriors. Just as dawn was breaking, the Persians came behind Leonidas's rear (Keaveney 2011: 59–61). Leonidas's force retreated to the hillock where they took their final position. The missile weapons of the Achaemenid light infantry finally wiped them out (Lazenby 1989: 62).

Historian Kurt Raaflaub claims that a military revolution occurred in Greece during the fifth century BCE. The principal feature of this revolution, according to him, was the emergence of large-scale naval combat. This in turn made warfare permanent, professional, brutal and total (Raaflaub 1999: 132–41). Besides Raaflaub, H.T. Wallinga also asserts that a naval revolution occurred in Greece just before Xerxes's invasion. For Wallinga, this revolution was an amalgam of managerial and technological inputs. The naval revolution that occurred at Corinth and especially at Athens had two features. First, the introduction of the trireme (a three-banked vessel) as the standard warship, which replaced the penteconter (two-banked ship) and merchant galley, constituted the principal feature of this revolution. Second, rather than privately owned ships that were commandeered during wartime, the poleis started building their own navies. Wallinga asserts that the principal causative factor behind the naval revolution in Greece was the rise of the Persian navy. Once Cambyses created a navy for conquering Egypt in 525 BCE, the Ionians and the citizens of mainland Greece felt threatened. Unlike the Assyrians, Cambyses in fact laid down the foundations of a state-owned navy. Therefore, the Greeks of Ionia and mainland Greece in turn started building public navies. The Persians responded by launching two pre-emptive invasions under Darius I and Xerxes before the Greek naval programme reached its logical culmination. And this in turn resulted in a naval race and war in the Eastern Mediterranean (Wallinga 2005: 7–33).

This book argues that, rather than any naval revolution, a slow and steady transformation in naval warfare occurred due to certain geopolitical factors. The rise of the Persian threat was only one of them. Athens developed a navy because sea communication was easier than land communication. Athens unlike Sparta had a export oriented maritime economy. If the Persian Empire was able to establish control over the Dardanelles Strait, then Athens's lifeline would have choked. Further, Athenian galleys were used to guard against the pirates. The metics (resident aliens) and slaves functioned as rowers in the Athenian ships.

The Spartiates (male citizens of Sparta) were full time hoplites and looked down upon rowing. The emancipated helots served as rowers (Bertosa 2005: 1–20). Therefore, we can say that war functioned as a vehicle for upward mobility of certain groups. Unlike Athens, Sparta did not become a premier sea power because it had its big heartland with hostile serfs in the Peloponnese. Continuous interstate war resulted in the technological development of the warship, which was reflected in the transition from penteconter to trireme.

Around 800 BCE, the Greeks fixed bronze rams to the prows of their ships. These prows were used for ramming enemy ships during battle. Around 700 BCE, Tyre, which was under Assyrian control, came up with the bireme. This vessel had 50 oars arranged in two superimposed banks. By late sixth century BCE, a specialised vessel for combat emerged, and this was known as the trireme. A trireme was a long and narrow warship some 115 feet in length with a crew of 200 persons, and of them, 170 were rowers, who were tightly packed in three banks. Besides oars, each trireme had two sails. These ships were highly manoeuvrable. The propulsive power of a trireme was three times that of a penteconter. During battle, the trireme could achieve a speed of 9.5 km per hour. A trireme required a depth of 12.7 feet of water for sailing (Raaflaub 1999: 141; Dawson 2001: 166–9; Wallinga 2005: 56, 64; Billows 2011: 281). As a point of comparison, the length of a standard battleship of the Axis powers during the Second World War came to about 820 feet, and it required a depth of minimum 35 feet of water for operating. After two or three months of sailing, the hulls of the wooden ships were fouled by marine growth (barnacles and sea grass). Roughness caused by such growth reduced speed. The hulls needed to be scraped (careened) regularly. Minor repairs, caulking and recoating with pitch (in order to make the hulls watertight) were required of the ships. This sort of refurbishing required the hull to be dried, and this required several days. Generally, the triremes were routinely hauled onto the beach overnight to prevent waterlogging. The trireme being narrow, there was no space for cooking, washing etc., by the crew. It was customary for the triremes to anchor next to the shore almost every night so that the crew could disembark to cook, eat and sleep (Harrison 1999: 168–71). Hence, triremes were fit for operations in an inland sea like Mediterranean with a lot of islands but not for sustained operations in the open sea. For operation, the triremes required friendly bases along the numerous islands of Aegean and along the coast of Asia Minor. This in turn explains the Greco-Persian rivalry for controlling the coastline of Asia Minor and also the numerous small islands scattered in the Aegean Sea. Thus, most of the trireme battles occurred near the coasts and in the environs of the islands. The technology and geography of naval warfare in the Mediterranean, as Chapter 7 shows, remained constant till the sixteenth century.

Xerxes had ordered ships to be built along the coastal regions of Egypt, Phoenicia, Cyprus and Anatolia. Xerxes built an armada of 1,200 warships. As a point of comparison, in 513 BCE, Darius I mobilised 600 ships for the Scythian campaign (Wallinga 2005: 14). The Achaemenid fleet under Darius I was manned mostly by the Phoenicians. However, Xerxes ordered his other subjects to make naval contributions. The Greek polities allied with Xerxes (like the cities along the coast of Anatolia, Rhodes etc.) supplied 320 ships. The Phoenicians supplied 300 vessels, Egypt 200 ships and Cyprus 150 ships. In addition, 850 horse transport ships also accompanied the warships (Munro 1960a: 229; Diodorus 2010: 11–14). This was probably the first time in history a polity was using horse transport ships in large numbers to carry a large cavalry force for power projection across the sea. The horse transports were essential because cavalry constituted the core of the Achaemenid army. Each horse transport had 60 rowers (Wallinga 2005: 40). In 492 BCE, a Persian fleet met disaster while trying to round the rough waters around Mount Athos in the Chalcidian Peninsula in North-Eastern Greece. To prevent the recurrence of such a tragedy, Xerxes ordered the building of a canal. The digging of the Athos Canal started in 483 BCE and was completed three years later (Wallinga 2005: 24–7). The building of this canal, along with the military roads constructed for the march of Xerxes's army was a great feat on the part of the Achaemenid military engineers. This was further proof that the Asian militaries were technologically on par with if not superior compared to the much-lauded *techne*-oriented Classical Greek militaries.

As Xerxes's fleet sailed around Attica, the non-Athenian Greeks were not eager to confront the numerically superior Achaemenid fleet. However, Themistocles (b. 524–d. 459 BCE), the principal politician of Athens, threatened that in such a case, the Athenians would make for Sicily and establish a new city there. This threat forced the Greeks to fight the Achaemenid fleet. The result was the Battle of Salamis in September 480 BCE (Keaveney 2011: 79).

The Narrows along the north-east side of Salamis Island have two parts. The first is the Salamis Strait that stretches east–west between the coast of Attica north-west of the Piraeus port in the east and the harbour of Salamis named Ayios Yeoryios at the western end. The second part continues north from Ayios Yeoryios and runs into Eleusis Bay. Xerxes deployed 1,000 ships for this battle. Of them, 207 were fast attack ships. Each of the ships in Xerxes's armada besides the rowers had 30 Persian, Medes and Saka mariners for boarding the hostile ships. The Persian aim was to attack the Greek fleet at its base near the city of Salamis. If the Persian ships could reach this harbour before the Greek fleet could deploy, then the latter would be fenced in the harbour of Salamis and the rest of the Greek ships could be pushed back to the coast of Salamis. If this plan succeeded, then the Persians planned to land troops in Salamis harbour to cause panic among the Greek ranks (How 1923: 130; Wallinga 2005: 55, 57–8).

In the Achaemenid order of battle, the Phoenicians formed the western wing and the Ionians constituted the eastern wing, followed by the Cypriots, Cilicians, Pamphylians and Lycians. The Greek ships were formed in two lines, with the vessels in the second line covering the gaps between the ships in the first line (Keaveney 2011: 81, 84). In the confined space of Salamis, neither the numerical superiority of the Achaemenid navy nor superior Phoenician seamanship could be used against the Greek ships. Due to lack of sea space, the Persian first line comprising 207 ships clashed with 375 Greek triremes. The Persian second line of ships, in their haste to support their first line, caused confusion and mayhem among the Persian ships in the narrow strait (Wallinga 2005: 60). The Phoenician tactics in general was to outrow the Greek ships and then turn and board the hostile ships. But implementation of this tactic required wide space. For the Greek navy, with slower ships and suffering from numerical disadvantage, the best option was to fight in the confined area around Salamis rather than in the open sea. In the narrow confines of Salamis, the Greek tactic of ramming the enemy ships proved effective (Cawkwell 2006: 96, 109). Again, the Greek mariners, being local, were accustomed to the varying depth, reefs and shoreline of the Salamis Channel, about which the crews of the Achaemenid navy were ignorant. The Achaemenid naval crew fought valiantly, but they soon lost their formation. Once their formation became chaotic, the Persian fleet suffered huge losses. The Athenian and Aeginetan naval contingents were mainly responsible for destroying large numbers of Achaemenid ships. Artemisia, the Queen of Halicarnassus, Xerxes's mistress and advisor, displayed great vigour in this battle. When her ship was pursued by a Greek ship, she rammed a Persian ship and then escaped (Herodotus 1972: 552). Xerxes, who was witnessing the battle, according to Herodotus, remarked: 'My men have turned into women, my women into men' (quoted from Herodotus 1972: 553). Finally, the veteran Egyptian contingent was absent. Xerxes had sent it to blockade the Kynosura exit. The Egyptian ships were blockading the strait between Kynosura, Keos and Megara. The Persians feared that the Greek fleet might escape through this region (Wallinga 2005: 71). Had the Egyptian ships been used to press the attack against the Greek fleet at Salamis, then the end result might have been different.

Instead of attempting to destroy the Greek fleet in the narrow Strait of Salamis, if Xerxes had just blockaded the former, history might have taken a different turn. The Persian army, with its base at Attica, could have sat down for a long blockade. And the Persian navy could have been resupplied and repaired at Piraeus harbour in Attica. The Greeks, with their base at the small unproductive island of Salamis, could not have sustained a long blockade. However, Xerxes feared that the Greek fleet might attempt to escape westward of Salamis (Wallinga 2005: 69). So the overconfident king decided to force the issue at Salamis.

After the destruction of the bulk of the Achaemenid fleet at Salamis, the Persians could not maintain their huge army in mainland Greece. Besides problems of supplying the large number of soldiers, the security of the coastline of Asia Minor was threatened. The Greek fleet was free to conduct amphibious expeditions along the shoreline of Asia Minor and was in a position to cut the Achaemenid invasion force's line of communications across the Hellespont. Further, Xerxes had staked his political fortunes on the successful invasion of Greece. The defeat at Salamis sent shock waves across the Achaemenid Empire and encouraged rebellion against the Great King. So Xerxes, with a part of the force, was required to be present in Asia. Then another naval contingent was necessary to maintain the army's line of communications across the Dardanelles Strait. Still, the Achaemenid Empire was able to maintain a sizeable ground force in mainland Greece. Xerxes withdrew with a part of his army through Boeotia, following the same route through which he had earlier advanced (Herodotus 1972: 562). Mardonius was left with a rump army in mainland Greece.

According to J.A.R. Munro, even after Xerxes's withdrawal, Mardonius could concentrate 120,000 men (including non-combatants and the 20,000 Greeks who were collaborating with Persia). Of the non-Greek 100,000 men, Mardonius himself commanded a contingent of 60,000 men, and the remaining 40,000 were drawn from the 60,000 strong contingent of Artabazus at Potidaea, guarding the line of communications with Asia. Perhaps Munro exaggerates the size of the Persian army in Greece after the withdrawal of Xerxes. The Spartan army defended the wall at the narrow neck of the Isthmus. It could not be carried by frontal attack, and the Persians at this stage (especially after Salamis) lacked the naval power to sail around the wall to outflank its defence (Munro 1960b: 317). Overall, Mardonius enjoyed cavalry superiority over the heavy infantry-centric Greek force.

In June 479 BCE, Mardonius, with his cavalry, occupied Attica and burnt Athens. Then he withdrew to Boeotia. where his cavalry could be deployed effectively in the open terrain and, also, friendly Thebes provided supplies for his force (Cawkwell 2006: 110). While Aristides commanded the Athenian army, Xanthippus was in charge of the Athenian fleet (Keaveney 2011: 91). At Plataea (August 479 BCE), Mardonius's force comprised Persians, Medes, Bactrians, Indians and Sakas (Briant 1999: 118). Mardonius ordered Masistus to launch a spoiling attack against the Greeks with a cavalry detachment. After Masistus fell, Mardonius moved his cavalry force near the Asopus River. The Persians were stationed opposite to the Spartans. On the right of the Persians were the Medians, and then came the Bactrians and the Sakas (Herodotus 1972: 467, 589). At Plataea, Mardonius was able to disorganise the Greek heavy infantry by harassing them from a distance. Had he continued this policy for a longer period, the Greeks, due to lack of supply and utter frustration, would have retreated. However, Mardonius could not resist the temptation when he saw that the Greek army was retreating and its right wing was in disarray. The Achaemenid cavalry charge was contained by 5,000 Spartan hoplites. Mardonius's death settled the issue in favour of the Greeks (Lazenby 1989: 62–3).

The Persian cavalry failed to launch a coordinated charge with their infantry. Further, if the cavalrymen and their mounts had been covered with armour, then the cavalry charge would have been more effective. Again, the terrain of Plataea was not suitable for fast moving cavalry operation. The flanks of the Greek heavy infantry were partly protected from Persian cavalry by rough ground. The Greek army was innovating due to pressure of war. Instead of depending merely on heavy infantry, the Greeks also deployed light infantry equipped with javelins and bows. Further, the Persian light infantry also failed to coordinate their movement with their cavalry. When the Persian cavalry was slowed down in the pebble-strewn rocky bank of the Asopus River, the Greek light infantry inflicted casualties on them. Persian light infantry failed to support Persian cavalry at this critical juncture. Lastly, there was no chain of command in the Persian army. When Mardonius died, the battle was by no means lost for the Persians. Under a skilful commander, their infantry and cavalry could have retreated to fight another day somewhere in Central Greece. However, the death of Mardonius resulted in breakdown of command and morale in the Persian army, and there was no designated commander to replace him (Jones 2001: 17–19).

At Plataea, the Achaemenid field army in mainland Greece was destroyed. Simultaneously, the remnants of the Achaemenid fleet (what was left of it after Salamis) was annihilated at the Battle of Mycale (27 August 479 BCE). The Greek fleet comprised 110 ships under the Spartan King Leotychides (r. 491–76 BCE). Remembering Salamis, the Persian commander Tigranes was not in a mood to face the Greek fleet at sea again. He sent the best ships of the Achaemenid fleet (the Phoenician ships) to their home ports and disembarked the crews in the promontory of Mycale opposite Samos. Tigranes fortified his encampment and waited for the Greeks to attack. Leotychides encouraged the Ionians to rebel against their Persian masters. Tigranes in response disarmed the Samians. Leotychides's men attacked from the east, and the Persian stockades were breached. The Samians inside the Persian camp rose up in rebellion. Tigranes was killed, but most of the Achaemenid soldiers were able to escape to Sardis (de Souza 2003a: 70–4). Tigranes's passive defensive strategy had failed. After the destruction of the Achaemenid fleet, the Greco-Persian War came to an end. Soon, Greece was embroiled in a series of civil wars.

The Peloponnesian War: 431–404 BCE

The First Peloponnesian War (461 to 446 BCE) was actually a confrontation between Athens and an alliance of Corinth and Thebes. This war resulted in increase of Athenian power that made Sparta uncomfortable. Here, we are concerned with the Second Peloponnesian War. The Second Peloponnesian War was actually a Greek civil war between the Athenian Confederacy and Sparta and its allies, with Persia standing on the margin but at times playing a crucial role. Both the Greek parties' military systems were more or less similar. Victory for Sparta was decided by extraneous financial support (the Achaemenid Empire) and the faulty strategy of Athens (the Sicilian Expedition). For this war, our main source is Thucydides (c. 460–395 BCE). He was an Athenian admiral who fought at Amphipolis and was exiled due to underperformance in 423 BCE. Sparta was afraid of the rising Athenian maritime imperialism. Sparta feared that Athens might attempt to spread her influence in Southern Greece. So Sparta launched a pre-emptive strike in the spring of 431 BCE (Hanson 2006: 7–15).

The rise of Athenian power could be traced back to the formation of the Delian League, which later was transformed into an Athenian maritime empire. In 478 BCE, the Delian League was formed to carry on war (especially raiding the coastline of the Persian Empire) against the Great King. The Ionians and the Athenians joined forces together to form this league to check any possible

Achaemenid counterattack in the near future. Sparta was a land power, and it was suffering from acute manpower shortage. So it was neither interested nor in a position to carry out offensive operations in the Aegean. However, Athens, with its rising manpower and interest in controlling the trade route to the Black Sea through the Dardanelles Strait, had economic and strategic interest in destroying the remnants of Achaemenid power along the eastern seaboard of the Aegean. During 454/453 BCE, the treasury of the Delian League was transferred from Delos to Athens and the Delian League became the Athenian maritime empire. In 449 BCE, the Achaemenid Empire and the Athenian Empire made peace with the signing of the Peace of Callias treaty. This treaty slightly favoured the Athenians vis-à-vis the Persians. All the Greek cities of Asia Minor were declared autonomous, and the Persian force was to stay away at a distance of three days' journey from the coast, and no Persian ship was to sail near the Strait of Dardanelles. In return, Athens promised not to attack any dominion of the Achaemenid Empire (Cawkwell 2006: 127; Kagan 2010: 35, 41–2; Keaveney 2011: 114–15). This treaty showed that even at the height of Athenian naval supremacy, Persian power could be reduced only along the coastline of Asia Minor. The core inland areas of the vast Achaemenid Empire remained outside the range of power projection capabilities of the Athenian navy. In other words, the Achaemenid Empire could not be brought to its knees merely with sea power.

Sparta was ruled by two kings, whose power was tempered by five ephors who were elected. While Athens was a sea power, Sparta was a land power. The number of Spartiates declined from 8,000 adult males in 480 BCE to 3,000 in 431 BCE and to 1,100 in 371 BCE. The helot population varied between 170,000 and 224,000, respectively. In the Spartan army, seven helots supported one Spartiate (male citizen of Sparta who functioned as a hoplite). The latter functioned as lightly armed infantry during war. The bulk of the Spartan army was tied up in policing the helots engaged in agricultural tasks in Messenia. At the beginning of the war, landlocked Sparta, with a primitive barter economy, had 10,000 heavy infantry, and Athens had 300 ships, a population of over 300,000 residents and a lot of coined money. Further, 200 foreign subject states paid annual tribute to Athens. When the war started, Athens had 6,000 talents of coined silver and 500 talents worth of precious metals. The annual revenue of the Athenian Empire varied from 600 to 1,000 talents (Talbert 1989: 23–4; Hanson 2006: 6, 15–16; Lazenby 2007: 14; Trundle 2010: 240).

The principal warship of Athens was the trireme. The maximum speed was nine knots (16.5 km per hour). As a point of comparison, a German battleship during World War II could sail at a speed of 54 km per hour. Ramming and boarding constituted the principal naval battle tactics. The Athenians mainly depended on ramming the hostile ships. The Athenian ships used speed and manoeuvrability in order to ram the enemy ships on their sides or sterns. This was achieved by rowing through the enemy line and by rowing round their flanks. This naval tactic would remain constant till the advent of naval guns in the sixteenth century. Thus, compared to Salamis, naval tactics had not altered much. The two biggest naval powers of Greece at that time were Athens and then Corcyra (an Athenian ally), with 120 warships. The financial reserve in Athens, 6,000 talents, could keep the 300 triremes operational for 20 months. Further, Athens had 29,000 hoplites (Lazenby 2007: 12–14).

Barley was Attica's main crop. And the principal export of Athens was olive oil and wine (Wees 2013: 230–1). Since only 22 per cent of the land in Greece was cultivable, destruction of cereals, vines and olive trees could cause disaster (Kagan and Viggiano 2013b: 8). Hanson persuasively argues that in the long run the effect of raiding on the agriculture of Attica was negligible (Hanson 1999a). However, in the short run, the effect was horrible. In the spring of 431 BCE, the Spartans marched into Attica assuming that a year or two of ravaging would bring Athens to its knees. The Spartans were the best hoplites in Greece. Unlike the hoplites of other poleis, the Spartan hoplites were not farmers but regular professional soldiers. This was possible because the helots tilled the land of Sparta, allowing the Spartiates time and opportunity to carry on military training throughout the year. So Pericles (b. 494–d. 429 BCE), the principal strategist of Athens, rightly decided not to face the Spartans in a land battle, something which the Spartans wanted. Pericles's strategy was based on the assumption that the Spartans, lacking siege materials, would not be able to capture the fortified city of Athens with its long walls and would retreat after devastating Attica. Moreover, in the long run, the lack of financial resources would prevent Sparta from maintaining its army in the field for a long period. The Spartan army ravaged the Athenian countryside but lacked the siege engines to demolish the walls of Athens. In terms of conducting siege warfare, Classical Greek armies were far behind the Assyrian army. Incapable of laying siege to Athens, Sparta established a landward blockade of Athens, and then disease played the crucial role in shaping the contours of war. The Periclean strategy of evacuating the Athenians from the countryside and concentrating them in the fortified city of Athens resulted in overcrowding and subsequently unhygienic conditions, which in turn facilitated the spread of the plague. Pericles himself died of plague in the city in 429 BCE. In 428 BCE, the Spartans invaded Attica again, for the third time since the beginning of the war. The Spartan army arrived when the grain was ripening in Attica. However, even after seven years of continual Spartan invasions of Attica and 80,000 Athenian deaths due to plague, there was no prospect of Spartan victory (Hanson 2006: 5, 11; Lazenby 2007: 49, 51). The plague was the unintended consequence of the Periclean defensive strategy, which in itself was a response to the superior Spartan land power. The plague, however, in the end weakened Athens and encouraged the Athenian politicians to pursue an offensive strategy by invading Sicily.

Imperial overstretch on the part of Athens led to strategic exhaustion, and this contributed to Athens's downfall in the end. Frustrated by regular Spartan raids into Attica, Athens decided to use its sea power to carry the war to Sparta indirectly by launching an amphibious expedition. Athens decided to invade Sicily in order to stop the export of grain from this island to the Peloponnese. Further, Athens feared that Syracuse, the principal power in Sicily, might aid Sparta

(Lazenby 2007: 57–8, 131–2). In 415 BCE, the Athenians sent an expeditionary force to Sicily. The Athenian navy, like Xerxes's navy earlier, used horse transport ships, and each of them was able to carry 30 horses (Evans 1986–87: 101). Divided command among the Athenians was one of the principal factors behind their defeat. Athens should not have opened a distant second front while Sparta remained in the game. The ships lost due to the failed Sicilian campaign could be replaced, but not the trained rowers, mariners and the heavy infantry. The defeat and destruction of the Athenian expeditionary force showed the Mediterranean world that the Athenian navy was not unbeatable. The Athenian defeat at Syracuse in 413 BCE was a sort of Athenian 'naval Stalingrad'. Psychologically and materially it was a turning point. Athens would launch attacks after 413 BCE, but it was not the mighty Athens of Pericles. Hanson compares the trauma suffered by Athens due to the failed Sicilian expedition (415–13 BCE) with that inflicted upon the Americans by the Civil War. Athens sent over 40,000 troops to Sicily, and all of them were lost. Some 600,000 Union and Confederate troops died during the American Civil War due to disease and combat, out of a population of 32,000,000. So 1 in 50 was lost. In two years, the fatalities in Sicily for Athens came to about 1 for every 25 people in the Athenian Empire (Hanson 2006: 10–11).

Sparta realised that the Athenian navy, already badly mauled due to the Sicilian misadventure, had to be destroyed in order to bring that city under its heel. However, maritime warfare was costly. The cost of paying a trireme's crew came to about 6,000 drachmae per month (Lazenby 2007: 3). Sparta lacked financial assets, and its maritime tradition was not as strong as that of the Athenians. So Sparta looked towards Persia to defeat their common enemy. Since 431 BCE, the Spartans had sent several embassies to the Achaemenid Court for a joint front against the Athenians (Eddy 1973: 256). In May 425 BCE, the Athenians established a base at Pylos which raised the spectre of helot rebellion in the Peloponnese. Instead of attacking Pylos, the Spartans sent 420 hoplites in the nearby island of Sphacteria. Their objective was that by garrisoning this island and deploying a fleet, they would be able to cut off the Athenian base at Pylos and starve the Athenian defenders to submission. However, the Athenian fleet returned, drove off the Spartan fleet and besieged Sphacteria. Some 70 Athenian ships were engaged in blockading Sphacteria. Thucydides writes: 'The Athenians always had two ships on patrol by day, sailing round the island in opposite directions, and at night the whole fleet was anchored all round it' (quoted from Thucydides 1988: 278). In the rugged uneven bushy terrain of Sphacteria, the Athenian light troops were able to conduct hit-and-run attacks and ambushes against the heavily armed clumsy Spartan hoplites. Some 128 Spartan hoplites died at the hands of Athenian light infantry. During August 425 BCE, the rest of the 292 Spartan hoplites in this small island, suffering from food shortages, surrendered (Talbert 1989: 24; Hanson 2006: 113–15).

In 424/3 BCE, Athens sent an embassy to Susa, and both parties agreed to renew the Peace of Callias. In 413 BCE, Tissapharnes was sent by the Great King to formulate and implement Achaemenid policy towards the warring Greeks. He was the *satrap* governing Lydia and Ionia from Sardis (Fox 2004a: 12). Tissapharnes's strategy was to support both the Athenians and the Spartans so that they would continue to fight till both of them cracked. In contrast, the Persian prince, Cyrus the Younger, when he came to Sardis, decided to support the Spartans to the hilt. Cyrus the Younger had his own hidden agenda: his objective was to use the Spartan mercenaries later in order to make a bid for the Achaemenid throne. With Persian finance, Sparta developed a navy. The non-Athenian rowers were attracted to the Spartan fleet due to the higher pay, which in turn was possible due to Persian largesse. At that critical juncture, Athens was suffering from the financial and military disaster of the failed Sicilian expedition. After the Athenian defeat at Aegospotami in 405 BCE, most of Athens's allies, except Samos, changed sides. In 404 BCE, Sparta ultimately won due to Persian financial support. And when Athens surrendered in that year, its democratic constitution was overthrown and an oligarchy came into power (de Souza 2003b: 85–91; Waters 2014: 14).

Despite being victorious at the end of the Second Peloponnesian War, Sparta could not maintain its hegemony in Greece due to its manpower shortage and the prospect of helot rebellion in Messenia and Laconia. The Spartiates constituted less than 2 per cent of the adult male population of Sparta (Lazenby 2007: 5; Strauss 2010: 188). The Second Peloponnesian War exhausted the Greek poleis, which in turn enabled the Achaemenid Empire to remain as the principal power in the Eastern Aegean. The Second Peloponnesian War showed the rising importance of light infantry (an important element of the Achaemenid military system) and the importance of mercenaries who were professionals and more combat effective than the part-time hoplite citizen militia of most of the Greek states.

Xenophon's ten thousand

Xenophon the Athenian, along with 11,200 Greek mercenaries, took service under Cyrus the Younger, when the latter hatched a plan to overthrow his brother the Great King Artaxerxes II, who had inherited the throne in 404 BCE. Besides Xenophon, there was an important Arcadian general named Sophaenetus with the expeditionary force. Various groups of Greek soldiers were under the command of different Greek generals, and all of them were paid by Cyrus the Younger. And the whole group collectively was called the Ten Thousand. Cyrus was brave and overambitious; he loved hunting and excelled in horsemanship, archery and javelin throwing (Braun 2004: 97, 107–8; Stylianou 2004: 69). The Greek mercenaries (hoplites and peltasts [light infantry] in the ratio of 5:1) set off from Western Asia on a campaign for their paymaster, the Persian prince Cyrus. Our main source for this historical event is Xenophon's *Anabasis* (March Upcountry), which was written in the 360s (Fox 2004a: 1; Whitby 2004: 217; Cawkwell 2006: 47–8; Waters 2014: 15). Xenophon's *Anabasis* is probably the first autobiographical account by a commander. The failed expedition of Xenophon shows the inconclusive clash between two diverging military traditions in West Asia.

Cyrus the Younger and the Great King's armies confronted each other at the battlefield of Cunaxa in 401 BCE. The Greek mercenaries numbered 10,400 hoplites and 2,500 peltasts. Cyrus, with 600 heavy armoured cavalry, was in the centre of the line. Artaxerxes II (r. 404–358 BCE), unlike Xerxes, possessed heavy cavalry to tackle the hoplites. The cavalrymen wore breastplates, armour to cover their thighs and helmets, and the horses' foreheads and breasts were covered with plates of armour. Hence, we see that the Persian cavalry had learnt their lesson from Plataea and had developed heavy armoured cavalry (quasi-cataphracts). The Great King deployed 200 scythe chariots in front, and 6,000 cavalry under Artagerses guarded the Great King. The scythe chariot was a new weapon introduced by the Persians in an attempt to dislocate the orderly advance of the Greek heavy infantry towards the Achaemenid battle line. This chariot had two long curved knives attached to the wheels on either side. The Persian objective was once the hoplites were disorganised by the charge of the scythe chariots, Persian cavalry and light infantry would complete their defeat. According to one author, the scythe chariot emerged due to the innovativeness of the Persian high command (Nefiodkin 2007: 493–502).

There is a possibility, we would argue, that the scythe chariots first emerged in Magadha (Bihar) in North India, and then this novel technology passed into the Achaemenid Empire through Punjab, which was under Persian occupation. Punjabi mercenaries who served in the Achaemenid army probably functioned as the medium of transfer of this piece of military technology. Cambyses first started the tradition of recruiting mercenaries from Punjab, and Xerxes continued this tradition (Srivastava 1985: 10).

Between 1200 and 400 BCE, the Aryans moved from Punjab into North India. With the introduction of iron technology, the Aryans were able to cut down the dense forest of the Ganga Valley and introduced agriculture. The monsoon-fed rivers deposited silt on the land, and the Aryan agriculturists were able to produce bumper crops. The pastoral nomadic society was transformed into an agrarian society, and territorial states emerged which were known as *mahajanapadas*. The *rajan* (tribal chief) became *raja* (king/ruler of a territorial unit). Instead of launching raids for capturing cattle (a method of warfare practised by the pastoral nomadic Aryan tribes), the *mahajanapadas* started fighting for land that was annexed permanently and utilised for agriculture, which was the principal source of wealth for these territorial units. North India during the sixth century BCE witnessed continuous warfare among the polities, a situation somewhat similar to the Warring States era (475–221 BCE) of ancient China. Continuous warfare backed by the agrarian wealth of the *mahajanapadas* resulted in innovations in the theory and practice of organised violence. Under Bimbisara (r. 544–492 BCE), Magadha became the principal power in North India. He introduced the first standing army in India. The *chaturangabala* (four-limbed) armies of the *mahajanapadas* comprised infantry, cavalry, chariots and elephants. By roughly 800 BCE, the Aryans had domesticated elephants and used them like main battle tanks in battles. Elephants were also used as command platforms for the generals. And the Aryans, aware of the importance of *asvas* (horses) in warfare, learnt to ride by 500 BCE. In fact, the Indian armies emerged as a well-rounded force with a logistical apparatus. The epics (the *Ramayana* and *Mahabharata*) tell us that the army was accompanied by doctors, nurses and labourers (Srivastava 1985: 9).

Under Bimbisara's successor and son Ajatasatru (r. 492–461/460 BCE), Magadha came up with *rathamusala* (scythe chariots) and stone-throwing machines. The scythe chariots were an advance upon the two-wheeled chariots introduced by the Aryans in around 1200 BCE, an issue discussed earlier. The stone-throwing machines (*sataghnis*, i.e. hundred killers) were used in siege warfare to batter down the defenders deployed along the walls of the forts. Use of these machines was necessary because the rulers of the *mahajanapadas* pursued the policy of constructing forts to protect the principal cities and the capitals of their kingdoms. Magadha had access to timber and iron deposits for constructing chariots and *sataghnis*. Further, a large number of elephants was available in the nearby forests. This gave Magadha battlefield supremacy over other *mahajanapadas* like Anga, Kosala etc. Later, Magadha evolved into the Nanda Empire (345–21 BCE) which ruled North India when Alexander invaded Punjab (Thapar 1990; Jha 2004: 81–95). The core of the Magadhan army comprised scythe chariots and elephants, and the Achaemenids somewhat ineptly tried to accommodate these two elements in their own military system.

Let us glance back to the battlefield of Cunaxa. Armoured cavalry in the Persian left wing was commanded by Tissapharnes. Next to them were Persian light infantry, and then came the Egyptian infantry, with wooden shields and archers. The Greeks sang the paean, and the phalanx started advancing. All the Greeks shouted their battle cry, '*Eleleu*', and they clashed their shields and spears together in order to frighten the enemy horses. Then the Achaemenids launched their scythe chariots. As the chariots came towards the Greek infantry, the latter opened ranks and shot at the chariot drivers. Cyrus was rightly afraid that due to quantitative superiority in cavalry on part of Artaxerxes, the latter's cavalrymen sooner or later would be able to overcome Cyrus' cavalry and then attack the flanks and rear of the Greek hoplites. A hoplite force could only attack forward but was extremely vulnerable to attack on its flanks and rear. So a desperate Cyrus decided to dislocate the Achaemenid high command, by charging towards Artaxerxes with a special unit comprising 600 cavalry. As Cyrus rushed towards the Great King, the latter's bodyguard cavalry made mincemeat of Cyrus's cavalry detachment (Xenophon 1972: 83–90). At Cunaxa, Cyrus personally wounded Artaxerxes, and Persian chariots had failed to throw the Greek heavy infantry formation into disarray (Whitby 2004: 225). It was a forerunner of the Battle of Gaugemela/Arbela (1 October 331 BCE). At Gaugemela, Darius III's scythe chariots failed to disorganise the Macedonian phalangites, and Alexander also led a cavalry detachment for attacking the Great King.

After the death of Cyrus the Younger, owing to his failed charge towards Artaxerxes, the Greek mercenaries lost their paymaster. They decided to retreat to the coast of Asia Minor. While retreating, the mercenaries engaged in widespread rape and plunder. In accordance with Greek military culture, raping and enslaving women of foreign territories was allowed (Fox 2004b: 185). While retreating, for seven days the 'Ten Thousand' met opposition from the Carduchis. They were a semi-autonomous tribe who inhabited the mountains of Northern Iraq (Armenia). The Carduchis were excellent archers who shot long arrows at the Greeks. Such arrows were able to pierce the Greek hoplites' shields and breastplates. The Greeks in turn used these arrows as javelins against the Carduchis (Stylianou 2004: 80). Xenophon ordered the Greek infantry to form a hollow square, and one general was in charge of the van while another guarded the rear (Xenophon 1972: 19). The Duke of Wellington would form such hollow squares of infantry to confront Marshal Ney's cavalry charges at the battlefield of Waterloo (18 June 1815). Xenophon commanded the rearguard, the most dangerous position, especially during retreat through enemy country. Chirisophus, the Spartan general, commanded the van. Logistical difficulties, more than the Persian attacks, proved to be the problem for the retreat of the Ten Thousand. Xenophon notes in his memoirs: 'without supplies neither a general nor a private is good for anything' (quoted from Xenophon 1972: 67). The Ten Thousand retreated through Kurdistan and Armenia into Asia Minor. In 400 BCE, the Greek mercenaries were able to reach the Hellespont (Xenophon 1972: 11). The overjoyed soldiers shouted 'the sea, the sea', and the Greek misadventure was finally over.

In 387/386 BCE, the Spartans signed a peace treaty with the Achaemenid Empire known as the King's Peace. Sparta, with its weak demographic and economic base already exhausted despite winning the Second Peloponnesian War, was attempting to establish its hegemony within Greece, and for that reason, it required benevolent Achaemenid neutrality. It proclaimed autonomy for all the Greeks except those cities in Asia Minor that were supposed to have been returned to Persia under the terms of the 411 BCE treaty. If anyone broke the terms of this common peace among the Greeks, emphasised one of the clauses of this treaty, then the Great King would make war on them (de Souza 2003b: 92). Thus, neither Athens nor Sparta but the Achaemenid emperor became the guarantor of peace among the Greeks. The situation in a way was similar to what it was during Darius I's time. Persia had made a comeback, and the mainland powers of Greece were exhausted. A new power was rising north of Greece: Macedonia, a theme that we will tackle in the next chapter.

Conclusion

Hanson asserts that civic spirit and the concept of equality gave greater cohesion to the Classical Greek soldiers as compared to the Persians during decisive battles (Hanson 2001: 1–59). Our analysis has shown that both in the naval and land battles, the Achaemenid soldiers fought valiantly, though their weapons, tactics and command failed them at times. We find divergence in the ground forces of Persia and Greece but convergence in naval matters. The Persians expanded upon the cavalry-centric force and logistical apparatus of the Neo-Assyrian military system. Although the Greek and Persian methods of conducting land warfare were different, there was a striking similarity in naval warfare. The Greek and the Persian fleets varied marginally only in matters of minor tactics. This was because both the Greeks and the Persians developed their mode of naval warfare from the Eastern Mediterranean maritime culture. However, it would be too much to argue that naval developments in the fifth century BCE resulted in a military/naval revolution. Further, we have seen that both the Achaemenid and the Vedic Indian systems of warfare were undergoing innovations with time. Therefore, the non-European military cultures were not static, as the proponents of the Western way of warfare would have us believe.

We can say that the hoplite warfare was not superior compared to the Persian cavalry-centric warfare. Xerxes's army had failed to conquer Greece, but it came all the way from Asia to threaten the Greek homeland. It was the Achaemenids who were projecting power into Greece. This was possible due to the superior logistical (which included economic and demographic resources) and engineering capabilities of the Achaemenids. After the defeat of Mardonius, the Athenian-controlled Delian League made pinprick attacks along the coastline of Asia Minor, which proved to be a nuisance for the Achaemenid Empire. The Athenian navy could not project power inland far away from the shoreline. And during the latter part of the Second Peloponnesian War, the Achaemenid Empire emerged as an important factor in shaping the trajectory of history of the Eastern Mediterranean. Xenophon's march showed that a disciplined Greek infantry with Persian cavalry support could penetrate into the core of the Achaemenid Empire. This was a forerunner of what Alexander would do in the near future. However, when the pretender Cyrus the Younger's Persian cavalry support was withdrawn, a stalemate followed. While Achaemenid cavalry and their light infantry were unable to overwhelm disciplined heavy Greek infantry, armoured heavy Greek infantry was also unable to destroy the Achaemenid cavalry. Ultimately, the Greek infantry had to withdraw due to logistical issues. As a result of battlefield interactions, both the Greek and Persian military systems were changing. The Second Peloponnesian War accelerated the rise of light infantry. The Persians came up with armoured cavalry, and the Greeks realised the importance of light infantry acting in conjunction with heavy infantry. This trend of light and heavy infantry cooperating in the battlefield against Persian cavalry had already started at Plataea. In fact, it was left to Philip and his son Alexander to synthesise the Greek and Persian methods of land combat and to create a hybrid military machine. This hybrid military machine proved superior to both the hoplites and Persian cavalry.

References

Adcock, F.E., 'The Reform of the Athenian State', in J.B. Bury, S.A. Cook and F.E. Adcock (eds.), *The Cambridge Ancient History*, vol. 4, *The Persian Empire and the West* (1926, reprint, London/New York: Cambridge University Press, 1960), pp. 26–58.

Archer, Robin, 'Chariotry to Cavalry: Developments in the Early First Millennium', in Garrett G. Fagan and Matthew Trundle (eds.), *New Perspectives on Ancient Warfare* (Leiden/Boston: Brill, 2010), pp. 57–79.

Bertosa, Brian, 'The Social Status and Ethnic Origin of the Rowers of Spartan Triremes', *War & Society*, vol. 23, no. 1 (May 2005), pp. 1–20.

Billows, Richard, *Marathon: How One Battle Changed Western Civilization* (2010, reprint, New York: Overlook Duckworth, 2011).

Braun, Thomas, 'Xenophon's Dangerous Liaisons', in Robin Lane Fox (ed.), *The Long March: Xenophon and the Ten Thousand* (New Haven/London: Yale University Press, 2004), pp. 97–130.

Briant, Pierre, 'The Achaemenid Empire', in Kurt Raaflaub and Nathan Rosenstein (eds.), *War and Society in the Ancient and Medieval Worlds: Asia, the Mediterranean, Europe, and Mesoamerica* (Washington, DC: Center for Hellenic Studies, distributed by Harvard University Press, 1999), pp. 105–28.

Cawkwell, George, *The Greek Wars: The Failure of Persia* (2005, reprint, Oxford: Oxford University Press, 2006).

Dawson, Doyne, *The First Armies* (London: Cassell, 2001).

de Souza, Philip, *The Greek and Persian Wars 499–386 BC* (London/New York: Routledge, 2003a).

de Souza, Philip, *The Peloponnesian War 431–404 BC* (2002, reprint, New York/London: Routledge, 2003b).

Diodorus Siculus, *The Persian Wars to the Fall of Athens, Books 11–14.34 (480–401 BC)*, translated with an Introduction and Notes by Peter Green (Austin, TX: University of Texas Press, 2010).

Eddy, Samuel K., 'The Cold War between Athens and Persia, ca 448–412 BC', *Classical Philology*, vol. 68, no. 4 (1973), pp. 241–58.

Evans, J.A.S., 'Cavalry about the Time of the Persian Wars: A Speculative Essay', *Classical Journal*, vol. 82, no. 2 (Dec. 1986–Jan. 1987), pp. 97–106.

Fagan, Garrett G., '"I Fell Upon Him Like a Furious Arrow": Toward a Reconstruction of the Assyrian Tactical System', in Garrett G. Fagan and Matthew Trundle (eds.), *New Perspectives on Ancient Warfare* (Leiden/Boston: Brill, 2010), pp. 81–100.

Fales, Frederick Mario, 'Grain Reserves, Daily Rations, and the Size of the Assyrian Army: A Quantitative Study', *State Archives of Assyria Bulletin*, vol. 4, no. 1 (1990), pp. 23–34.

Fox, Robin Lane, 'Introduction', in Robin Lane Fox (ed.), *The Long March: Xenophon and the Ten Thousand* (New Haven/London: Yale University Press, 2004a), pp. 1–46.

Fox, Robin Lane, 'Sex, Gender and the Other in Xenophon's Anabasis', in Robin Lane Fox (ed.), *The Long March: Xenophon and the Ten Thousand* (New Haven/London: Yale University Press, 2004b), pp. 184–214.

Gladitz, Charles, *Horse Breeding in the Medieval World* (Portland, OR: Four Courts Press, 1997).

Gray, G.B. and Cary, M., 'The Reign of Darius', in J.B. Bury, S.A. Cook and F.E. Adcock (eds.), *The Cambridge Ancient History*, vol. 4, *The Persian Empire and the West* (1926, reprint, London/New York: Cambridge University Press, 1960), pp. 173–228.

Hale, John R., 'Not Patriots, Not Farmers, Not Amateurs: Greek Soldiers of Fortune and Origins of Hoplite Warfare', in Donald Kagan and Gregory Viggiano (eds.), *Men of Bronze: Hoplite Warfare in Ancient Greece* (Princeton/Oxford: Princeton University Press, 2013), pp. 176–93.

Hanson, Victor Davis, *Warfare and Agriculture in Classical Greece* (1998, reprint, Berkeley/Los Angeles: University of California Press, 1999a).

Hanson, Victor Davis, 'Hoplite Obliteration: The Case of the Town of Thespiai', in John Carman and Anthony Harding (eds.), *Ancient Warfare: Archaeological Perspectives* (Gloucestershire: Stroud, 1999b), pp. 203–17.

Hanson, Victor Davis, *Carnage and Culture: Landmark Battles in the Rise of Western Power* (New York: Doubleday, 2001).

Hanson, Victor Davis, *A War Like No Other: How the Athenians and Spartans Fought the Peloponnesian War* (2005, reprint, New York: Random House, 2006).

Hanson, Victor Davis, 'The Modern Historiography of Ancient Warfare', in Philip Sabin, Hans Van Wees and Michael Whitby (eds.), *The Cambridge History of Greek and Roman Warfare*, vol. 1, *Greece, the Hellenistic World and the Rise of Rome* (Cambridge: Cambridge University Press, 2007), pp. 3–21.

Hanson, Victor Davis, 'The Hoplite Narrative', in Donald Kagan and Gregory Viggiano (eds.), *Men of Bronze: Hoplite Warfare in Ancient Greece* (Princeton/Oxford: Princeton University Press, 2013), pp. 256–75.

Harrison, Cynthia M., 'Triremes at Rest: On the Beach or in the Water?', *Journal of Hellenic Studies*, vol. 119 (1999), pp. 168–71.

Herodotus, *The Histories*, translated by Aubrey de Selincourt, Revised with an Introduction and notes by A.R. Burn (1954, reprint, London: Penguin, 1972).

Herzog, Chaim and Gichon, Mordechai, *Battles of the Bible: A Military History of Ancient Israel* (1978, reprint, London: Greenhill, 1997).

Hill, G.F., 'Coinage from Its Origin to the Persian Wars', in J.B. Bury, S.A. Cook and F.E. Adcock (eds.), *The Cambridge Ancient History*, vol. 4, *The Persian Empire and the West* (1926, reprint, London/New York: Cambridge University Press, 1960), pp. 124–36.

Hirsch, Steven W., 'Cyrus' Parable of the Fish: Sea Power in the Early Relations of Greece and Persia', *Classical Journal*, vol. 81, no. 3 (1986), pp. 222–9.

How, W.W., 'Arms, Tactics and Strategy in the Persian War', *Journal of Hellenic Studies*, vol. 43, Part 2 (1923), pp. 117–32.

Jha, D.N., *Early India: A Concise History* (New Delhi: Manohar, 2004).

Jones, Archer, *The Art of War in the Western World* (1987, reprint, Urbana/Chicago: University of Illinois Press, 2001).

Kagan, Donald, 'Pericles, Thucydides, and the Defense of Empire', in Victor Davis Hanson (ed.), *Makers of Ancient Strategy: From the Persian Wars to the Fall of Rome* (Princeton: Princeton University Press, 2010), pp. 31–57.

Kagan, Donald and Viggiano, Gregory F., 'Introduction', in Donald Kagan and Gregory Viggiano (eds.), *Men of Bronze: Hoplite Warfare in Ancient Greece* (Princeton/Oxford: Princeton University Press, 2013a), pp. xi–xxi.

Kagan, Donald and Viggiano, Gregory F., 'The Hoplite Debate', in Donald Kagan and Gregory Viggiano (eds.), *Men of Bronze: Hoplite Warfare in Ancient Greece* (Princeton/Oxford: Princeton University Press, 2013b), pp. 1–56.

Keaveney, Arthur, *The Persian Invasions of Greece* (Barnsley: Pen & Sword, 2011).

Krentz, Peter, *The Battle of Marathon* (New Haven/London: Yale University Press, 2010a).

Krentz, Peter, 'A Cup by Douris and the Battle of Marathon', in Garrett G. Fagan and Matthew Trundle (eds.), *New Perspectives on Ancient Warfare* (Leiden/Boston: Brill, 2010b), pp. 183–204.

Krentz, Peter, 'Hoplite Hell: How Hoplites Fought', in Donald Kagan and Gregory Viggiano (eds.), *Men of Bronze: Hoplite Warfare in Ancient Greece* (Princeton/Oxford: Princeton University Press, 2013), pp. 134–56.

Lazenby, John, 'Hoplite Warfare', in John Hackett (ed.), *Warfare in the Ancient World* (London: Sidgwick & Jackson, 1989), pp. 54–81.

Lazenby, J.F., *The Peloponnesian War: A Military Study* (2004, reprint, London/New York: Routledge, 2007).

Maurice, F., 'The Size of the Army of Xerxes in the Invasion of Greece', *Journal of Hellenic Studies*, vol. 50, Part 2 (1930), pp. 210–35.

Melville, Sarah C., 'The Last Campaign: The Assyrian Way of War and the Collapse of the Empire', in Wayne E. Lee (ed.), *Warfare and Culture in World History* (New York/London: New York University Press, 2011), pp. 13–33.

Melville, Sarah C., *The Campaigns of Sargon II: King of Assyria, 721–705 BC* (Norman: University of Oklahoma Press, 2016).

Munro, J.A.R., 'Some Observations on the Persian Wars', *Journal of Hellenic Studies*, vol. 19 (1899), pp. 185–97.

Munro, J.A.R., 'Marathon', in J.B. Bury, S.A. Cook and F.E. Adcock (eds.), *The Cambridge Ancient History*, vol. 4, *The Persian Empire and the West* (1926, reprint, London/New York: Cambridge University Press, 1960a), pp. 229–67.

Munro, J.A.R., 'The Deliverance of Greece', in J.B. Bury, S.A. Cook and F.E. Adcock (eds.), *The Cambridge Ancient History*, vol. 4, *The Persian Empire and the West* (1926, reprint, London/New York: Cambridge University Press, 1960b), pp. 317–46.

Nefiodkin, Alexander K., 'On the Origin of the Scythed Chariots', in Everett L. Wheeler (ed.), *The Armies of Classical Greece* (Aldershot: Ashgate, 2007), pp. 493–502.

Parker, Robert, 'One Man's Piety: The Religious Dimension of the Anabasis', in Robin Lane Fox (ed.), *The Long March: Xenophon and the Ten Thousand* (New Haven/London: Yale University Press, 2004), pp. 131–53.

Raaflaub, Kurt, 'Archaic and Classical Greece', in Kurt Raaflaub and Nathan Rosenstein (eds.), *War and Society in the Ancient and Medieval Worlds: Asia, The Mediterranean, Europe, and Mesoamerica* (Washington, DC: Center for Hellenic Studies, distributed by Harvard University Press, 1999), pp. 129–61.

Raaflaub, Kurt, 'Early Greek Infantry Fighting in a Mediterranean Context', in Donald Kagan and Gregory Viggiano (eds.), *Men of Bronze: Hoplite Warfare in Ancient Greece* (Princeton/Oxford: Princeton University Press, 2013), pp. 95–111.

Randsborg, Klavs, 'Into the Iron Age: A Discourse on War and Society', in John Carman and Anthony Harding (eds.), *Ancient Warfare: Archaeological Perspectives* (Gloucestershire: Sutton, 1999), pp. 191–202.

Rosenstein, Nathan, 'Phalanges in Rome?', in Garrett G. Fagan and Matthew Trundle (eds.), *New Perspectives on Ancient Warfare* (Leiden/Boston: Brill, 2010), pp. 289–303.

Schwartz, Adam, 'Large Weapons, Small Greeks: The Practical Limitations of Hoplite Weapons and Equipment', in Donald Kagan and Gregory Viggiano (eds.), *Men of Bronze: Hoplite Warfare in Ancient Greece* (Princeton/Oxford: Princeton University Press, 2013), pp. 157–75.

Srivastava, A.K., *Ancient Indian Army: Its Administration and Organization* (Delhi: Ajanta Publications, 1985).

Strauss, Barry, 'Slave Wars of Greece and Rome', in Victor Davis Hanson (ed.), *Makers of Ancient Strategy: From the Persian Wars to the Fall of Rome* (Princeton/Oxford: Princeton University Press, 2010), pp. 185–205.

Stylianou, P.J., 'One Anabasis or Two?', in Robin Lane Fox (ed.), *The Long March: Xenophon and the Ten Thousand* (New Haven/London: Yale University Press, 2004), pp. 68–96.

Talbert, Richard J.A., 'The Role of the Helots in the Class Struggle at Sparta', *Historia*, vol. 38, no. 1 (1989), pp. 22–40.

Thapar, Romila, *From Lineage to State: Social Formations in the Mid-First Millennium BC in the Ganga Valley* (1984, reprint, Delhi: Oxford University Press, 1990).

Thommen, Lukas, *An Environmental History of Ancient Greece and Rome*, translated by Philip Hill (Cambridge: Cambridge University Press, 2012).

Thucydides, *History of the Peloponnesian War*, translated by Rex Warner with an Introduction and notes by M.I. Finley (1954, reprint, London: Penguin, 1988).

Trundle, Matthew, 'Coinage and the Transformation of Greek Warfare', in Garrett G. Fagan and Matthew Trundle (eds.), *New Perspectives on Ancient Warfare* (Leiden/Boston: Brill, 2010), pp. 227–52.

Tuplin, Christopher, 'All the King's Horse: In Search of Achaemenid Persian Cavalry', in Garrett G. Fagan and Matthew Trundle (eds.), *New Perspectives on Ancient Warfare* (Leiden/Boston: Brill, 2010), pp. 101–82.

Ure, P.N., 'The Outer Greek World in the Sixth Century', in J.B. Bury, S.A. Cook and F.E. Adcock (eds.), *The Cambridge Ancient History*, vol. 4, *The Persian Empire and the West* (1926, reprint, London/New York: Cambridge University Press, 1960), pp. 83–123.

Viggiano, Gregory F., 'The Hoplite Revolution and the Rise of the Polis', in Donald Kagan and Gregory Viggiano (eds.), *Men of Bronze: Hoplite Warfare in Ancient Greece* (Princeton/Oxford: Princeton University Press, 2013), pp. 112–33.

Wallinga, H.T., *Xerxes' Greek Adventure: The Naval Perspective* (Leiden/Boston: Brill, 2005).

Waters, Matt, *Ancient Persia: A Concise History of the Achaemenid Empire, 550–330 BCE* (Cambridge: Cambridge University Press, 2014).

Wees, Hans van, 'Farmers and Hoplites: Models of Historical Development', in Donald Kagan and Gregory Viggiano (eds.), *Men of Bronze: Hoplite Warfare in Ancient Greece* (Princeton/Oxford: Princeton University Press, 2013), pp. 222–55.

Whitby, Michael, 'Xenophon's Ten Thousand as a Fighting Force', in Robin Lane Fox (ed.), *The Long March: Xenophon and the Ten Thousand* (New Haven/London: Yale University Press, 2004), pp. 215–42.

Xenophon, *The Persian Expedition*, translated by Rex Warner, With an Introduction and Notes by George Cawkwell (1949, reprint, London: Penguin, 1972).

4

THE ERA OF HYBRID MILITARY MACHINES

Macedonia, the Hellenistic monarchies, India and China

Introduction

The rise of Macedonia as a world power occurred to a great extent due to the burning political ambitions and military capabilities of two rulers: Philip II and his son, Alexander (b. 356–d. 323 BCE). The father–son case shows how macro changes occur due to micro (personal/individual) factors. As this book argues, despite the operation of long-term structural factors, selected individuals are important, and their actions often change the trajectory of history. This chapter traces how Philip II of Macedonia created a hybrid military machine by synthesising military elements from both the Achaemenid and the Classical Greek military systems. This composite military system enabled Philip and his son, Alexander, to conquer both Classical Greece and the Achaemenid Empire. In this chapter, the terms Persian and Achaemenid are used interchangeably, and Alexander's force is referred to as the Macedonian army, which also included allied Greek contingents. In India and China, the military systems which came into existence were in many ways dissimilar to Macedonian warfare. However, there were some commonalities between the Asian and Greco-Roman theoretical corpus. This was because both sides were learning from interactions with each other. Alexander, while campaigning in India, had to adapt his military machine in line with local conditions. Alexander's successors further refined their military system by absorbing elements from the Indian military tradition. And the dialectics of learning also operated in the case of the Indian military culture, which absorbed certain elements from the Macedonian military machine.

Macedonia rising

This section shows that by absorbing selected elements of the diverging military cultures of West Asia and Classical Greece, Philip II of Macedonia constructed

DOI: 10.4324/9780429284069-5

a war-winning military machine. In the fourth century BCE, Macedonia was a tribal state. The monarchy was central to the Macedonian polity and its most important institution. The king was the commander-in-chief and led the army in times of battle. The monarch was expected to display personal bravery in battle. Philip II came to the throne in 359 BCE. Philip captured the gold mines of Mount Pangaeus. Exploitation of the gold and silver mines in Damastium further strengthened the Macedonian economy. The marshy plain of Crenides was drained and put under cultivation as part of a land reclamation policy. The royal monopolies in the export of timber from the Balkan hinterland and on the mining of precious metals provided Philip money to raise, train and equip his army (Fuller 1989: 28; Hamilton 1999: 166–71; Worthington 2016: 39–43).

Before Philip, Macedonia's army comprised a tribal levy. The personnel were mainly peasants who provided their own armour and fought as light infantry only when the harvesting season was over. He took steps to create a professional standing army and introduced regular pay and promotion structures. Cash bonuses and land grants were also introduced to raise the motivation of the soldiers. The poor infantry personnel were provided with arms and armour (Worthington 2016: 33). Regular pay initiated by Philip enabled the Macedonian soldiers to fight throughout the year and also when required outside their home territory. Training is the lifeblood of the army, and a rigorous training program was what Philip introduced.

He introduced the tradition of disciplined heavy infantry from Greece into Macedonia but modified the hoplite panoply. He armed his heavy infantry soldiers with sarissas. A sarissa was about 16 feet in length, compared to the 9-foot-long pike of the hoplite (Waterfield 2011: 60). The heavy infantry was arranged in the rectangular formation with several rows known as phalanx. In the sarissa-armed formation, pikes from the first five rows projected forward of the first rank between each pair of soldiers (Markle 1978: 484). In the hoplite phalanx, pikes from the first three rows (due to their shorter length) projected forward beyond the first rank of each pair of soldiers. The Macedonian phalangite (heavy infantry soldier of the phalanx) had a smaller shield compared to a hoplite. Since both the hands of a phalangite were required to hold the longer sarissa, weighing 6.3 kg, he, unlike a hoplite, could not carry a bigger round shield. The sarissa being longer than the pike of the hoplite, the Macedonian phalangites were able to rupture the hoplite phalanx long before the hoplites' pikes could penetrate the former's frontal rank. Since the sarissa came into action before the hoplite pike could make contact with the Macedonian phalanx, the phalangite did not require a large round shield like the hoplite. The phalangites were organised in a phalanx on a territorial basis. The phalangites were Macedonian peasants, and every free man was under the obligation to serve (Worthington 2016: 34–5).

The tactical effectiveness of the Macedonian heavy infantry was further increased due to the introduction of certain managerial changes. This involved disposition of the infantry into several hierarchical formations and the introduction of officers at various levels of command. The basic tactical unit of the infantry was the *dekas*. The *dekas* constituted a file of 10–16 men drawn in deep battle order within a phalanx. Each *dekas* was commanded by a *dekadarchos*, who led from the front rank. Each file was subdivided into two half-files, with each commanded by a *dimoirites* and two *dekastateroi*. Above the *dekas* was the *lochos*. Each *lochos* had 512 men assembled in 32 files. The *lochos* were organised in phalanx divisions which were also known as *taxeis*. A phalanx had about 1,500 men and was generally 16 ranks deep. During combat, the first four or five ranks used the sarissas in couched position, and the rest kept their weapons in a vertical position. Generally, Persian archers shot from a sitting position. And their arrow following parabolic motion hit the hostile infantry at the rear of the phalanx. The series of long pikes held by the rear ranks were projected upward and prevented the arrows (following a parabolic trajectory) shot by the enemy archers from hitting the phalangites at the rear of the phalanx. The latter ranks used their body weight to push the former and thus increase the momentum of the front line. In accordance with the circumstances of battle, the depth of the phalanx varied. For instance, the *dekas* could be doubled to increase the depth of the phalanx to 32 ranks or could be reduced to eight ranks. During the Battle of Issue (discussed later in this chapter in greater detail), the phalanx initially was 32 ranks deep. But as the army came to the plain, the phalanx expanded to cover the ground in order to prevent outflanking moves by Persian cavalry. As a result, the depth of the phalanx was reduced to eight-deep ranks (Bosworth 1993: 259–61; Sekunda 2007: 326–31). Such variations in depth required transfer of the *dekas* from one position to another like chess pieces on a chessboard. The men had to keep their position within a particular *dekas* to prevent confusion in the battlefield. This was possible because of the presence of non-commissioned officers among the infantry and high training inculcated among the common soldiers.

The light infantry known as peltasts in Greece became important during the later stages of the Second Peloponnesian War. Peltasts wore no armour and were equipped with slings, javelins and bows. Peltasts only became common in Greece after the Second Peloponnesian War, and their employment in the battlefield was credited to the Athenian General Iphicrates (b. 415–d. 353 BCE). The Achaemenid army had used them long before. In the Sicilian expedition, the Athenians used Rhodian slingers. During Xenophon's (b. 430–d. 354 BCE) retreat, the slingers played an important role in holding off Persian cavalry. Philip II used slingers and javelin men (Tarn 1930: 6–21) in conjunction with heavy infantry. The slingers of the Macedonian army used marble shots that were more effective than shots made from baked clay. Slingers shooting stones weighing 30–40 grams could be as deadly as modern army shooters in the battlefield equipped with telescopic rifles. During the Battle of Plataea, Mardonius was brought down by a Spartan slinger (Rihll 2013: 2–3, 10). As a result of these reforms, the army of Philip was much more combat effective than the untrained traditional Macedonian tribal infantry levy equipped haphazardly with swords, spears and wicker shields (Fuller 1989: 42–8).

Further, Philip raised heavy cavalry, aping the Achaemenids. Macedonians were mainly pastoralists, and the country, unlike mainland Greece, was suited to raising a large number of horses (Hammond 2001: 8, 12). The Macedonian aristocracy traditionally provided the cavalry, which fought under the king's personal leadership. Their honorific title was the Companions. Each Companion was armed with a spear about 14.7 feet in length. Hence, the Companions were able to penetrate hoplite infantry formations equipped with 9-foot-long spears. Traditionally Greek cavalry fought in a square or rectangular formation. In contrast, the Scythians (Sakas) and the Thracian cavalry fought in a wedge or triangle formation. Such tactics influenced Philip. The Companions attacked in a delta-shaped wedge formation (Devine 1989: 104; Lendon 2005: 98). This formation raised the penetrative power of the cavalry in the battlefield.

Philip reduced the number of non-combatants accompanying the army drastically, which increased the marching speed of his army. He permitted each cavalryman to have one attendant, and there was only one attendant for 16 infantry soldiers (Markle 1978: 82). The slow-moving carts drawn by oxen used for carrying equipment and provisions were replaced with faster-moving pack animals (Worthington 2016: 37).

Classical Greek hoplite battles were indeed 'limited'. The hoplites of two warring parties clashed with each other, and after the *osthismos* (the push), one side's hoplites broke ranks and fled. However, there was no pursuit by the victorious hoplites because the heavily armed infantry soldiers arranged in phalanx formation were only capable of marching slowly and were incapable of quick pursuit. And if the victorious hoplites broke ranks to pursue the fleeing hoplites, then they became vulnerable to a counterattack by the hostile forces. In addition, the hoplites, lacking siege engines, were incapable of capturing the enemy's fortified places.

However, Philip's new army changed the paradigm of land warfare by using siege engines and waging battles of annihilation. The reformed Macedonian army, thanks to its logistical reforms, could march rapidly to the enemy country and force the enemy to accept a decisive battle. The phalanx made short work of the hoplite force, and then the peltasts and heavy cavalry of Philip were able to destroy the defeated hoplites. And if the defeated enemy took recourse to holding a fortified place, the Macedonian siege engines demolished them.

Classical Greek siege warfare was quite backward considering Near Eastern standards. During the Trojan War, the Achaeans, Homer tells us, tried to storm the walls of Troy but failed miserably. They captured the city later by a ruse. During the early Classical era, for defensive purposes the Greeks constructed walls made of mud brick upon stone plinths. The experience of the Persian occupation of Athens, with the availability of more money and manpower when the Athenian maritime empire was at its height, resulted in developments in the sphere of siege warfare. The Greek walls tended to become stronger, with towers, gateways and sally ports for launching assaults on the besiegers by the defenders. If the walls of a defending fort had protruding crenellated wall towers and balconies, then it provided a wide arc of fire to the defenders. In such a scenario, it became doubly difficult for the assault infantry to climb up the walls with the aid of scaling ladders. In the Second Peloponnesian War, the Athenians and the Spartans learnt the Assyrian techniques of building mounds and using rams, tortoises (covered sheds to protect the siege workers), sapping (to undermine the enemy walls) and counter-sapping (mining to intercept enemy sappers). The attacking soldiers from the top of the artificial mound drove away the defenders from the battlements of the walls of the city, and the rams were used for battering the walls. During this war, both Athens and Sparta resorted to the expedient of establishing a fortified post in enemy territory. From this post, assault parties were launched to destroy the economic infrastructure of the enemy (Strauss 2007: 237–40; Rihll 2013: 30).

Philip, by adopting siege techniques that were perfected by the Neo-Assyrians and the Carthaginians, made the Macedonian army capable of capturing fortified hostile cities. The movable siege towers and rams used by the Assyrians were integrated in the Macedonian siege train. An example of an Assyrian siege will illustrate the sophistication achieved by them in the field of siege warfare. One of the most famous sieges undertaken by the Assyrians was the Siege of Lachish by Sennacherib in 701 BCE. Lachish was a stronghold of Judah which guarded the south-west approaches of Jerusalem. The Assyrian army surrounded the city and then constructed a ramp. A ramp or a causeway of piled-up earth, rubble and wood enabled the attackers to strike the thinner and fragile upper portion of the walls. The ramp enabled them to bring forward the siege engines. The latter were placed on top of artificial earthen mounds. Mobile battering rams on wheels, propelled by muscle power, aimed to make a breach at the thinner upper part of the wall near the gates. The battering ram was metal-tipped wood housed in a wooden framework. The Assyrians also used wooden siege towers from where their sharpshooters poured fire on the defenders on the top of the walls. The defenders responded by constructing wooden barriers to protect themselves from the arrow storm unleashed by the besieging army from the wooden towers. The defenders shot flaming arrows, poured hot oil and threw torches to burn down the rams. In turn, the Assyrian engineers covered the rams with heavy oiled leather canopies. The Assyrians deployed archers covered by a shield wall for protecting the rams. Further, the assault infantry, wearing body armour, attempted to get on top of the walls with scaling ladders. While undertaking a siege, the Assyrians made it a point to devastate the surrounding countryside for acquiring food and siege materials (wood, stone, etc.) for themselves and to prevent supplies from reaching the enemy camp (Wiseman 1989: 46–9).

Catapults (pre-modern missile-throwing machines) became an integral part of the Macedonian army. Two types of torsion catapult came into existence: an arrow shooting machine and a stone thrower. Both types were stand mounted and spanned by winch, supplemented in larger machines by pulleys, and they used a slider in a grooved stock. Tracey Rihll, in her book on the catapult, asserts that all types of field and handheld weapons shooting bolts and darts as well as the city breakers (huge stone throwing machines), etc., in the ancient period were invented in the region stretching from Sicily in the west to Crete in the east. Further, she credits Philip of Macedonia with several inventions in the sphere of field and siege artillery. She argues that the mechanisation of the sling occurred in the middle of the fourth century BCE with the invention of the *onager* – the one-armed torsion catapult. And the two-armed torsion catapult was the fusion of two-armed bow or tension catapult and the one-armed torsion catapult. She asserts that in 405 BCE Dionysius I (b. 432 BCE–d. 367 BCE), the ruler of Syracuse in Sicily, invented the catapults in response to the Carthaginian invasion (Rihll 2013: xx, 1–27). Rihll's work provides us with a lot of information about the various siege engines (both heavy and portable) used in Mediterranean Europe. Nevertheless, there is a serious limitation in her approach. Her book is a rabid example of the Eurocentric approach. It seems she neither knows nor has consulted the sources east of the Near East, and she may be unaware of the development of military technology in China and India and neglectful of the development of siege warfare techniques in the Middle East.

The torsion catapult was probably a Phoenician invention, and these machines were used efficiently by Alexander during his siege of Tyre (333–32 BCE). These machines from Phoenicia reached their kinsmen, the founders of Carthage. The use of such machines by the Carthaginians in Sicily resulted in the adoption of such knowledge by the Sicilians. From Sicilian Greeks, such ideas reached Athens and finally Macedonia. However, in a humid climate, the torsion catapult was a disadvantageous machine to use because the tension of the skeins varied with changes in atmospheric humidity. Hence, torsion catapults did not become common in South and South-East Asia. The torsion catapult machines were not big enough to throw large stones which could batter down massive walls, but they provided effective covering fire that enabled the besiegers to advance towards the walls of the defending cities and use other techniques, such as rams and scaling ladders, to overcome the defence. The torsion catapults were replaced by trebuchets during the Middle Ages (Hacker 1968: 34–50; Fuller 1989: 43–5). In the East, the traction trebuchet came into use.

Philip followed combined arms tactics. During battles, his aim was to crush the enemy cavalry with his own heavy cavalry and then attack the enemy's infantry in one of its flanks. Simultaneously, the Macedonian phalanx attacked the enemy infantry frontally. In a decisive battle, his phalangites pinned the enemy formation while his heavy cavalry dealt a body blow at one of the flanks of the enemy line. After the enemy force was defeated, the Macedonian cavalry, by going around their rear and riding down the fleeing enemy force, destroyed them.

In 338 BCE, at the Battle of Chaeronea, Philip had 30,000 infantry and 2,000 cavalry. He gave command of the Macedonian left flank and the Companion cavalry to his 18-year-old son. Philip commanded the Macedonian right wing. The Greek allied force had Athenians in the left wing, Thebans in the right wing and the soldiers of the lesser Greek states in the centre. Philip deliberately initiated a tactical retreat in order to draw the Athenians forward of the allied Greek centre. As the Athenian troops surged forward and got separated from their allies, Philip counterattacked. Simultaneously, Alexander's Companions crushed the Thebans (Fuller 1989: 35; Worthington 2010: 119). Close coordination between father and son commanding the two wings and thorough discipline of the troops enabled Philip to conduct a tactical retreat and then launch a counterattack, giving victory to the Macedonians. A deliberate tactical retreat in front of the enemy force and then launching a planned counterattack was a technique that Hannibal later perfected.

Chaeronea established Macedonian supremacy in Greece. The age of the poleis was over. Philip was planning an invasion of the Achaemenid Empire. For transporting his troops across the Strait of the Dardanelles, Philip for the first time introduced the navy in Macedonia. He built upon the traditional naval tradition of Athens and Persia. Each vessel was 120 feet in length and carried 200 men. The vessel was propelled by a square-rigged sail and by a single row of oars on each side, with three rowers to each oar. He sent an advanced force of 10,000 men under General Parmenion to establish a bridgehead in Asia (Fuller 1989: 38, 53–4). Philip was assassinated (an act in which one of his wives named Queen Olympias had a hand) in late summer 336 BCE, and his son, Alexander, succeeded him.

Alexander's campaigns in Asia: 334–323 BCE

The principal sources for Alexander's military activities are Roman authors such as Quintus Curtius Rufus (first century CE) and Arrian (second century CE). In 334 BCE, 12,000 Macedonian infantry together with 1,500 Companion cavalry were left behind in Macedonia under Alexander's regent, Antipater (Devine 1989: 105), to guard against any possible Persian amphibious expedition and to deter the principal Greek states from rebelling against Macedonian overlordship. Alexander crossed into Asia Minor with an army numbering 48,000 infantry and 6,000 cavalry supported by a fleet of 120 warships (Worthington 2010: 119). At that time, Macedonia's population was 500,000 (Worthington 2016: 14). The total cavalry strength (the core of the Persian military system) of the Achaemenid Empire came to about 50,000 (Tarn 1930: 60). In addition, the Achaemenid Empire had about 25,000 Greek mercenary hoplite infantry. The battle doctrine of both the Achaemenids and the Macedonians was to fight decisive battles in order to decide the issue. But while the Achaemenids depended mainly on cavalry, with the infantry in an auxiliary role, the Macedonian battle doctrine focused on combined arms tactics. Three battles and one siege (Tyre) effectively destroyed the Achaemenid Empire.

The first battle that Alexander fought with the Persians was Granicus. The Achaemenids did not enjoy numerical superiority either in infantry or in cavalry vis-à-vis Alexander's army in this battle. It was an erroneous strategy on the part of the Achaemenid commanders to fight a decisive set piece battle with the Macedonian army. Rather, the Persian commanders should have conducted hit-and-run attacks to harass Alexander's line of communications and waited for reinforcements from the Great King Darius III (r. 336–323 BCE). This would have distracted Alexander by increasing his supply problems. The Persians could have traded space for time. In the meantime, the Achaemenid fleet should have carried on amphibious attacks along the Greek coastline and attempted to cut Alexander's lifeline with Macedonia. The Achaemenid fleet had been rebuilt and after the Peloponnesian War was the strongest naval force in the Mediterranean. The Achaemenid navy, with 300 ships and 60,000 crew, was almost three times larger than Alexander's fleet (Hammond 2001: 90). Memnon the Rhodian, the Greek commander on Persian payroll, had actually argued for such an attritional strategy. However, most of the Persian commanders wanted to fight a pitched battle with Alexander as early as possible, and in the end the latter group prevailed. This was probably because it was politically degrading for the Achaemenid monarchy to lose territory even temporarily to an external invader.

At Granicus in 334 BCE, the Persians assembled 20,000 infantry (including Greek mercenaries) and cavalry. The Achamenids were aware of their lack of superiority in cavalry. So instead of choosing a wide open plain, they stationed their cavalry behind the Granicus River. The river was 80 feet wide. Though the river was fordable at that time, the Persians hoped that attacking uphill along the steep rocky bank of the Granicus would hobble the cavalry charge of Alexander. Alexander deployed his phalangites in the centre and cavalry with light infantry at both the flanks. Alexander himself like his father commanded the right flank of the Macedonian army. The position of honour in the battlefield according to Macedonian military culture was the command of the right wing where the Macedonian king always took his place. The right flank under Alexander (who displayed heroic leadership) crossed the river first and attacked the stationary Persian cavalry. Momentum (mass × velocity) partly enabled them to break through the static Persian cavalry stationed at the bank of the Granicus. Moreover, cooperation between the Macedonian cavalry and infantry enabled them to overcome the Persian cavalry that fought alone. Further, the Achaemenid cavalry had the added disadvantage of fighting with the low evening sun in their eyes while the Macedonians had the sun behind them. In close-quarter combat, better discipline and the superior cornel wood spears of the Macedonian cavalry against light javelin-equipped Persian cavalry decided the day. As soon as the Persian cavalry fled the battlefield, the Macedonian phalanx frontally attacked the Greek mercenary infantry, while Alexander's cavalry attacked them at both the flanks and their rear. The Greek mercenaries were all slaughtered. Alexander's force suffered 115 killed and some 1,000 wounded (Delbruck 1990: 187–9; Badian 2012: 224–43).

Achaemenid tactics in this battle were faulty. The Persian cavalry failed to coordinate its actions with light infantry and the Greek mercenary infantry. Persian cavalry and Greek mercenary infantry fought separately and collapsed against the combined arms tactics of the Macedonian army. During the battle, the Persian cavalry formed the first line and the Greek mercenary infantry constituted the second line. There was no cooperation and coordination among the formations deployed in these two lines. In hindsight, we can say that the Persians should have deployed light infantry armed with javelins and bows with the Greek mercenary infantry along the bank of the Granicus to harass Alexander's cavalry and the phalanx as they slogged up along the steep bank of the river. Then the Persian light infantry should have retreated, and the Persian cavalry ought to have charged the tired and harassed Alexander's cavalry just as it had reached the riverbank. Tension between Memnon, the general of the mercenary Greek infantry, and the Persian *satraps* commanding the cavalry prevented coordination between these two arms in this battle.

At Granicus, the combined forces of the regional *satraps* were destroyed. Meanwhile Darius mobilised the royal army and advanced from Babylonia to Syria and then to Cilicia in Asia Minor. In October 333, Darius, with his army, appeared at Alexander's rear near the Pinarus River. It was a brilliant operational move on the part of Darius, who has generally been considered an incompetent leader by the historians. At Issus, Darius had 12,000 Greek mercenary infantry plus Persian infantry and cavalry. Against more than 50,000 Persians, Alexander had 40,000 men with him (Tarn 1930: 18; Delbruck 1990: 210; Worthington 2016: 165–6). Darius held the north bank of the river, and he placed a strong detachment on the mountains to his left. Alexander, coming from the south, advanced northwards. The Persian battle order comprised cavalry on the right near the sea, then the Greek mercenaries along the river in the centre, and then came the *kardakes*. Darius himself stayed in the centre. His plan was that his centre should remain on the defensive and contain the Macedonian phalanx while the bulk of the Achaemenid cavalry massed on his right flank under Nabarzanes would defeat and envelop the Greek left wing and then attack the phalanx on its left flank and rear. The Macedonian battle line comprised the phalanx in the centre, a cavalry detachment under Parmenion constituted the left wing, and Alexander commanded the right wing comprising Companion cavalry. Thus, the 5,000 cavalry with Alexander were roughly equally divided among his right and left flanks (Hammond 2001: 100–9; Badian 2012: 417). More or less, the same battle order was to be followed by Alexander in the next great battle, Gaugemela/Arbela.

Meanwhile, back at the battlefield of Issus, the phalanx advanced and punched a hole into the *kardakes*. Simultaneously, the Macedonian cavalry on the right wing advanced but were attacked by Hyrcanian and Median cavalry. The latter in turn were counterattacked by Greek light infantry equipped with javelins and archers. As at Granicus, the Macedonian light infantry–cavalry combination won against the Achaemenid cavalry. Coordination and cooperation among various arms had reached a high state in the Macedonian army thanks to the reforms and training of Philip II. Further, the Persians had weakened their left flank and massed most of their cavalry in their right flank. Once the Persian cavalry on their left flank disintegrated, the Macedonian cavalry attacked the Persian centre on its left flank and rear. Meanwhile, the Macedonian left wing was in trouble.

Heavy armoured Persian cavalry attacked the Thessalian cavalry and pushed them back. Despite the collapse of the Macedonian left wing, the situation was saved because the Persian left wing and centre had disintegrated and Darius fled the battlefield. Only then did the victorious Persian cavalry on their right wing start retreating through the Amanus Range (Hammond 2001: 106–7). Greek casualties in this battle amounted to 450 killed and 4,000 wounded (Delbruck 1990: 201). Darius could not take advantage of his numerical superiority in cavalry because he deployed his force in a narrow plain. The battlefield was not suitable for the deployment of a large number of cavalry. Rufus had written: 'Every accessible area in the defile was packed with his [Darius's] troops, and the wings stood on the mountains on one side and on the sea shore on the other' (quoted from Rufus 2004: 40).

Let us pose a counterfactual to see whether Alexander's victory was inevitable. It will throw light on the thorny issue whether the Macedonian military machine was inherently superior compared to the Achaemenid ground force. If Darius had selected a wide open plain, then the heavy Persian cavalry would have been able to overwhelm the Macedonian left wing earlier and the battle might have gone the other way. Both at Granicus and in Issus, Alexander gambled for high stakes, risking his personal life repeatedly. Alexander's style of leading from the front and taking risk was similar to that of his father, Philip. Alexander's command style reminds one of the personalised risky daredevil command style of the Nazi panzer commanders like Joachim Peiper and 'Panzer Graf' Hyazinth von Strachwitz. Darius collected the survivors and then retreated to the Euphrates (Hammond 2001: 107).

In the aftermath of Issus, Alexander conquered Palestine, Syria and Egypt to capture the Achaemenid naval bases in order to neutralise the Persian navy. Alexander's diversion in the south and south-west of the Achaemenid Empire gave time to Darius to raise another field army for challenging the Macedonian incursion. However, the capture of the Achaemenid Empire's Mediterranean coastline was necessary to prevent any Persian amphibious attack against Greece and also to ensure protection of Alexander's supply lines stretching from Anatolia to Macedonia. A Persian expeditionary force operating in Greece would have encouraged Athens, Sparta and other poleis to rebel against Antipater. And continuous inflow of recruits from Macedonia required safety of the umbilical cord across the Strait of Dardanelles.

This task required Alexander to conduct the tiresome Siege of Tyre in 332 BCE. The city of Tyre was situated on an island half a mile ashore, and its 4.5 km circuit of walls rose to a height of 150 feet (Heckel 2013: 67). Alexander's own ships supported by the vessels provided by the Cypriots defeated the ships of Tyre. Only after the defeat of Tyre's naval power could the Macedonian army lay siege to this city. Alexander's troops acquired wood for the siege engines from Lebanon. Alexander requisitioned food and other materials from the coastal plain of Palestine, and his ships at Tyre and Gaza also brought supplies from farther afield. Besides the Macedonian engineers, Alexander also enlisted specialists of siege warfare from Cyprus and Phoenicia. He deployed catapults in his horse transport ships and on his triremes to bombard the city. The Hellenistic generals (especially Demetrius) would later emulate this practice. Once the rams had battered the wall on the southern section, Alexander himself led his elite Hypaspists into the breach (de Souza 2007b: 440–58).

After being defeated at Issus, Darius again started raising a second army, and this time he drew heavily on the resources from the eastern *satrapies* of his empire. At Gaugemela, Alexander had 47,000 men including 7,000 cavalry. Darius did not contest the crossing of the Euphrates and Tigris rivers by the Macedonians but waited for them in the great plain east of these two rivers, not far from the ruins of the Assyrian capital Nineveh. Darius's plan at Gaugemela was to conduct a sort of *Kettelschlacht* (cauldron battle) with his 12,000 cavalry. The plan was to attack both the flanks of the Macedonian army with cavalry and using cavalry superiority to envelop the latter. At Issus, the Persian high command weakened one of its flanks (left flank) and concentrated the bulk of the cavalry on its right flank. The objective was to attack one flank of the Macedonian army (its left flank) and then engulf it. This resulted in the Macedonians attacking the weakened Persian flank (its left flank). So this time, the Persians strengthened both of the flanks and aimed for a double envelopment: every general's dream till the present time. Envelopment, assumed the Persian high command, would hopefully lead to encirclement and finally annihilation of the Greeks. Since Darius was mainly depending on his cavalry for defeating the Greeks, he planned to fight the battle in a wide plain. So Darius ordered his men to artificially level the terrain by removing all possible obstacles. However, there was a caveat. Until the Persian cavalry overwhelmed both the flanks of the Macedonian army, the Persian centre had to hold its own against the formidable phalanx. At this point, due to casualties in the previous two battles, the Achaemenid Empire did not have a substantial number of Greek mercenary infantry. Darius tried to counter the Macedonian phalanx by ordering his infantry to adopt long spears instead of javelins. However, inadequate drill and training due to lack of time prevented Darius's newly minted infantry to emerge as a combat effective force. Darius also depended on a few Indian elephants and scythe chariots to disrupt the Macedonian phalanx (Tarn 1930: 59; Delbruck 1990: 210–21).

On 30 September 331 BCE, the day before the great battle, Darius feared a nocturnal attack by Alexander on his camp. So he kept the men under arms (Badian 2012: 467). Next day, when the battle started under the scorching sun at a temperature of about 35 degrees Celsius, the Persian soldiers were suffering from a lack of sleep. Darius opened the battle by launching the scythe chariots towards the Greek line. Just as at Cunaxa, the chariot charge failed. As the scythe chariots advanced towards the Greek infantry, the Greek sharpshooters easily killed the drivers, and many of the chariots overturned (Delbruck 1990: 214). Again, Darius had too few elephants and they were not used effectively. In the next great battle that Alexander would fight in Punjab, the elephants would give him a run for his money. The Macedonian phalanx was not disrupted by the chariots and elephants, and it slowly advanced towards the Persian centre. Meanwhile, Persian cavalry on the right and left flanks under Mazaeus and Bessus attacked the left wing of the Macedonian line under Parmenion and the right wing comprising the Companion cavalry, respectively. Parmenion's cavalry was hard-pressed by Mazaeus. Two phalanx battalions had to support Parmenion's faltering cavalry (Tarn 1930: 64). And on Alexander's right, the Companion cavalry was slowly pressed back by Bessus's Saka cavalry.

At midday, the battle was finely balanced. Alexander's left wing was pushed back slowly but steadily, and his right wing was faltering too. Persian cavalry might break both of his wings before the Macedonian phalanx could come to grip with the Persian infantry in their centre. If this happened, then all would be lost. Alexander with his reserve cavalry could aid either his left or his right wings. If he went to reinforce the Macedonian left wing, his right wing would break. On the other hand, if he reinforced his right wing, the left wing under Parmenion would disintegrate. In contrast, the Persians had enough cavalry to sustain advance on either of their wings. It seemed that it was Darius's high noon.

Judging by normal standards, Alexander had lost the battle. However, despite the criticism of certain recent historians, Alexander was a military genius. Amidst the braying of horses, dust storms, shouting and howling of wounded men and groans of the wounded beasts with the sun shining brightly, Alexander made a momentous decision. He decided to attack the nerve centre of the Persian command, where Darius was stationed. God helps those who help themselves. The Achaemenid line between their centre and left wing stretched out to link up with Bessus's cavalry contingent, which was trying to outflank the Companions, and consequently a gap opened between Achaemenid centre and the left wing. Alexander at once charged through this gap towards Darius, and the latter fled. Seeing the Great King fleeing, the Persian soldiers in the centre of their line also started deserting. This in turn broke the Persian line at their centre (Tarn 1930: 64). When the Persian centre collapsed, Bessus, to escape encirclement by the Companions and the phalanx, withdrew with his hitherto victorious cavalry force from the Achaemenid left wing. In contrast to Darius's passive battlefield leadership, Alexander was a virtual living dynamo. After charging towards Darius, Alexander and his reserve cavalry hastened back to aid Parmenion. Alexander was always rushing to the threatened points along the forward edge of the battle area like Jon Snow at the Battle of Winterfell in the *Game of Thrones* HBO series. Mazaeus saw that the Persian centre and left wing had vanished from the battlefield. In response, he also withdrew with his force from the battlefield.

However, the advance of the main body of the phalanx towards Persian centre and the diversion of two phalanx battalions to aid Parmenion also opened a gap in the Macedonian heavy infantry line. Through this gap, the Punjabi cavalry charged. The reserve Punjabi cavalry was posted in the centre of the Persian line specially to guard Darius. However, the Punjabi cavalry, instead of attacking the Macedonian phalanx from the rear, went for the Macedonian baggage train in order to rescue the family of Darius (Tarn 1930: 64–5). Had the Punjabi cavalry charged at the rear of the Macedonian phalanx, then the battle could have gone the other way, and Alexander would have remained as a footnote in history. And even if the reserve Punjabi cavalry had remained at their post, Alexander's rash charge towards the centre of the Persian line where Darius was stationed might not have succeeded. It was a probable turning point of the battle. At Cunaxa, Cyrus the Younger, like Alexander at Gaugemela, had launched a cavalry charge towards the Great King in an attempt to dismantle the Persian high command. However, Cyrus the Younger's charge was defeated by the Great King's bodyguard cavalry. The Persian commanders, before starting the Battle of Gaugemela, rightly assumed that, like Cyrus the Younger, Alexander might also make a mad dash towards their Great King. For blunting such a desperate charge, they had stationed the Punjabi cavalry contingent to protect Darius. Nevertheless, at Gaugemela, Darius sent his bodyguard cavalry (Punjabi horsemen) on a different errand – to liberate the Achaemenid royal family held captive at Alexander's camp after the Persian defeat at Issus. Darius's defeat at Gaugemela was not a foregone conclusion. Had Darius played his cards right and kept his nerve a bit longer, Bessus with the Saka (Scythian) cavalry (the best in the world at that time) would have defeated the Companions (the elite cavalry of Alexander) and rolled up the phalanx (which was also coming under serious threat from Mazaeus's cavalry on its left flank) from its right side. Alexander understood the combat effectiveness of the Central Asian cavalry, and during his campaigns in Bactria and Sogdiana (327 BCE) took care to incorporate such horsemen in his hybrid force. At Gaugemela, Alexander's army lost 500 killed and many more wounded (Delbruck 1990: 214). It must be noted that the Greco-Roman historians deliberately downplay all the figures about the size and casualties of the Greek army and overestimate the size and casualties of the Macedonian 'superhero's' enemies in order to give a bloated picture of his heroic military activities. Further, the victorious Greeks controlled the battlefield. So unlike the Persians, they could recover many of their wounded soldiers.

The Persian high command disintegrated after Gaugemela. Bessus, the *Satrap* of Bactria commanding the largest and most militarily effective contingent in the hitherto defunct Achaemenid army, was seething with anger for Darius's not-so-good performance in the battlefield. Darius was murdered by Bessus, but the latter did not emerge as the centre of unified Persian resistance. Bessus retreated to his own *satrapy* of Bactria. And many Persian nobles voluntarily surrendered to Alexander. The Achaemenid Empire as a unified politico-military entity was finished. To sum up, military defeat and fractious political moves resulted in the collapse of the Achaemenid Empire.

Alexander's infantry faced its toughest opponent in Paurava's (Porus for the Greeks) elephants in the Battle of Hydaspes/Jhelum (May 326 BCE). As Alexander moved into Punjab, in May 326 BCE, a regional potentate known as Paurava confronted him with the *chaturanga bala*. Since Paurava had deployed at the eastern side of the Jhelum River with his force, Alexander could not cross the river in open sight. Moreover, the Greek horses were afraid of the elephants that stood like 'black mountains' on the east of the river. For several days, Alexander marched his army up and down, which generated a false sense of security among Paurava's men. Then one night when it was raining heavily, Alexander with a specially chosen force (6,000 infantry and 5,000 cavalry) marched upstream and crossed the river in rafts at midnight. In the early morning, Paurava sent his son with 2,000 infantry and 60 chariots to check the Macedonian force which had landed further upstream. Paurava believed that the main Greek force with Alexander was still in the camp on the west of the river opposite to the Indian camp in the east side of the river. Alexander's crossing of the river was a masterstroke. His decoy operation has certain similarities with the *maskirovka* (deception) operations conducted by the Red or Army during the Second World War. Sending a small detachment to contain the force which had crossed the river under Alexander was an erroneous move on part of Paurava. Alexander sent his horse archers, who made short work of the force under Paurava's son, named younger Paurava (McCrindle 1983: 100–1). This was the first time the horse archers played an important combat role in Indian history. And the horse archers would continue to dominate the subcontinent's military landscape till the early eighteenth century. As Alexander marched south, Paurava prepared to move his force north to prepare for the climactic battle.

Paurava deployed 85 elephants in the front, and behind them were stationed 300 chariots. These four-horse chariots (somewhat akin to the Hittite chariots) were bigger than the two-horse chariots introduced by the Aryans in India in around 1200 BCE. The third line comprised 30,000 infantry, and some of them were equipped with heavy bows. The arrows shot by such bows were able to penetrate Greek body armour, and one such arrow actually punctured Alexander's lungs during the siege of Malli in Punjab during 325 BCE. Paurava opened the battle by launching the chariot charge. As in Gaugemela, Alexander's disciplined light infantry made short work of the chariots. Then the elephants and the infantry advanced to the beating of the drums. The infantry archers had to rest their heavy bows on the ground while pulling the strings. Due to rainfall during the night, the ground was slippery, and the Indian bowmen could not discharge their long arrows (McCrindle 1983: 203–10, 325). However, the elephants created mayhem among the Macedonian infantry. Rufus portrays the face of the encounter in the following words: 'Thus the battle was doubtful, the Macedonians sometimes pursuing and sometimes fleeing from the elephants, so that the struggle was prolonged till the day was far spent' (quoted from McCrindle 1983: 211). The Greek phalangites hacked at the limbs and trunks of the beasts with axes and swords. The wounded elephants ran madly, creating disorder in the ranks of both Alexander's and Paurava's force. Many phalangites were trampled to death by the angry beasts, and several were flung upwards, caught by the elephants' trunks. As the Indian elephants and the Greek infantry were continuing their bloody slogging match, the battle was decided by cavalry encounter. The Dahae horse archers of Alexander's army attacked the Indian cavalry at Paurava's left wing. Simultaneously, Demetrius and Coenus with their cavalry detachments drove away Paurava's cavalry at his right wing. Superior discipline combined with quantitative superiority gave victory to the Greek cavalry against Paurava's horsemen. In contrast to Darius, Paurava, though wounded, remained in the battlefield. Paurava himself, riding an elephant, was heroically urging his men forward when he was captured. He thus started the Indian tradition of generals commanding their armies from the back of the elephants. In the preceding Bronze Age, military commanders directed the battle from chariots. And this tradition of commanding from the back of elephant would continue till the British conquest of India during the late eighteenth century. The Indian infantry and elephants now surrounded gave up the struggle just before evening as the news of Paurava's capture spread through the rank and file of the Indian soldiery (McCrindle 1983: 211, 308).

After the battle, a wounded Paurava was brought before Alexander. The latter when asked the former how he wanted to be treated; Paurava replied 'like a king'. Alexander made his valorous ex-opponent his client ruler. Alexander was in need of client rulers to manage the outlying parts of his expanding empire. The Indian casualties amounted to 12,000 killed and 9,000 as prisoners of war. The Macedonian losses numbered 280 cavalry and more than 700 infantry dead and several hundred wounded (McCrindle 1983: 107–9). In this battle, Alexander's infantry suffered its greatest losses, especially in combat against the elephants. If a small regional king with some 85 elephants could cause so many casualties to the Macedonian army, then fighting the much larger Nanda army in the Ganga Valley was indeed a dangerous proposition for Alexander's men. The Nanda army included 20,000 cavalry, 2,000 four-horse chariots, 200,000 infantry and 3,000 elephants (Jha 2004: 91). India did not produce good-quality horses. The Indian monarchs throughout history obtained horses mainly from Central Asia and Afghanistan through Punjab. Since the Nanda Empire did not extend up to Punjab, the Nanda military establishment lacked substantial numbers of cavalry. The Macedonian army would have made short work of the Nanda infantry, cavalry and chariots. But the large number of war elephants would possibly have torn the Macedonian phalanx apart. No doubt the terrific Battle of Hydaspes and the fear of the Nanda war elephants to a great extent created the Beas Mutiny in the summer of 326 BCE, which discouraged Alexander from moving further eastwards into India. Moving downward from Punjab into Sind, Alexander went back to Persia along the Makran Coast.

Hellenistic militaries

Alexander died at Babylon around 3.30 pm on 11 June 323 BCE at the age of 32. When Alexander died, his empire extended from the Indus River in the east to the Danube River in the west and in the south up to Egypt. It was bigger than Darius I's empire because the whole of Greece was part of the Macedonian Empire. Alexander's empire stretched over 5 million sq km (Waterfield 2011: 1–11). After Alexander's death, his generals started fighting for his empire, and this ensuing conflict is known as the Wars of the Diadochi (Successors). Out of the ashes of this war emerged the Hellenistic monarchies. While Antigonus, the one eyed (nicknamed Cyclops), and his son, Demetrius I, wanted to re-establish the unified empire of Alexander, the other generals such as Lysimachus, Ptolemy and Seleucus wanted to divide the empire among them.

Let us have a look at the hybrid armies of Alexander's generals. Infantry was of three types: light, medium and heavy. The sarissa-equipped Macedonian phalangites constituted the heavy infantry. The Greek hoplite mercenaries constituted the medium infantry. The light infantry equipped with bows, slings and javelins was mainly non-Greeks (Asians), and their job was to screen the main force as it was deploying. Light infantry functioned as missile troops and were effective in rough terrain. Light cavalry functioned as scouts. Both light cavalry and light infantry functioned as skirmishers during the battle. The Macedonian upper classes provided the heavy cavalry (Waterfield 2011: 60–2). The heavy cavalry functioned as shock troops, and the cavalrymen were armoured and were equipped with short spears (Sekunda 1989: 130). While the Seleucids used Indian elephants, the Ptolemies used the smaller African forest elephants. Generally, the elephants were placed at the front of the army and were spaced 50–150 feet apart to discourage enemy cavalry charges (Carey et al. 2005: 86). In contrast, the Indian and Chinese generals used elephants to disorganise the hostile heavy infantry.

Overall, the generals of Alexander improved upon the hybrid military system of their master. If anything, they further adapted their military machines by adopting certain Asian military techniques such as chariots, light infantry and elephants. This was because Alexander's marshals understood that in the different terrains of Eurasia for combat against different types of enemies, they required different types of force structures. In the semi-desert region of Palestine and swampy riverbed of lower Euphrates and Tigris, cavalry was non-operational and light infantry was at a premium. Not all the parties would oblige the Hellenistic rulers by fighting set piece battles. For sieges of the cities (for instance in Judea/Judaea), infantry and elephants were useful. And even when hostile parties offered set piece battle, then elephants along with heavy infantry could function as a force multiplier. For crossing the large distances of Asia, light cavalry was efficient. It remained so until the advent of internal combustion engines in the early twentieth century. For conducting set piece battles and guarding the far-flung frontiers along the Oxus and Jaxartes rivers in Sogdiana and Bactria against Central Eurasian nomads, inclusion of the mounted nomads with their skill in long-distance archery was essential.

While Persia provided the money and cavalry and India supplied elephants, from Phoenicia and Egypt came the shipbuilders and sailors, and Greece and Macedonia were the sources of heavy infantry. The model for using chariots was provided by both the Achaemenid and Paurava's armies. Light infantry was recruited from Asia and the Eastern Mediterranean lands. For instance, Crete provided archers and Achaians supplied slingers. In fact, a mercenary market operated in the Peloponnese which was able to feed the 'hungry' armies of Alexander's successors. There were two types of mercenaries: long-term volunteers who served in the royal field armies and short-term soldiers raised to tide over emergencies (Chaniotis 2005: 78–9; Bennett and Roberts 2008: 135). The Macedonian army was a national army, but the Seleucids and the Ptolemies depended on the military settlers. These two monarchies settled Macedonians and Thracians in the various military settlements across their empires. These men were given land, and in exchange they had to perform military service (Tarn 1930: 24–5). According to one estimate, the military settlements provided the Seleucid monarchy with 44,000 heavy infantry, 3,000 medium infantry and 8,500 cavalry. They were further supplemented by 16,000 mercenaries (Bar-Kochva 1979: 9).

Compared to Alexander's times, the size of the armies rose. While Alexander's field army comprised about 40,000 infantry and 6,000 cavalry, during the late third century BCE, the Seleucids could deploy about 70,000–80,000 men in the battlefield. However, the Macedonian kingdom's field army was smaller: between 25,000 and 35,000 men (Carey et al. 2005: 84). In 319 BCE, Antigonus, who ruled Asia Minor and Syria, had 60,000 infantry, 10,000 cavalry and 30 elephants. In contrast, the size of Ptolemy's army was smaller. He could field 18,000 infantry and 4,000 cavalry. Most of the personnel were Hellenistic mercenaries, and some were recruited from Egypt (Bennett and Roberts 2008: 59, 100). Antigonus in that year, with an army comprising 40,000 infantry, 7,000 cavalry and elephants, covered 462 km in a forced march of seven days and nights in order to destroy the army of his opponent Alcetas (Sekunda 1989: 130).

Ptolemy, son of Lagus (later known as Ptolemy I), got Egypt, where Alexander had left a war chest that contained 8,000 talents. This enabled Ptolemy to start recruiting soldiers. In 322 BCE Ptolemy, by annexing parts of North-East Libya, was able to control the caravan trade of that region. In 320 BCE, while Ptolemy was controlling Egypt and Cyrenaica, Seleucus was in charge of Babylonia. In the autumn of that year, Ptolemy occupied Palestine and Phoenicia. He was attempting to create a Greater Egypt like the great *pharaohs*. Ptolemy created a navy for controlling the Eastern Mediterranean (Waterfield 2011: 27–70).

Naval innovation occurred in the Hellenistic era. In 315 CE, Antigonus commanded 240 ships in Phoenicia. Of them, 210 were big heavy ships fitted with siege machinery. They had fighting decks for marines and catapults above the oarsmen. Further, these ships had heavy screens along their sides for protection of the oarsmen. For constructing warships, in 315 BCE, Antigonus used the cedars and cypresses of Lebanon and wood from the Taurus Mountains. In the same year, Ptolemy sent 100 warships and 10,000 soldiers to conquer Cyprus, which was to function as a bastion of Ptolemaic naval power in the Eastern Mediterranean. Early in 315 BCE, Seleucus took charge of a fleet comprising 100 warships and cruised along the Levantine shore, disrupting Antigonus's shipbuilding activities and wooing away his allies and dependants. Then he raided the coastline of Asia Minor, especially along Lydia and Ionia, the core areas of Antigonus's realm. By 305 BCE, Demetrius I started using very large and heavy ships known as polyremes. These were not used for ship versus ship combat. Rather, these polyremes had siege towers. They were floating siege machines and were used for capturing the coastal cities of Asia Minor. Antiochus III (r. 222–187 BCE) had a fleet geared for conducting amphibious warfare. In his fleet, there were light/fast triremes, each of which carried only 14 soldiers. In addition, he had decked vessels (quinquiremes) and trireme troopships. A quinquireme was able to transport 300 men and the trireme about 100 men (de Souza 2007a: 358, 360; Bennett and Roberts 2008: 78, 99, 117).

In 316 BCE, when Eudamus and Peithon were drawn into the vortex of the wars of the Diadochi, Chandragupta Maurya (340–297 BCE), an Indian warlord, conquered the Indian provinces under Greek rule. Chandragupta for a time had served in Alexander's army and had experience of the Macedonian art of war. Further, Chandragupta also recruited mercenaries from the north-west frontier of India. These mercenaries had experiences of fighting the Greeks and also occasionally had served in the armies of Alexander's governors stationed at the eastern edge of the Macedonian Empire. Therefore, Chandragupta's army was also to an extent an amalgam of Greco-Indian military traditions. He established the Maurya Empire in 322 BCE, which lasted till 187 BCE. While Antigonus ruled over Asia Minor, Seleucus, ruler of Syria in 306/305 BCE, took the royal title. In 305 BCE, Seleucus advanced towards Indus in order to reconquer the provinces lost to the Maurya Empire. Chandragupta, who was campaigning in the Narmada Valley, on hearing of the Greek advance turned north-westwards with an army of about 400,000 men. The 9,000 elephants in Chandragupta's army struck fear in the Greek force (Jha 2004: 100; Bennett and Roberts 2008: 127). The Greco-Roman chroniclers definitely exaggerated the size of Mauryan army, but there is no doubt that it was huge by the then-global standard. According to Indian tradition, Chandragupta in a series of encounters around Indus badly mauled Seleucus, who had ironically taken the title *Nicator* (the victor). The treaty between Seleucus and Chandragupta favoured the latter. In return for 500 war elephants (which the Maurya Empire could easily allow due to the vast elephant herds of India), Chandragupta gained Eastern Afghanistan and present-day Pakistan and married a daughter of Seleucus named Helen (not to be confused with Helen of Troy). Later, a Greek ambassador named Megasthenes was stationed at Pataliputra, the capital of the Maurya Empire (Roy 2012: 62). The eastward expansion of the Greeks was over. But this treaty was mutually beneficial. Seleucus, a young infantry captain during the Battle of Hydaspes, was impressed by Porus's war elephants. He would use the war elephants acquired from Chandragupta with great effect in his wars with other generals of Alexander. The West's acquisition of knowledge from the East is still to be studied in detail. The Indian doctors impressed Alexander himself. So much so that many Indian doctors were enlisted in the Macedonian army to cure the soldiers of diseases and pain. They were especially esteemed for curing snake bites (Habib and Jha 2004: 157).

Antigonus, for his attack on Egypt, massed 80,000 infantry and 8,000 cavalry. In response, Ptolemy and Seleucus banded together against the common danger. At the Battle of Ipsus (301 BCE), both sides deployed about 80,000 soldiers each (Bar-Kochva 1979: 7). The Seleucid elephants were battle ready thanks to the efficient logistical apparatus of the empire. It is to be noted that each adult elephant consumed 272 kg of green fodder per day. These elephants had marched from Baluchistan through Iran to Babylon and finally reached Ipsus in Asia Minor. At Cappadocia, Seleucus had 480 elephants, and at Ipsus he deployed 400 of these beasts. Against them, Antigonus could only deploy 75 elephants. To a great extent, the Seleucid victory at Ipsus was due to a superior number of Indian elephants. After the battle, Seleucus constructed a stable for his elephants at Apamea on the Orontes (Glover 1944: 257). In 217 BCE, at the Battle of Raphia, Antiochus III's Indian war elephants battled with Ptolemy IV's African elephants. On the Egyptian right wing, the smaller African war elephants refused to charge the bigger Indian beasts (Carey et al. 2005: 86–7). As part of the action–reaction dialectic, even the Macedonian army acquired elephants. For instance, Antipater's successor Polyperchon also possessed 67 war elephants (Habib and Jha 2004: 16).

The Seleucid Empire was beset with constant internal rebellions. The Maccabean Revolt started in 167 BCE. At that time, Judea was a small province of the Syrian-based Seleucid Empire. Antiochus III took Palestine from the Ptolemies of Egypt and incorporated Judea in his empire in 198 BCE. Judea is mostly mountainous territory broken by *wadis* (dry riverbeds), with the Mediterranean Sea in the west and the Jordan Valley and the Dead Sea in the east. The population of this region was about 250,000 people. Antiochus IV came to power in 175 BCE. In the east, he faced a threat from the Medes and the Persians, and in the west, Rome posed the principal threat. In the south-west, Egypt posed a minor threat. Antiochus IV's policy of imposing Hellenic culture as part of the policy to unify his empire partly gave rise to the Maccabean rebellion (Herzog and Gichon 1997: 265–7). At the Battle of Ammaus (165 BCE), the Seleucids deployed 40,000 infantry and 7,000 cavalry. Of them, 20,000 were royal troops, and the rest were provided by the Philistines and Edomaeans. Crushing the Jewish insurgency proved costly for the Seleucids from the military point of view. Besides the insurgent ethnic groups, Seleucid *satraps* like those of the Achaemenid Empire at times proved to be disloyal and raised the banner of rebellion. In April 160 BCE, the suppression of the rebellion led by Timarchus in the eastern *satrapies* involved the participation of half the available royal troops of the Seleucid Empire: 20,000 infantry (including 10,000 phalangites) and 2,000 cavalry (Bar-Kochva 1979: 13–15).

Threats from the other successor states of Alexander's empire, internal rebellions, the Romans in the west and the steppe nomads in the north were the four security problems faced by the Seleucid monarchs. Antiochus III, in an attempt to reestablish his power in the eastern *satrapies* of his empire, marched through Media, Hyrcania, Bactria and Punjab with 100,000 infantry and 20,000 cavalry. Of these troops, 10,000 infantry and 2,000 cavalry constituted the standing army of his empire. The rest were mercenaries and troops supplied by his allies. After securing the eastern *satrapies*, Antiochus mobilised a fleet to conquer Greece. Antiochus landed in Greece at the beginning of the winter of 192–91 BCE. In 191 BCE, at the Second Battle of Thermopylae (I call the battle between Leonidas's force and Xerxes's army the First Battle of Thermopylae), Antiochus III deployed 12,000 cavalry, 45,200 infantry, 1,000 light infantry and 54 elephants, each of which carried four archers on their backs. The hybrid military system of the Seleucids had finally reached the Balkans. In 190 BCE, at the Battle of Magnesia, the Seleucid force comprising 60,000 infantry and 12,000 cavalry was defeated by a Roman force of 30,000 infantry (Bar-Kochva 1979: 8–18).

In 129 BCE, Antiochus VII made a final attempt to resurrect Seleucid power in the eastern part of his collapsing empire. He collected 80,000 soldiers and 200,000 civilian followers. Of these soldiers, the military settlements of Asia Minor provided 8,000 infantry and 500 cavalry. As Antiochus VII crossed the Tigris, most of the military settlers from Media and Susiana also joined his ranks, thus swelling his force (Bar-Kochva 1979: 10–11). In 129 BCE, the Parthians, by defeating Antiochus VII, destroyed Seleucid rule in Iran. And in 63 BCE, the Romans ended Seleucid rule in Syria. The Parthians were able to defeat the Seleucids because the latter had no counter to the former's mounted archery. In fact, in the flat open plains of Mesopotamia, the Parthians were later able to defeat even the Roman legionaries several times. And in Greece and Asia Minor, the superior Roman infantry arranged in a flexible maniple system (discussed in the next chapter) proved superior to the cumbersome hoplite-phalanx infantry organisation of the Hellenistic monarchies. The next section turns the spotlight on the evolution of theory and practice of warfare east of the Hellenistic monarchies.

India and China

In both China and India, professional bureaucratic armies emerged after 400 BCE. During the Spring and Autumn Period (771–481 BCE), the Chinese armies were small, inefficiently arrayed, ineptly led, poorly equipped, badly trained and haphazardly supplied. However, during the Warring States era (403–221 BCE), a big polity could maintain 1,00,000 troops who were well armed, properly supplied and commanded by professional commanders. Such armies, unlike the undrilled peasant levies of the Spring and Autumn Period, were capable of conducting tactical manoeuvres using gongs, drums, flags, banners and bells. Post-500 BCE Chinese armies comprised infantry and cavalry. The chariots were outdated (Sun Tzu 1963: 7). In Mauryan India, the royal standing army comprised infantry, cavalry and elephants. The Indian armies in fact displayed a learning curve. After Hydaspes, the chariots were marginalised in the Maurya order of battle. Further, the Indian strategic managers emphasised the necessity of drilling and training the troops and recommended use of combined arms formation too. For instance, elephants carrying archers on their *howdahs*, cavalry and infantry armed with spears, bows and swords were trained to cooperate with each other (Kautilya 1992: 676–84).

India to a greater extent and China to a lesser extent continued to use elephants in warfare. In 506 BCE, when the maritime state of Wu attacked Ch'u, the latter's commander sent elephants with torches tied to their tails into the ranks of Wu soldiery in order to dislocate the latter's formation. However, the use of elephants in warfare died out. This was because the extensive forests of South China which were the natural habitat of the elephant herds were cut down by an expanding population practising paddy cultivation. And due to ecological conditions, elephants that require warm tropical conditions did not thrive in arid North China. The forests of Central India, East India and Orissa were the chief sources of elephants for the Hindu rulers of the ancient period. In fact, the monarchs took care to preserve forests and grasslands for raising elephants. A professional class known as *hastipa* came into existence, charged with the duty of upkeep of the war elephants. Further, India invented the *ankush* (a rod with curved end). It was used by the *mahout* (driver) to control the elephant. Further, the Indians came up with *howdah*, a boxlike structure made of wood and tied to the back of the elephant. The *howdah* functioned as a platform from which the *mahout* controlled the elephant, and warriors fought with bows and javelins. The Ptolemids and Carthaginians used Indian *mahouts* along with *ankush* and *howdahs* (Schafer 1957: 290; Habib and Jha 2004: 29; Chakraborty 2015: 75–6; Trautmann 2015: 1–10).

Regarding cavalry warfare, a sort of convergence occurred between the Chinese sedentary polities and the steppe nomads inhabiting the north of the Gobi Desert. Horse riding was introduced into North China from the neighbouring steppe nomads in or about the fourth century BCE (Goodrich 1984: 280–1). In 320 BCE, King Wu Ling of Chao State made cavalry an integral part of the army and introduced trousers (a Central Eurasian dress) for the cavalrymen (Sun Tzu 1963: 11). Trousers were more effective for riding horses than the skirts of the Macedonian cavalrymen. Select units of Chinese cavalry were equipped with iron swords during this period (Barnard 2005: 80). Further, Wu Ling established stud farms for horses in the border areas. Chinese cavalry forces were trained in the style of the nomads. Furthermore, nomadic people were also recruited in the cavalry branch. China acquired horses from the nomads in return for tea (Dreyer 2012: 24). In India, horses were introduced by the Aryans in around 1200 BCE. The earliest representation of the stirrup in the world comes from the Sanchi sculptures of India, which can be dated to about the late second century BCE (Dien 2005b: 185). It is to be noted that cavalrymen under Alexander and in the Hellenistic polities rode their mounts without saddles. Strange as it may seem, one could argue that the military effectiveness of the Indian cavalry was higher than the much-talked-about cavalry force of Alexander's successors.

During the Warring States-era China and Mauryan India, infantry became important on the battlefields due to drilling and adoption of armour and better weapons. Indian steel was better than that produced in Achaemenid Europe or in Greece. Charcoal from the extensive forests of North and Central India was used for smelting the iron ores. The iron used for making swords had 0.7 per cent carbon (Habib and Jha 2004: 122). In China, iron-tipped spears became common. The adoption of long swords that replaced short daggers increased the combat effectiveness of the Chinese infantry. The long swords raised the combat effectiveness of the Chinese infantry vis-à-vis the chariots (Sun Tzu 1994: 35–6). This transformation had occurred in the Near East around 1200 BCE, and it occurred in China during the fourth and third centuries BCE. During the fourth century BCE, iron plates of armour laced together through pairs of holes placed at the edges came into use in China. In addition, lamellar leather armour continued to be used (Dien 2005a: 85, 92). By the third century BCE, the Chinese infantry was using crossbows and the volley fire technique in order to counter the attacks by the horse nomads using mounted archery. The crossbow was adopted by the Chinese several centuries before its adoption by the Roman army. Probably crossbow technology spread from China to Mediterranean Europe (Kierman, Jr. 1974: 64; Wilbur 2005: 51–9).

A fusion occurred in the field of siege artillery. India had developed some siege machines before the invasion of Alexander. The Greek mercenaries in Mauryan army introduced the scientific techniques regarding Greek siege machines. The result was that siege engines under the Maurya Empire were upgraded to world standards. The Mauryan army's siege artillery comprised wooden siege towers, battering rams and big stone-throwing machines. These engines during campaigns were transported by bullock trains and river boats (Habib and Jha 2004: 30).

Due to the vast demographic resources and advanced agrarian economies of India and China, polities in these two countries were able to maintain big armies. The Maurya Empire maintained 600,000 infantry, 30,000 cavalry and 9,000 elephants. The horses and the elephants were maintained at the king's expense. We know that in the Maurya Empire, taxes were paid partly in cash. From the sixth century BCE, punch-marked silver coins were in use in India. In the Maurya Empire, the state mints mass-produced punch-marked coins. Silver and lead were extracted from the mines near Aravalli Hills. South of the Maurya Empire was the Andhra Kingdom. It had 30 well-fortified towns and maintained 100,000 infantry, 2,000 cavalry and 1,000 elephants (Habib and Jha 2004: 21, 28, 38, 121). The forests of south India and Sri Lanka were suitable for breeding elephants. However, good horses cannot be bred in the hot and humid climate of South India, where the landscape was dotted with extensive paddy cultivation. This explains the comparatively small size of cavalry force of the Andhras. At times, war was bloody in ancient India. Asoka's Kalinga Campaign (262–261 BCE) resulted in the death of more than 100,000 persons. The large size of the ancient Chinese armies and the lethal nature of warfare could be inferred from the huge number of battle casualties. For example, between 356 and 236 BCE, the Qin armies killed more than 1.5 million soldiers of the enemy states (Jha 2003:

96–7; Hui 2005: 87). Only during First World War would Europe experience battle casualties of such magnitude.

Ancient China and India also developed sophisticated theories of war. Mencius (371–289 BCE), a Confucian scholar, advocated punitive military expeditions to chastise evil rulers. However, unlike Achaemenid Persia and Hellenistic Greece, Chinese bureaucracy and the Hindu intellectuals of ancient India looked down upon the profession of arms and military force (Sun Tzu 1994: 32).

The most famous ancient Chinese theoretician of war is Sun Tzu/Sun Zi/Master Sun, who composed *The Art of War*, and his equivalent in ancient India was the famous theoretician Kautilya, author of the *Arthashastra* (Science of Politics). Scholars still debate the authorship and date of both works. Victor H. Mair asserts that *The Art of War* is not the product of a single author who supposedly lived during late sixth and early fifth century BCE. Other scholars claim that Sun Tzu lived in the fourth century BCE. Ancient Chinese historian Ssu-ma Chien, writing in around 100 BCE, notes that Sun Tzu, an inhabitant of Chi, offered his text to King Ho-lu of Wu around 500 BCE. Sun Tzu probably conceptualised his theory based on combat which occurred during the Warring States era. We can surmise that his followers developed his thoughts as *The Art of War* (Yates 2005: 278–9; *Art of War* 2007: LI, 1).

As far as Kautilya is concerned, while one group of scholars argues that the *Arthashastra* is the product of a group of learned individuals, the other view is that Kautilya/Chanakya/Vishnugupta is the author of this compendium. Generally, there is an agreement that the core of the work focuses on the Maurya Empire. One interpretation is that Kautilya was a learned Brahmin from Taxila or Gandhara who initially served the Nanda Empire but had to leave the Nanda Court at Pataliputra due to personal problems. He became an inveterate enemy of the Nanda Empire and aided the young Chandragupta in overthrowing the Nandas. And when Chandragupta established the Maurya Empire, Kautilya became his principal minister. Probably Kautilya composed the *Arthashastra* during retirement after his stint in the Mauryan Court was over (Roy 2012: 58–9).

Kautilya, in *Arthashastra*, attempts to put forward the timeless principles of politics, economics, diplomacy and war. Kautilya's work does not refer to any particular historical event. This is because, in accordance with the tradition of the *shastras* (Hindu genre of learned works), great works expound timeless principles. Any reference to fleeting historical individuals and events would reduce the value of such a work. However, no theorist can remain unaffected by the surrounding historical reality and pre-existing scholarly traditions. Kautilya was certainly influenced by the conditions existing before and during the formation of the Maurya Empire. *Arthashastra* is divided into 15 books. Book 7 deals with foreign policy and the use of force to gain strategic objectives. Books 9 and 10 deal with military preparations by the *Vijigishu* (ruler with hegemonic ambition). Book 12 shows how defence at times can triumph over offence. In fact, this book examines how a weak power can survive against a stronger power. Book 13 describes siege warfare in details (Roy 2012: 60–1).

Arthashastra and *The Art of War* focus on indirect strategy. Initially, Chandragupta attempted a direct attack on the Nanda capital and failed. Then Chandragupta, in accordance with the advice of Kautilya, practised indirect strategy. He first won over the peripheral areas and then moved towards the core of the Nanda Empire. Punjab was in a state of chaos at that time. The Greek officers murdered Paurava. Chandragupta, after hiring mercenaries from the outlying tribes along the north-west frontier of India, defeated the Greek officials left behind by Alexander and then moved towards the final kill by advancing towards the Nanda capital, Pataliputra, in 321 BCE. In the final battle with the Nandas, more than 1,000 war chariots along with elephants covered with iron armour (each beast carried a *mahout* and three archers in the *howdah*) and infantry clashed. In the end, the Nanda army under its general, Bhaddasala, was defeated. By 312 BCE, Chandragupta's conquest of North and North-West India was complete (Mookerji 1960: 35; Sinha 1965: 131; Roy 2012: 61–2). There were several reasons for Chandragupta's victory over the Nandas. He had access to the mercenaries from North-West India who had experience of fighting with and against the Greeks. Thus, these military personnel were veterans of the latest combat techniques. Further, Chandragupta, unlike the Seleucids and Carthaginians, had started the practice of covering the elephants with iron armour, which increased their military effectiveness. Further, Chandragupta had access to warhorses through Punjab. Lastly, the Nanda Empire was seething with internal rebellions due to the heavy taxation introduced by the monarch.

Sun Tzu, like Kautilya, is in favour of achieving victory without fighting, if possible. Sun Tzu and Kautilya prescribe psychological measures for unnerving the enemy rather than crushing him in a decisive battle. Their techniques are to demoralise, misdirect, mislead and outmanoeuvre the enemy. The objective is to break down the hostile party's social cohesion and generate chaos and disorder in the enemy's camp. Force is to be used only in the last resort and in a careful, graduated manner. Deception is at the heart of Kautilya and Sun Tzu's military method. Both Sun Tzu and Kautilya give overt importance to intelligence matters. Sun Tzu speaks of the need for efficient signalling during the heat of battle (*Art of War* 2007: xiv–xxii, 2). Sun Tzu speaks of fixing the enemy and then bringing a manoeuvring detachment to surprise and topple the enemy army (Wallacker 2005: 299).

Besides Sun Tzu, there is another theoretician named Sun Bin/Pin (*c.* 380–320 BCE). He served the state of Qi. At the famous Battle of Maling (341 BCE), he defeated the Wei army under Pang Juan (*Art of War* 2007: 6–7). Sun Pin had placed 10,000 crossbowmen in an ambush position, which annihilated the enemy (Sun Tzu 1963: 9). Both Sun Tzu and Kautilya would have been very happy with Sun Pin's use of an unorthodox method for winning victory. However, Sun Tzu II or Sun Bin is less aggressive than Kautilya. Sun Bin warns: 'Those who enjoy militarism, however, will perish. . . . War is not something to enjoy, victory is not to be an object of ambition' (quoted from Sun Tzu II 1996: 21). In contrast, Kautilya urges the *Vijigishu* to achieve paramountcy over the Indian subcontinent by acting aggressively.

Arthashastra argues for the establishment of a centralised bureaucratic empire. To an extent, the Maurya Empire was a centralised entity. The Mauryas were great road builders like the Romans, and all the roads built by them in the subcontinent lead to the Maurya capital Pataliputra. *Arthashastra* speaks of the necessity of maintaining a standing army and the importance of paying the military personnel in cash regularly. In fact, the Mauryan economy was a cash economy, and Kautilya speaks of die striking (punch-marked) coins. Kautilya also speaks of a royal monopoly over the production of weapons (Roy 2012: 61–3).

Ajatasatru, the ruler of Magadha (r. 493–462 BCE) had fortified Rajagriha (his capital) with a 40 km-long wall. This wall was much longer than the famed 33.7 km long walls of Athens which linked the city with Piraeus Harbour. Then Chandragupta laid siege to Kusumapura in 320 BCE. All these sieges influenced Kautilya's theory of siege warfare. Rather than undertaking a costly and time-consuming siege operation, Kautilya is for launching commando attacks in order to capture enemy forts by stealth. While laying siege to a fort, Kautlya advises that the besieging army should lay waste to the countryside to prevent supplies reaching the besieged garrison (Kautilya 1992: 537, 564). Similarly, between 240 and 220 BCE, an engineer cum theorist named Philon/Philo of Byzantium submitted a detailed report to his patron, a general named Ariston. Philo is the author of work on mechanics and artillery. Philo discusses siege warfare in detail and how to use a navy to conduct sieges of harbours and coastal cities. Philo was obviously influenced by Demetrius I's large polyremes. However, Philo is more technical compared to Kautilya as regards the discussion of siege machines. Though *Arthashastra* speaks of *yantras* (machines) for conducting siege warfare, Kautilya, unlike Philo, does not give precise mechanical details about these siege machines. Philo, for instance, discusses a catapult powered by bronze springs for conducting siege warfare (Murray 2014: 129–31). However, ancient India also produced a technical treatise. *Arthashastra* offers a lot of technical data about the construction of defensive fortifications in the forts. Another example of military-technical literature is Visvakarmiya's *Rathalaksanam*. This treatise provides measurements and other technical details (i.e. types of wood to be used for manufacturing different parts of the vehicle, such as axles, spokes etc.) for actual construction of various types of chariots (*Visvakarmiya Rathalaksanam* 1994).

Use of deception and treachery in warfare are part and parcel of both Eastern and Western ways of warfare. Kautilya advocates that *sama* (diplomacy, conciliation), *dana* (bribery) and *bheda* (treachery, deception and trickery for creating dissension among the enemy ranks before attacking) should be used with *danda* (military force) as part of the grand strategy to weaken and destroy the enemy (Roy 2012: 61). Kautilya was probably influenced by past historical incidents. Ajatasatru became the ruler of Magadha by killing his father Bimbisara. And *Arthashastra* warns the *Vijigishu* against treachery by sons, queens and close relatives. Further, Ajatasatru created dissensions among the Vajji Confederacy before overthrowing them. Like Sun Tzu and Kautilya, Philo notes that the best way to capture a hostile city is through stealth or deceit when the enemy is least prepared for an attack, as this technique results in minimum casualties. *Arthashastra* notes: 'Miraculous results can be achieved by practicing the method of subversion' (quoted from Kautilya 1992: 519). Sextus Julius Frontinus (*c.* 130–80 CE) a Roman magistrate cum theorist is the author on the water supply of Rome and on stratagems (*Strategematica*), and Polyaenus from Macedonia, a theoretician of the second century CE, compiled a collection of 900 historical anecdotes about stratagems published at the beginning of Parthian War in 162 CE. Successful tricks called *strategemata* are given weight by theorists like Polyaenus and Frontinus (Murray 2014: 133).

Kautilya is not merely a military theorist but also a realist theorist of power. For him the interstate system is a *mandala* (circle of state) which is characterised by *matsanya* (anarchy), where the big fish swallows the smaller fries. *Arthashastra* speaks that all states seek to maximise power in the anarchic international system in order to ensure security. The *Arthashastra* says: 'Power is possession of strength and success is obtaining happiness' (quoted from Roy 2012: 68). Power, which is a combination of political, economic and military assets, is the only security in the big bad world characterised by disorder and chaos. The power balance among the states is a fluid one and is constantly changing, thus creating opportunities as well as danger for the *Vijigishu*. The operating principle is 'Might is right'. As part of the survival strategies of the state, Kautilya asks the *Vijigishu* to remain ready to launch a pre-emptive strike or to start a preventive war in order to prevent a permanent imbalance of power in the *mandala*. A preventive war is waged to arrest the growth of hostile military power through bold and timely action, exploiting one's advantage while one can. And a pre-emptive strike is an attack launched against the enemy for self-defence when a massive enemy strike is almost certain to come and cannot be effectively checked by mere passive defence (Roy 2012: 66–7).

Similarly, the Greek historian Thucydides also uses his historical analysis of the Second Peloponnesian War to bring out the timeless principles about the origins of interstate conflict. The principal factors for going to war, according to him, are hegemonic ambitions, past injuries and land disputes (Robinson 2006: 250). Thucydides implies that fear on the part of Sparta due to rising power of Athens was the principal reason for the Spartan attack on Attica which started the Second Peloponnesian War. So Thucydides, like Kautilya, also harps on preventive war.

Conclusion

Philip created a hybrid army by adopting elements from both the Classical Greek military tradition and the Assyrian-Achaemenid military system. This composite army was the tool used by Alexander to overthrow the Achaemenid Empire. Later, Alexander and his generals took further the process of hybridisation by including Asian elements in their military machines. Geographical imperatives and differing security concerns forced them to accelerate the process of hybridisation of their military machines. The military machines of both East and West experienced innovations and hybridisation. While the Achaemenids tried to increase the combat effectiveness of their force by integrating heavy cavalry with mounted bowmen, scythe chariots and elephants, Macedonia was able to build a much more combat-effective balanced force. In all the battles with the Macedonian army, infantry–cavalry combined arms tactics plus Alexander's daredevil heroic generalship won the day. The scope and intensity of warfare in Eurasia increased in the post-Alexander era. The unique Indian military technique of using war elephants in battle also influenced the Seleucids to adopt this military practice and implement it in the battles fought in the Near East. The use of Indian war elephants by the Seleucids forced the Ptolemids to use African war elephants too, which in turn influenced Macedonia, Carthage and Rome. Further, this chapter has shown that presence of trained infantry and evolution of secular military theories were not unique to Classical Europe but were also present in China and South Asia.

As regards the theory of war, there is striking similarity between Sun Tzu of China and Kautilya of India. Superficially, it seems that the military theories of East and West are poles apart. The assumption of John Keegan and Geoffrey Parker that while the Asian (Chinese and Indian) military theories seem to focus on guile and treachery, the Western (Greek) militaries focus on decisive set piece battles is erroneous. Actually, there is convergence in the military theories and practices of the East and West. We have seen that both the Macedonian and the Achaemenid as well as the Indian doctrines of war emphasised fighting decisive battles. However, the Greeks were also not above using treachery when it was required. The biggest example is use of the so-called Trojan Horse and later Frontinus's work. Greco-Roman theorists, like their Asian counterparts, at times stressed treachery and deception. And despite Kautilya and Sun Tzu's injunctions of not fighting battles, in reality, ancient Chinese and Indian rulers during the Warring States era and in Mauryan India frequently engaged in battles and protracted sieges. It must not be forgotten that the greatest test of Alexander's army was during the Battle of Hydaspes. And China experienced several decisive battles during the Warring States era. Both Kautilya and Thucydides could be categorised as realist theorists of war. In fact, we could speak of a common pool of Eurasian body of military thought in Classical Antiquity. The next chapter focuses on the rise of the Roman superpower, its enemies, and military divergence in China and India.

References

Anon, *Visvakarmiya Rathalaksanam: A Study of Ancient Indian Chariot (With Sanskrit Text, English Translation and Historical Notes)*, translated by R.P. Kulkarni (Delhi: Kanishka Publishers, 1994).

Badian, Ernst, *Collected Papers on Alexander the Great* (London/New York: Routledge, 2012).

Bar-Kochva, B., *The Seleucid Army: Organization and Tactics in the Great Campaigns* (1976, reprint, Cambridge: Cambridge University Press, 1979).

Barnard, Noel, 'Did the Swords Exist?', in Peter Lorge (ed.), *Warfare in China to 1600* (Aldershot: Ashgate, 2005), pp. 71–82.

Bennett, Bob and Roberts, Mike, *The Wars of Alexander's Successors 323–281 BC*, vol. 1, *Commanders & Campaigns* (Barnsley: Pen & Sword, 2008).

Bosworth, A.B., *Conquest and Empire: The Reign of Alexander the Great* (1988, reprint, Cambridge: Canto, 1993).

Carey, Brian Todd, Allfree, Joshua B. and Cairns, John, *Warfare in the Ancient World* (Barnsley: Pen & Sword, 2005).

Chakraborty, Debraj, 'Elephants in Pre-modern India', in Kaushik Roy and Peter Lorge (eds.), *Chinese and Indian Warfare: From the Classical Age to 1870* (London/New York: Routledge, 2015), pp. 74–90.

Chaniotis, Angelos, *War in the Hellenistic World: A Social and Cultural History* (Oxford: Blackwell, 2005).

Delbruck, Hans, *Warfare in Antiquity: History of the Art of War*, vol. 1, translated from the German by Walter J. Renfroe, Jr. (Lincoln/London: University of Nebraska Press, 1990).

de Souza, Philip, 'The Hellenistic World and the Roman Republic, Military Forces, B. Naval Forces', in Philip Sabin, Hans Van Wees and Michael Whitby (eds.), *The Cambridge History of Greek and Roman Warfare*, vol. 1, *Greece, the Hellenistic World and the Rise of Rome* (Cambridge: Cambridge University Press, 2007a), pp. 357–67.

de Souza, Philip, 'The Hellenistic World and the Roman Republic, Battle, B. Naval Battles and Sieges', in Philip Sabin, Hans Van Wees and Michael Whitby (eds.), *The Cambridge History of Greek and Roman Warfare*, vol. 1, *Greece, the Hellenistic World and the Rise of Rome* (Cambridge: Cambridge University Press, 2007b), pp. 434–60.

Devine, Albert, 'Alexander the Great', in John Hackett (ed.), *Warfare in the Ancient World* (London: Sidgwick & Jackson, 1989), pp. 104–29.

Dien, Albert E., 'A Study of Early Chinese Armor', in Peter Lorge (ed.), *Warfare in China to 1600* (Aldershot: Ashgate, 2005a), pp. 83–144.

Dien, Albert E., 'The Stirrup and Its Effect on Chinese Military History', in Peter Lorge (ed.), *Warfare in China to 1600* (Aldershot: Ashgate, 2005b), pp. 185–208.

Dreyer, Edward L., 'Continuity and Change', in David A. Graff and Robin Higham (eds.), *A Military History of China* (Lexington, KY: University Press of Kentucky, 2012), pp. 19–38.

Fuller, J.F.C., *The Generalship of Alexander the Great* (1960, reprint, New York: Da Capo, 1989).

Glover, Richard, 'The Elephant in Ancient War', *Classical Journal*, vol. 39, no. 5 (1944), pp. 257–69.

Goodrich, Chauncey S., 'Riding Astride and the Saddle in Ancient China', *Harvard Journal of Asiatic Studies*, vol. 44, no. 2 (1984), pp. 279–305.

Habib, Irfan and Jha, Vivekanand, *A People's History of India*, vol. 4, *Mauryan India* (New Delhi: Tulika, 2004).

Hacker, Barton C., 'Greek Catapults and Catapult Technology: Science, Technology, and War in the Ancient World', *Technology and Culture*, vol. 9, no. 1 (1968), pp. 34–50.

Hamilton, Charles D., 'The Hellenistic World', in Kurt Raaflaub and Nathan Rosenstein (eds.), *War and Society in the Ancient and Medieval Worlds: Asia, The Mediterranean, Europe and Mesoamerica* (Washington, DC: Center for Hellenic Studies, distributed by Harvard University Press, 1999), pp. 163–91.

Hammond, N.G.L., *Alexander the Great: King, Commander and Statesman* (1980, reprint, London: Duckworth, 2001).

Heckel, Waldemar, *The Conquests of Alexander the Great* (2008, reprint, New Delhi: Cambridge University Press, 2013).

Herzog, Chaim and Gichon, Mordechai, *Battles of the Bible: A Military History of Ancient Israel* (1978, reprint, London: Greenhill, 1997).

Hui, Victoria Tin-bor, *War and State Formation in Ancient China and Early Modern Europe* (Cambridge/New York: Cambridge University Press, 2005).

Jha, D.N., *Ancient India in Historical Outline* (1977, reprint, New Delhi: Manohar, 2003).

Jha, D.N., *Early India: A Concise History* (New Delhi: Manohar, 2004).

Kautilya, *The Arthashastra*, edited, rearranged, translated and introduced by L.N. Rangarajan (1987, reprint, New Delhi: Penguin, 1992).

Kierman, Jr., Frank A., 'Phases and Modes of Combat in Early China', in Frank A. Kierman, Jr. and John K. Fairbank (eds.), *Chinese Ways in Warfare* (Cambridge, MA: Harvard University Press, 1974), pp. 27–66.

Lendon, J.E., *Soldiers & Ghosts: A History of Battle in Classical Antiquity* (New Haven/London: Yale University Press, 2005).

Markle, III Minor M., 'Use of Sarissa by Philip and Alexander of Macedon', *American Journal of Archaeology*, vol. 82, no. 4 (1978), pp. 483–97.

McCrindle, J.W., *The Invasion of India by Alexander the Great* (1896, reprint, New Delhi: Cosmo, 1983).

Mookerji, R.K., *Chandragupta Maurya and His Times* (1943, reprint, Delhi: Motilal Banarasidass, 1960).

Murray, William M., *The Age of Titans: The Rise and Fall of the Great Hellenistic Navies* (2012, reprint, New York: Oxford University Press, 2014).

Rihll, Tracey, *The Catapult: A History* (2007, reprint, Yardley, PA: Westholme, 2013).

Robinson, Eric, 'Thucydides and Democratic Peace', *Journal of Military Ethics*, vol. 5, no. 4 (2006), pp. 243–53.

Roy, Kaushik, *Hinduism and the Ethics of Warfare in South Asia: From Antiquity to the Present* (Cambridge/New York: Cambridge University Press, 2012).

Rufus, Quintus Curtius, *The History of Alexander*, translated by John Yardley with an Introduction and Notes by Waldemar Heckel (1984, reprint, London: Penguin, 2004).

Schafer, Edward H., 'War Elephants in Ancient and Medieval China', *Oriens*, vol. 10, no. 2 (1957), pp. 289–91.

Sekunda, Nicholas, 'Military Forces, in Philip Sabin, Hans Van Wees and Michael Whitby (eds.), *The Cambridge History of Greek and Roman Warfare*, vol. 1, *Greece, the Hellenistic World and the Rise of Rome* (Cambridge: Cambridge University Press, 2007), pp. 325–57.

Sekunda, Nick, 'Hellenistic Warfare', in John Hackett (ed.), *Warfare in the Ancient World* (London: Sidgwick & Jackson, 1989), pp. 130–5.

Sinha, B.P., 'Art of War in Ancient India (600 BC-300 AD)', in Guy S. Metraux and Francois Crouzet (eds.), *Studies in the Cultural History of India* (Agra: Shiva Lal Agarwal and Company, 1965), pp. 115–63.

Strauss, Barry, 'Archaic and Classical Greece, Battle, B. Naval Battles and Sieges', in Philip Sabin, Hans Van Wees and Michael Whitby (eds.), *The Cambridge History of Greek and Roman Warfare*, vol. 1, *Greece, the Hellenistic World and the Rise of Rome* (Cambridge: Cambridge University Press, 2007), pp. 223–47.

Sun Tzu, *The Art of War*, translated with an introduction by Samuel B. Griffith (London: Clarendon Press, 1963).

Sun Tzu, *The Art of War*, translated with introductions and commentary by Ralph D. Sawyer with the collaboration of Mei-chun Sawyer (Boulder: Westview, 1994).

Sun Tzu II, *The Lost Art of War*, translated with commentary by Thomas Cleary (New York: HarperCollins, 1996).

Sun Zi, *The Art of War: Sun Zi's Military Methods*, translated by Victor H. Mair (New York: Columbia University Press, 2007).

Tarn, W.W., *Hellenistic Military & Naval Developments* (London: Cambridge University Press, 1930).

Trautmann, Thomas R., *Elephants and Kings: An Environmental History* (Ranikhet: Permanent Black, 2015).

Wallacker, Benjamin E., 'Two Concepts in Early Chinese Military Thought', in Peter Lorge (ed.), *Warfare in China to 1600* (Aldershot: Ashgate, 2005), pp. 235–9.

Waterfield, Robin, *Dividing the Spoils: The War for Alexander the Great's Empire* (Oxford: Oxford University Press, 2011).

Wilbur, C. Martin, 'The History of the Crossbow, Illustrated from Specimens in the United States National Museum', in Peter Lorge (ed.), *Warfare in China to 1600* (Aldershot: Ashgate, 2005), pp. 51–62.

Wiseman, D.J., 'The Assyrians', in John Hackett (ed.), *Warfare in the Ancient World* (London: Sidgwick & Jackson, 1989), pp. 36–53.

Worthington, Ian, 'Alexander the Great, Nation Building, and the Creation and Maintenance of Empire', in Victor Davis Hanson (ed.), *Makers of Ancient Strategy: From the Persian Wars to the Fall of Rome* (Princeton/Oxford: Princeton University Press, 2010), pp. 118–37.

Worthington, Ian, *By the Spear: Philip II, Alexander the Great, and the Rise and Fall of the Macedonian Empire* (2014, reprint, Oxford: Oxford University Press, 2016).

Yates, Robin D.S., 'New Light on Ancient Chinese Military Texts: Notes on Their Nature and Evolution, and the Development of Military Specialization in Warring States China', in Peter Lorge (ed.), *Warfare in China to 1600* (Aldershot: Ashgate, 2005), pp. 273–310.

5

MILITARY DIVERGENCE AGAIN

Rome, Carthage, Parthia, India and the Far East

Introduction

The period under review in this chapter shows the military divergence between the eastern and western parts of Eurasia. While in Western Eurasia, the Roman military machine was centred on heavy infantry, in Eastern Eurasia, cavalry played the dominant role. There were certain variations within this broad division. For instance, though the Romans continued the Classical Greek tradition of heavy infantry, the former came up with a more nimble and manageable form of infantry organisation. This was exemplified in the transition from hoplite format to the legionary formation. In the eastern half of Eurasia, certain variations in the military practices of different regions were present too. While elephants and cavalry held sway in South Asia, cavalry supported by auxiliary infantry constituted the dominant military paradigm in China. The two contrasting military cultures – one based on heavy infantry and another based on cavalry – confronted each other in the Middle East, as exemplified in the Roman–Parthian Wars. Further, massive and intricate military fortifications emerged in South Asia during this period which dwarfed the fortifications of the Roman Mediterranean world. Ancient Chinese siege artillery was probably superior to the machines used by the Greco-Roman military engineers. Now, we shift the focus to the rise of Roman military power in the Mediterranean, where sea power played an important role. Unlike the ground war, which was characterised by divergence, naval warfare in the Mediterranean exhibited convergence. This was because both Carthage and Rome adopted and upgraded the Greek–Phoenician–Persian maritime tradition.

Military landscape of the Mediterranean region

The rise of Rome was the defining feature of the Mediterranean military landscape from the third century BCE onwards. Roman Republican imperialism

DOI: 10.4324/9780429284069-6

developed in three stages. The first stage involved the conquest of Italy, and the second stage was concerned with the projection of Roman power abroad: North Africa, Spain, Gaul (France) and Greece. The final stage involved the civil wars that transformed the Roman Republic into the Roman Empire. The traditional date for the foundation of the city of Rome on the bank of the Tiber River is 753 BCE. Like the Classical Greeks and the Macedonians, the military doctrine of the Romans emphasised fighting pitched battles in order to reach a quick political conclusion. During the seventh and sixth centuries BCE, the Etruscans adopted the hoplite formation from the Greeks (Rosenstein 2010: 290). Initially, the Romans also followed the hoplite organisation, but later they improved upon it. And this innovative military structure helped Rome to conquer Italy.

The infantry structure of the Roman army was an amalgamation of Greek and Celtic practices and innovations introduced by the Romans themselves. The Romans adopted military equipment from their neighbours. The large shield (*scutum*) and the short stabbing sword (*gladius*) were adopted from the Celts, and the barbed javelin with an iron shaft (*pilum*) was adopted from Spain (Fagan and Trundle 2010: 14). The Romans recruited slingers from the Balearic Islands. The slingers shot balls of stone and lead. The Roman heavy infantry was formed in legions. Each legion had 4,000 to 5,000 soldiers, which during the third century BCE was subdivided into 10 cohorts. A legionary was equipped with a short sword, two *pilums* and a shield. Each legion had 300 unarmoured cavalrymen. The cavalry was a mere adjunct of the heavy infantry (Bagnall 2003: 20, 25–6). The standard tactic of the Roman legionaries was to deliver a massed volley of *pilums* at the enemy from a distance of about 50 feet and then charge at the hostile troops with swords (Goldsworthy 2003: 22). The Roman infantry was deployed in three lines: *hastati* (first line), *principes* (second line) and the *triarii* (third line). While the first two lines charged at the enemy with swords and *pilums*, the third line, equipped with long thrusting spears, constituted the third and ultimate line of defence. The spear carried by a *triarii*, like a hoplite spear, was 7 feet long, with a bronze butt spike and a 30 cm-long socketed iron head. The use of *triarii* was actually an adoption of the Greek hoplite style of combat (Daly 2002: 64–5).

Most of the Romans in the fifth century BCE were farmers and grew wheat for their own consumption. The average size of their holdings was between 3.7 and 6 acres. To acquire political capital, the consuls had to win victories. War was actually Rome's most important public business. War offered scope for gaining manly excellence, glory and renown. Hence, the cultural value system of the ruling class encouraged a tendency for continuous warfare among the city's elites (Rosenstein 1999: 196, 200).

To an extent, the direction of Roman imperialism was partly a product of factional politics within the Roman assembly. The Claudii family in Rome favoured a southern expansion, in contrast to the Fabii family, which supported northern expansion (Bagnall 2003: 15). This was somewhat similar to the operation of two lobbies in interwar Japan; one of them favoured expansion in South-East Asia, and another lobby was for moving against Russia.

Carthage, a Phoenician colony, was situated along the Mediterranean coast of Tunisia. The city was probably founded in about 814 BCE. The city had a big naval base with defensive walls, arsenals and two harbours. The outer harbour was for commercial shipping and the inner harbour was reserved for the warships (Rawlings 2010: 262). Since Carthage derived the bulk of its revenue from overseas commerce, control over the Mediterranean was an absolute necessity for her. Carthage was very wealthy and largely depended on mercenaries for manning its army (Miles 2011: 179). The mercenary army's commanders came from the principal families of Carthage. The allied Numidian light cavalry, however, had their own leaders. Members of the wealthy families held the principal offices in Carthage. In fact, the top offices were sold to the highest bidders. The Romans called the dark-skinned Carthaginians Poeni, which led to the name Punic (Bagnall 2003: 12–21). The Romans fought three wars (known as the Punic Wars) with Carthage which involved land and sea battles. The naval force of both the parties was more or less similar. However, the ground forces of Rome and Carthage differed widely. While the Roman army was actually a conscript-based heavy infantry force, the Carthaginian army was a force of mercenaries, and its shock arm was light cavalry, with light infantry playing a subordinate role. And the Carthaginian army absorbed certain elements (like war elephants) from the Asian (Indian) militaries.

The First Punic War broke out in 264 BCE between Rome and Carthage over the issue of domination of the Central Mediterranean, and especially Sicily. This war forced Rome to build a navy for controlling the Mediterranean. In the fourth century BCE, Carthage was the principal naval power in the Central and Western Mediterranean. Historian Richard Miles claims that Carthage first came up with the quadrireme, and it was bigger than the trireme. It was replaced by the quinquereme that was developed by Carthage's enemy, Dionysius of Syracuse. Carthage adopted it and improved upon the original design. The improved hull of the Carthaginian quinquereme gave it extra protection against ramming. The initial performance of the Roman navy was not good. In 282 BCE, when a small Roman fleet was sent against Tarentum, the Romans were defeated. In 263 BCE, Carthage sent a force to Sicily. Acragas became the Carthaginian headquarters in Sicily. The Romans laid siege to this city, and the siege continued for five months. Then Hanno, the Carthaginian commander, with a force of 50,000 infantry, 6,000 cavalry and 60 elephants, came to relieve the Roman siege of Acragas. The siege dragged on for another two months. Hanno's leadership was passive. Further, the mercenaries of the Carthaginians were no match for the veteran Roman legions. Finally, the Carthaginian garrison pulled out, and Acragas surrendered to the Romans (Miles 2011: 177–80).

In 260 BCE, Rome initiated a massive shipbuilding programme. She decided to construct 100 quinqueremes and 20 triremes. This naval programme was in response to the Carthaginian policy of raiding Italy's coast with their fleet. Further, Carthaginian naval supremacy also discouraged many cities in Sicily from siding with the Romans. Rome got hold of a Carthaginian quinquereme that had ran aground and then copied its design while building its own fleet. The Roman rowers were mainly poorer citizens of Rome and its allies. They were trained to row by making them sit in the benches on dry land in the same order as on the benches of the ships themselves. At the command of the officers, the crews were accustomed to bringing their hands up and down at the same time. The Romans knew that in terms of seamanship and manoeuvring capacity they could not compete with the Carthaginian fleet. Therefore, the Romans innovated. They came up with the corvus. It was a boarding bridge that could be raised up at an angle with a pulley. During sea battle, the corvus was released so that it fell on the deck of the hostile ship. A heavy pointed spike at the underside of the corvus pierced the timber of the hostile ship's deck so that the enemy ship was firmly attached and could not make its escape. Then the Roman marines used the bridge to board the enemy vessel. And in close-quarter combat on the deck of the enemy ship, the Roman legionaries, due to their greater discipline, training, superior non-commissioned officers and cohesion, had an advantage over the Carthaginian mercenaries (Miles 2011: 181–2). Thanks to the corvus, the Romans transformed the sea battle into a land battle in which their drilled and disciplined heavy infantry enjoyed an advantage.

In 256 BCE, the Romans decided to bypass Sicily and attack North Africa, the hinterland of Carthage. From Rhegium, a Roman armada comprising 330 ships sailed towards North Africa. The Carthaginian fleet numbered 350 ships. The two fleets clashed at Cape Ecnomus. The Roman corvus emerged as the battle winner. The Carthaginian centre fled first. Against Roman losses of 24 ships, the Carthaginian fleet lost 94 vessels. The victorious Roman fleet landed the expeditionary force near Aspis in the Cape Bon Peninsula. Regulus, the Roman general, commanded 15,000 infantry and 5,000 cavalry in North Africa. Soon, Regulus was able to capture Tunes, a city near Carthage (Miles 2011: 185–7). In 255 BCE, Xanthippus, a Spartan mercenary commander on the payroll of Carthage with 12,000 infantry, 4,000 cavalry and elephants, offered battle to the Romans. In this battle, Xanthippus led a hybrid army which was a combination of Western and Eastern military elements (the Western element of disciplined heavy infantry was fused with two Eastern elements: light cavalry and elephants) against the heavy infantry-centric Roman force. The Carthaginian elephants mauled the legionaries. Simultaneously, Numidian cavalry attacked the hapless legionaries from the rear. Only 2,000 Romans escaped, and 500 Romans with Regulus himself were taken prisoners (Bagnall 2003: 41). In 241 BCE, due to general exhaustion of both sides, the First Punic War came to an end.

Hannibal (b. 247–d. 182 BCE) was the principal figure during the Second Punic War (218–201 BCE). Credit is due to him for fine-tuning the hybrid military machine by adopting and integrating military techniques both from the East (Asia) and the West. Hannibal's army included elephants, Numidian light cavalry and medium infantry. Hannibal adopted the use of war elephants from the Ptolemids, who in turn copied it from the Seleucids. Ptolemy III captured Indian elephants in the service of the Seleucid army when the former invaded the Seleucid Empire during the Third Syrian War (246–241 BCE). The Seleucids had acquired this military element from the Indian Maurya Empire. Hannibal used forest elephants, which were then found in North Africa. He also acquired elephants from Syria. The elephants acquired from the forests around Carthage, the Atlas Mountains and Morocco were smaller than the Indian elephants. In Syria, the Seleucids established an elephant-breeding centre. The Maurya Empire, which was on good terms with the Seleucid Empire, supplied several beasts to this elephant breeding centre in Syria. And the Seleucid monarchy, due to military pressure from the Romans, cooperated with Rome's enemy Carthage. The Carthaginians used Indians as elephant trainers and elephant drivers. Here, we find evidence of a long-distance mercenary service. Even earlier, Indian mercenaries from Punjab had fought in Darius and Xerxes's Persian force. In addition, Hannibal's elephant contingent adopted the use of the *howdah* from the Indians. For some time at least, the Roman generals toyed with the idea of adopting elephants in their military machine. At Magnesia (190 BCE), the Romans used African elephants (Bagnall 2003: 20; Charles and Rhodan 2007: 363–7). Nevertheless, gradually, the use of elephants faded among the Romans. This was because after the collapse of Carthage and the Hellenistic monarchies, the Romans did not have to face any hostile force structure deploying elephants. Further, the Romans were uneasy with the wholesale adoption of this foreign military element into their own military machine. Finally, the Romans fought most of their battles in Mediterranean Europe. The topography and climate of this region was not suited for breeding elephants, and none of the European enemies of Rome deployed elephants.

In November 218 BCE, Hannibal crossed the Alps and entered North Italy. On 2 August 216 BCE, about 50,000 Roman and allied soldiers were annihilated in the narrow plain north of the little town of Cannae. The Carthaginian army comprised infantry at their centre and Numidian cavalry at both the flanks. The Roman army mainly comprised legionaries tightly packed like a square. Hannibal executed double envelopment of the Roman army with his Numidian cavalry. His cavalry attacked both the flanks of the Roman army and its rear. Simultaneously, the frontal charge of the legionaries was contained by the infantry at the Carthaginian centre, which put into effect a slow withdrawal. The Roman heavy infantry, attacked by the Numidian cavalry both at the flanks and rear, in desperation surged forward and made a frontal rush towards the Carthaginian centre. The light and medium Carthaginian infantry (comprising Spanish and Gallic allies and mercenaries) could not meet them in a head-on struggle. The Carthaginian infantry retreated systematically. The Carthaginian centre was shaped like a concave mirror and drew the Roman legionaries inward. Once the Romans were attacked on their right, left and rear by Numidian cavalry, then the Carthaginian infantry in the centre started advancing and counterattacking the panicking Romans frontally. By this time, the Carthaginian centre, like the reflective surface of a convex mirror bulged outward and pushed the legionaries back into a smaller and smaller pocket, which was pressed both in the front and rear as well as along its two flanks. Double envelopment led to encirclement and annihilation of the Roman army (Goldsworthy 2004: 44; Hoyos 2008: 58–9).

The iron discipline of the Carthaginian infantry, who executed controlled retreat in the presence of the enemy force and then a timely advance, and close coordination between the Carthaginian infantry and Numidian cavalry under the overall direction one of the ancient world's greatest generals, Hannibal, gave victory to the Carthaginians. And this tactic (*Kettelschlacht*, i.e. cauldron battle) became the dream of all the generals after Hannibal. The German General Staff under Count Alfred von Schlieffen before the First World War raised double envelopment from the tactical to strategic level, and this became the famous Schlieffen Plan. Moltke the Younger, at the opening stage of the First World War, tried to implement this plan unsuccessfully. During the Second World War, the panzer generals executed double envelopment campaigns at the operational level.

After destroying the Roman field force, Hannibal could not capture Rome because he lacked a siege train. Post-Cannae, the Romans raised several armies that were bigger in size compared to the force that could be raised by Carthage. Rome's staying power was greater than its opponent's was. Thanks to the political system and social structure of the Roman Republic, this political entity could mobilise greater demographic and economic resources, thus generating greater military power than Carthage. Most of the Roman citizens were under obligation to serve as legionaries. And the Roman allies had to provide military manpower (Rich 1995: 1–3). During the Hannibalic War, some 29 per cent of all the adult Roman male citizens served as legionaries. And each Roman legionary who was conscripted proved cheaper for the Roman state compared to a mercenary hired by Carthage (Oakley 1995: 28–9). Political resilience and greater demographic resources enabled Rome to sustain higher casualties than Carthage and still maintain several large armies in Italy, Spain and North Africa. Thus, in the long run Rome finally won the war of attrition. Carthage, like Germany during the two World Wars, won battles but could not prevail in the long-drawn wars. The lesson we can derive from studying both Carthage and Imperial-Nazi Germany was that despite tactical chutzpah, war in most of the cases is determined not merely by battles but by politics, economy and demography.

The new Roman armies under Maximus Fabius (b. 280–d. 203 BCE) carried out a Fabian strategy. It meant avoiding pitched battle with Hannibal, which the latter was sure to win. Fabius followed a wait-and-watch policy and conducted hit-and-run attacks against Hannibal's supply lines. Due to the loss of naval supremacy, Carthage could not send any further reinforcements to Hannibal. It was a perfect case of Roman sea power influencing the grammar of land war in Italy. So the Fabian strategy could only succeed because the Romans had naval superiority. One can infer that the Romans had consciously chalked out a joint land–sea strategy. While Hannibal was caged in Italy, new Roman armies started destroying the Carthaginian base in Spain (Hoyos 2008: 60–89). The Fabian strategy, with its motto of 'slow and steady wins the race', suited attrition warfare by the Roman state, which was conducting a battle of materiel (*Materialschlacht*). Later, the Fabian strategy entered the Marxist lexicon and was followed by those Marxists who wanted to destroy the bourgeois state by a slow war of attrition instead of launching a Leninist revolutionary struggle.

In 206 BCE, just before the Battle of Ilipa, Hasdrubal, son of Gisco, joined force with his deputy Mago Barca. The combined field army of the two Carthaginian generals operating in Spain comprised 70,000 infantry and 4,500 cavalry (including superb Numidian horse led by Prince Masinissa) and 32 elephants. For the Romans, Scipio commanded 45,000 infantry (of them, 35,000 were Italians and the rest were unreliable Spanish levies) and 3,000 cavalry. During the Battle of Ilipa, Hasdrubal deployed his Libyan heavy infantry in the centre, and the Spaniards (who were fighting on both sides) and his cavalry at the flanks. Roman *velites*' (light infantry) missile weapons (javelins and sling shots) drove the elephants back into the Punic lines and spread confusion. Roman *velites* were actually a modified form of Greek peltasts. Then Scipio's infantry charged forward. The Numidian cavalry men equipped with light spears and javelins rode bareback, without saddles and bridles. Hence, they were not suited for launching a cavalry charge against disciplined infantry but were good for reconnaissance, scouting, pursuit of the defeated force etc. Hasdrubal failed to use the Numidians imaginatively against the flanks and rear of the Roman cohorts as they advanced towards the Carthaginian army. Once the Roman infantry reached the Carthaginian line, the former's drill and discipline won the day. Hasdrubal retreated with 6,000 soldiers. Soon, he left for Africa (Bagnall 2003: 20; Goldsworthy 2004: 68–74; Lendon 2005: 178; Hoyos 2015: 181–3). Next, a Roman army landed in North Africa under Publius Cornelius Scipio (later known as Africanus). At the Battle of Zama (October 202 BCE), even the 'genius for battle' Hannibal could not save the raw Carthaginian force against the battle-hardened Roman legions. Further, the Numidians changed sides and provided cavalry to the Romans. Without cavalry superiority and trained light infantry, Hannibal could not execute a second Cannae. At Zama, the Carthaginians suffered 40,000 casualties (Hoyos 2008: 104–16). This single battle destroyed Carthage's status as a great power.

After being defeated in the Second Punic War, Carthage became a third-grade power with no chance of defeating the Roman Republic. The Third Punic War (149–146 BCE) gave the final blow to a declining Carthage. In 146 BCE, under Scipio Aemilianus, the Roman army laid siege to the city of Carthage, which was surrounded with walls that were made of huge sandstone blocks. The Romans blocked the entrance to the harbours of Carthage by constructing a mole. Further, Carthage was also sealed from its North African hinterland. So the stricken city got no food from outside. The Carthaginian commander, Hasdrubal, assumed that the Romans would attack the commercial harbour first. However, the Romans outwitted him by attacking the military harbour first and then occupied the marketplace (Miles 2011: 1–3). After almost 700 years of existence, Carthage ceased to exist. The Mediterranean finally became *mare nostrum* (our sea) for the Romans. This also became Mussolini's dream in the early twentieth century. Next, Rome focused on the Eastern Mediterranean region controlled by the Hellenic polities.

The series of wars between the Hellenistic monarchies and the Roman Republic that occurred between 214 to 168 BCE was a clash between heavy infantry forces: the phalanx versus the legionaries. The Romans won because the legion was a more tractable and supple formation and the Roman political system could raise more conscripts compared to the Hellenistic royal polities, dependent on the limited number of costly mercenaries. The Romans first encountered a Hellenistic force in 280 BCE, when King Pyrrhus of Epirus supported the city of Tarentum against Rome. He defeated the Roman legions at Heraclea in the same year and at Asculum in 279 BCE, but was beaten at Malventum in 275 BCE. Pyrrhus' initial victories were due to the use of small number of elephants. The Romans found these unfamiliar beasts terrifying (Goldsworthy 2004: 88).

The Third Macedonian War lasted from 172 to 168 BCE. During this war the Romans used a small number of elephants supplied by their Numidian allies, but Perseus, the ruler of Macedonia, had no elephants. This was because Perseus controlled neither North Africa nor the Seleucid territories in Asia. The Battle of Pydna was fought on 22 June 168 BCE. The elephants deployed at the Roman right flank caused considerable damage among the Macedonians. The noise and smell of the elephants completely disoriented the Macedonian horses. Elephants and light infantry provided by Rome's allies completely defeated the Macedonian left wing. In close-quarter combat against the phalanx, the looser Roman legions triumphed because they were more manageable than the cumbersome, unwieldy phalanx. One reason for the superiority of the legionaries over the phalangites was the fact that the Roman legions had a higher proportion of officers compared to the Macedonian phalanx. Each legion had six tribunes and 60 centurions plus the *legati* and other members of the staff for managing the three lines of legionaries (*hastati*, *principes* and *triarii*). Further, in close-quarter infantry combat, Spanish swords in the hands of the legionaries proved more effective than the long unwieldy pikes in the hands of the phalangites. The Roman military theorist Vegetius recommended 3 feet of frontage per infantryman. So unlike the tightly packed phalangites, the legionaries were able to use their swords much more freely. The Spanish sword was a cut and thrust weapon with a tempered steel blade that caused horrible disfiguring wounds instead of merely punctured wounds caused by the pikes. In one hour of combat, 20,000 Macedonian infantrymen were killed and 6,000 were taken prisoner (Goldsworthy 2004: 88, 101–3; Gilliver 2007: 133). This battle marked the end of the reign of the phalanx and the rise of the legion in the Mediterranean military landscape. Nevertheless, there was a continuity between the Greek and Roman military tradition. The supremacy of heavy infantry for conducting land war in the Mediterranean littoral, however, continued. One form of heavy infantry was replaced by another. And the legion to an extent was an open-order, lighter phalanx.

The Roman army was constantly evolving. Ironically, at Pydna, the Romans fought with a quasi-hybrid force structure. However, after the annexation of Macedonia, the Roman army reverted to the heavy infantry-oriented force. A brief organisational overview of this force structure is necessary. Senator C. Marius, who held military command between 107 and 100 BCE, introduced certain reforms which came to be known as the Marian reforms. Each legion carried an eagle as the standard. This functioned as the rallying point and raised the cohesion of the legionaries in the battlefield during the heat of combat. The use of a wooden pin in the iron shank of the Roman javelin (*pilum*) became common. The iron shank with a pyramidal head was attached to the wooden shaft with a socket or a tang. When the Roman soldier threw the *pilum* towards an enemy soldier, and if it struck the hostile troop's shield, then the pin broke. As a result, the enemy could not throw the *pilum* back. Moreover, the shield of the enemy soldier with the *pilum* stuck in it became useless. Then the enemy infantry was easy meat against the sword-equipped Roman infantry covered with armour. The Roman infantrymen, like the Macedonian infantry soldiers of Philip, were ordered to carry their own equipment in order to limit the number of pack animals and camp followers. This in turn raised the infantry's rate of marching. The soldiers carried heavy backpacks and hence were called 'Marius's mules'. Besides their weapons and armour, the legionaries on the march carried other equipment suspended from a pole on their shoulders. This included a bag for their personal possessions, bronze saucepans for cooking, entrenching tools such as a pickaxe and stakes to construct barriers in front of ramparts. Further, a section of eight men carried a leather tent for them (Rankov 2007: 31–2, 61). Under the Republic, legionary service was restricted to the property owners who could arm and maintain themselves. However, due to the increase in the size of the army, landless labourers were enrolled in large numbers from the late third century BCE. Their recruitment was regularised in 107 BCE. The dominant ideology was that farmers made the best soldiers and discharged soldiers should return to farming, generate revenue for the state and produce future recruits for the army. This ideology is somewhat akin to the martial race ideology of the late nineteenth-century British Empire. A fixed daily sum was paid to the soldiers during active service. Command was exercised at four levels: centurions, tribunes, legates and consuls. The cohorts were commanded by tribunes, who were either promoted from the centurions or selected from the magistrates. The legion was commanded by a legate. Senior officers such as the prefects were drawn from the equestrian and senatorial orders. The centurions of the infantry and decurions of the cavalry were either promoted from the ranks or directly commissioned from the leading municipal families (Bagnall 2003: 26; Rathbone 2007: 158–9, 161). The presence of permanent battle-hardened junior officers like centurions gave the legion its staying power.

After conquering the Mediterranean Basin, the Roman Republic imploded. Ambitious generals jockeying for power fought a series of wars known as Civil Wars. These wars represented the differing parties fighting with similar instruments. Better military leadership decided victory. The struggle between Marius and Sulla known as the First Civil War seriously weakened the Roman Republic. And the series of battles between Gaius Julius Caesar (b. 13 July 100 BCE–d. 15 March 44 BCE) and Pompey the Great (Gnaeus Pompeius Magnus, b. 106–d. 48 BCE) constituted the Second Civil War, which resulted in Caesar's dictatorship and then establishment of the principate/empire by his successor Octavian/Augustus. The First Civil War started in 88 BCE, when Sulla (Pompey served under him) marched with his army against the Roman Assembly, which had removed him from the command of the expedition directed against Mithridates of Pontus in Anatolia. The army supported Sulla because they wanted a lucrative campaign against Mithridates and feared that Sulla's removal would result in Marius's legionaries taking their place (Alston 2007: 180–1). In order to acquire the loyalty of the soldiers, generals started clamouring for land in order to settle the veterans and discharged troops. The latter also realised that only their commanding generals could offer them lucrative economic packages (Goldsworthy 2003: 9). Hence, the different armies became loyal to their commanding generals and not to the Roman Assembly.

Julius Caesar rose to prominence by conquering Gaul and Britain for Rome. The Iron Age started around 800–700 BCE in North-West Europe. The tribal armies of North-West Europe consisted of small warrior bands maintained by individual chieftains and supplemented by free tribesmen. They were able to equip themselves and fought in clan and familial groupings. The biggest war leaders were able to maintain several thousands of military followers at the most. Tribal armies were clumsy in their movement. Due to logistical weakness, they generally dispersed after a few weeks of campaigning (Goldsworthy 2007: 81). Overall, the tribal hosts, comprising mainly undisciplined light infantry, who suffered from a shortage of iron weapons and armour and lacked logistical support, were no match for the trained legions led by Caesar. According to one calculation, in Gaul after the Roman conquest, the population fell by 14 to 28 per cent. This was due to large-scale enslavement and disastrous harvests for some years. The latter were partly due to loss of manpower caused by Caesar's campaigns. War, combined with bad harvests, resulted in widespread famine. One author claims that over a million people had been killed in Gaul in a decade due to Caesar's brutal campaigning. So the impact of warfare on the conquered society was massive in ancient times. When the Romans invaded Britain, the Britons were still using chariots. This was despite the fact that chariots had vanished much earlier from the mainland of Eurasia (Goldsworthy 2003: 29; Wileman 2014: 18, 48). The separation of Britain from the continental mainland explains this military divergence cum backwardness in the island's military system. Britain faced no external threat before the Roman invasion. Hence, there was no imperative to upgrade its pre-existing military machine.

After the end of the Gallic campaigns, Caesar commanded 10 legions. His auxiliaries came from the Gallic and Germanic tribes. In 49 BCE, Caesar crossed the Rubicon, which formed the boundary of his own area of command. And then civil war became inevitable. As he crossed the river, Caesar uttered the famous phrase: 'the die is cast'. Caesar advanced rapidly into Italy to prevent Pompey from raising soldiers. Pompey was forced to leave Italy and went to the Balkans. Despite the fact that Pompey had a superior navy, Caesar took the risk, crossed the Adriatic with his legions and landed in the Balkans. This feat is somewhat similar to Alexander crossing the Hellespont even though the Achaemenid Empire enjoyed naval superiority. Pompey raised troops from Spain and the eastern provinces of the Roman Republic. Pompey's new legions were not as combat effective as Caesar's legions, who had fought under his leadership for quite a long time. Therefore, Pompey depended partly on cavalry for defeating Caesar's infantry force. At the Battle of Dyrrachium (10 July 48 BCE), Pompey could not take advantage of his cavalry superiority due to the rugged ground (Goldsworthy 2003: 22, 29, 44; Alston 2007: 183). Pompey's failure to fight the battle in good cavalry country cost him victory.

At the decisive Battle of Pharsalus (9 August 48 BCE), Pompey, lacking confidence in his green legions, was for avoiding a pitched battle with Caesar's veteran infantry. Pompey was in favour of wearing down Caesar's infantry by denying the latter provisions with his superior cavalry force. However, pressure from the senators forced Pompey to settle the issue quickly. Caesar deployed 22,000 infantry and 1,000 cavalry. Pompey had 45,000 infantry and 7,000 cavalry. Pompey deployed his cavalry at his left flank under Labienus. His plan was to attack Caesar's legions on its right flank and then at the rear. The legions raised from Cilicia and Spain constituted Pompey's right flank, and the Syrian legions formed the centre of his army. They were to maintain a holding action against the left flank of Caesar's army under Mark Antony and the centre under Domitius Calvinus (Goldsworthy 2003: 48–51). So Pompey, like Darius III at Ipsus, concentrated his cavalry in one wing with the aim of smashing the right wing and centre of the opponent's infantry-oriented army. Caesar used reserve infantry to check the thrust by Pompey's cavalry. Caesar writes in his account of the Civil War:

> Caesar, keeping his previous order of battle, had stationed the Tenth Legion on the right wing. . . . At the same time, he observed the dispositions . . . and fearing that his right wing might be surrounded by the large numbers of Pompeian cavalry, he quickly took one cohort from each legion from his third line and formed them into a fourth line, which he stationed opposite the cavalry.
>
> *(quoted from Caesar 1984: 151)*

Caesar took the calculated risk of weakening his third line in order to create a reserve fourth line to check the charge of Pompey's cavalry, knowing that the untrained legions of Pompey would not be able to overwhelm the first and second lines of his infantry. Since Pompey's legions remained on the defensive, even the first and second lines of Caesar's infantry were not threatened. Pompey's generalship in this battle was not laudable. As soon as his cavalry fled, a distraught

Pompey, like Darius III, left the battlefield early. From this battle we can deduce that the handling of cavalry by one of the greatest Roman generals, Pompey, and the quality of cavalry used by the Romans were below standard. He should have used light infantry to back up his cavalry charge against Caesar's right flank. Pompey's Roman cavalry, enjoying numerical superiority, could not overcome Caesar's veteran infantry. The Roman cavalrymen armed with short swords were no match against the *pilum*–sword combination of disciplined legionaries. However, Parthian cavalry, which the legionaries would face in the near future in the Middle East, was a different matter altogether. On Pompey's side, 24,000 were taken prisoner at Pharsalus and 15,000 were killed. Pompey fled to Egypt and was shortly assassinated (Goldsworthy 2003: 52, 55).

Both Caesar and Pompey fought with the same force structure: heavy infantry. Pompey's attempt to integrate cavalry with heavy infantry during the heat of war failed. Acquisition of skill, training and doctrine for such an attempt to succeed required time which Pompey did not have. After the death of Pompey, his adherents carried out a fruitless struggle for some time in North Africa and in Spain but were finally overwhelmed, thanks to the military genius of Caesar combined with the greater resources at his command.

On 15 March (the Ides of March) 44 BCE, the senators under the leadership of Cassius and Brutus assassinated Caesar. Octavian (Caesar's nephew and adopted son) broke with Mark Antony in October of that year. The former started raising mercenaries in Italy, offering the men who would sign up a cash bounty worth two years' pay (Brice 2011: 43). After defeating Antony, Octavian became the first Roman emperor. The force structure of both Octavian and Antony was similar: heavy infantry. Victory went to the party that had greater combat experience and better leadership and training, plus good luck. Octavian's legionaries from Italy were better trained and veterans of past campaigns compared to the green recruits whom Antony raised from the eastern provinces. Further, Antony's own experienced troops to a large extent were destroyed during his disastrous campaign against the Parthians. Worse, Antony's intimate association with Cleopatra, whom the Italians hated, provided a good public relations exercise for Octavian. Octavian projected himself as one fighting for the Romans against the 'barbarian' queen and a rogue Roman general (Antony) who had gone native. The disgruntled Roman senators had murdered Caesar to end his military dictatorship. However, the ultimate result was a civil war ending with the establishment of a monarchy, a development about which the senators had not even dreamed.

During the late second and first century BCE, the Roman army was transformed from a citizens' militia into a professional force. Under Augustus, the legions became permanent structures. In this way, accumulated military experience could be passed on to successive generations of recruits (Goldsworthy 2007: 79–80). The Roman imperial army was a standing army comprising professional soldiers who normally served for 25 years (Campbell 1999: 217). The use of field artillery, a legacy of the Macedonian military system, became more systematic under the Romans. Mobile bolt-shooting scorpions were mounted on carts, and larger stone-throwing ballistas were also used in pitched battles (Gilliver 2007: 128). Augustus, after the Civil War, reduced the number of legions from 100 to 28 in 31 BCE (Brice 2011: 53). Still, during the first century CE, more than 40 per cent of the empire's income went to sustain the Roman army (Campbell 1999: 222).

Middle East

In the Middle East, Parthia emerged as the principal enemy of Rome. The Parthians were a Central Asian pastoral nomadic confederation who spoke an Indo-Iranian language. They were pushed southwards by the Sakas (Scythians), who were in turn pushed by the Kushanas (Yeuh-Chi), and the latter were in turn under pressure from the Huns (Xiongnu). The Parthians and the Sakas destroyed the Greek Kingdom of Bactria. The Parthians finally destroyed the Seleucid power in Persia and in 247 BCE established the Arsacid Kingdom at Parthia/Khorasan. In 128 BCE, the Parthians were making inroads into Mesopotamia (Tarn 1980). The Arsacid Kingdom emerged as the principal intermediary of the lucrative overland trade between China and the Roman Empire (the so-called Silk Road). Taxation of this long-distance trade enabled the Parthians to maintain a large cavalry force. At times, the physical geography as well as the hostile force structure shaped the configuration of the military system of a particular side. The Sakas were mostly mounted archers. Further, the Saka-Parthian battleground was North-East Iran, a region famous for breeding high quality warhorses. The flat alluvial plains were suited for large-scale cavalry encounters. And the large stretches of desert were unsuitable for campaigning by heavy infantry. In this region, a cavalry force supported by two-humped Bactrian camels could operate effectively. In the second century BCE, the Parthians employed Greek mercenaries as heavy infantry. However, the repeated defeats of the Greek heavy mercenary infantry in the Parthian army by the Saka mounted archers convinced the Parthian monarchs that for warfare in Persia, disciplined heavy infantry equipped with long spears were a liability (Farrokh 2007: 123–6). Hybridisation is not a formula for success in all circumstances. The context is important So, the Parthians came up with an all-cavalry force.

The Roman army relied mainly on heavy infantry against the Parthians, who depended on cavalry just as the Classical Greeks primarily deployed heavy infantry against Achaemenid Persia, which depended mostly on cavalry. However, both the cavalry and infantry underwent modifications with time. Unlike the hoplite phalanx of the Greeks, Roman heavy infantry was arrayed in maniples, which made the deployment and command of the legionaries more flexible, especially in broken ground. Further, the legionaries, unlike the hoplites, had the light throwing spear known as a *pilum*. Still, the reach of the *pilum* was less compared to the range of the compound bows carried by the Parthian cavalry. In fact, Parthian cavalry was more combat effective than Achaemenid cavalry.

While Achaemenid cavalry was mostly equipped with light spears, battle axes and swords, Parthia deployed both armoured heavy cavalry (cataphracts) for close-quarter combat and light cavalry for long-distance archery. Heavy cavalry developed among the Parthians from the third century BCE. And for several reasons which are discussed below, the Parthian heavy cavalry was more combat effective than the heavy cavalry which the Achaemenids tried to raise. The Parthian heavy cavalryman wore scale armour and a helmet with faceguard, and an armoured coat protected his mount. The armoured vest of the cavalrymen covered his torso and knees. It was made of rawhide with scales of iron and could withstand powerful blows from the enemy swords and javelins. Each heavy cavalryman carried a 12-foot lance with a butt spike. Since the cavalryman held the lance with both hands, he carried no shield. Some Parthian heavy cavalry was equipped with maces and long swords. The light cavalrymen were unarmoured and were equipped with compound bows. The use of horseshoes further raised the staying power and military effectiveness of Parthian cavalry. In the absence of horseshoes, carrying extra weight and rough terrain caused damage to the horses' hooves. In fact, the effectiveness of Parthian cataphracts (each cavalryman and the mount were covered with 57 kilos of armour) largely was due to the extensive use of horseshoes. Further, use of horseshoes increased the distance that a horse could cover without damaging its hoofs. Moreover, compared to the smaller Spanish horses used by the Roman cavalry, the Nisean breed gave birth to bigger horses, which were used by the Parthian cataphracts. Such horses had more stamina and were able to carry more weight, horse armour as well as the body armour of the riders. Thanks to the horseshoes, Parthian cavalry could cover 560 km in 3–4 days. In addition, the Parthians improved the Achaemenid carpet saddle with a saddle that was curved at both the front and the back. Such a saddle provided firm support to the thighs and base of the spine of the rider. This saddle kept the rider stable and allowed him to lean either way when holding his lance or when engaged in slashing action with his sword. So this new saddle further increased the combat effectiveness of Parthian heavy cavalry (Farrokh 2007: 131–2, 144). Thus, while the horseshoes raised the operating range of Parthian cavalry in general, the improved saddles increased the effectiveness of the light cavalry (horse archers) as well as shock power of the heavy cavalry in close-quarter pitched combat.

The Parthian battle tactics sought to disorganise the hostile infantry formation by launching light cavalry which shot arrows from a distance that could not be reached by the *pilums* and slingshots. Mobility and agility were the trademarks of the mounted archers. The Parthian horse archers were able to shoot arrows accurately even when their mounts were galloping. In fact, the Parthians were famous for the 'Parthian shot', which meant the archer could turn their body 180 degrees and shoot backwards even as the horse was in motion. Superb equestrian skill on part of the Parthian rider plus the use of the curved saddle enabled him to control the galloping horse with only knee pressure while he used both his hands to hold the bow and shoot backward, bending his body. And arrows from the compound bows were able to pierce Roman body armour. At times, the Roman infantry adopted the tortoise formation (as shown in several Asterix comics), which meant the legionaries covered their sides and tops with their rectangular shields. But in the hot dusty plain of Mesopotamia, the tortoise formation could not be maintained for long. And when the legionaries broke ranks due to the continuous barrage of arrows from the skirmishing light cavalry, heavy armoured Parthian cavalry provided the knockout blow by riding them down. First, the lancers charged and disorganised the tired and distracted legionaries, and then the latter were finished off by armoured cavalry equipped with maces and swords. For protecting the legions, the Roman army deployed German, Spanish and Gallic allied cavalry equipped with short swords and spears. In terms of stamina, horsemanship and equipment, Roman cavalry were no match for Parthian cavalry (How 1923: 118–19; Farrokh 2007: 133–7). The Roman military system had no credible answer to the Parthian way of war in the flat plains of Mesopotamia. In fact, the Turkish cavalry from the ninth century onwards more or less followed the Parthian tactics against the Rajputs of North India and against the Byzantine/Eastern Roman Empire in Asia Minor.

At Carrhae (53 BCE), a Roman legionary army under Crassus was wiped out. Crassus was assigned Syria and decided to win military laurels by invading Parthia. He advanced with 42,000 infantry (including 4,000 light infantry and 6,000 irregular Arab infantry) and 4,000 cavalry. Rustaham Surena, the Parthian general, confronted Crassus with 1,000 cataphracts (armoured heavy cavalry equipped with lances and long swords) and 10,000 mounted archers. Soon, the Parthian mounted archers surrounded Crassus's force. Each Parthian light cavalryman carried 30 arrows. The Parthians brought extra arrows on the backs of a string of camels, and this in turn enabled the horse archers to maintain a continuous barrage against the hapless legionaries. On average, each Parthian horse archer was able to deliver 8–10 arrows per minute. To give an idea about the barrage of the Parthian horse archers, at Carrhae, 10,000 mounted archers shot more than 1.5 million arrows in 20 minutes. So, Crassus's legionaries faced continuous arrow shots, an experience somewhat equivalent to what the British infantry encountered on the first day of the Somme (1 July 1916), when German machine-gun fire from all sides slaughtered the 'Tommies'. Camel trains continued to resupply the mounted archers. The Parthians had established a string of workshops throughout Luristan, Persis, Khorasan, Media etc. These workshops manufactured arrows which were supplied to the ammunition-hungry horse archers at Carrhae. The horse archers shot their arrows at a distance of 150 feet from the legionaries, and the latter's weapons (slings, javelins and *pilums*) could not reach their mounted tormentors. The lightly armoured Gallic cavalry's javelins had no effect on the Parthian cataphracts in close-quarter cavalry combat. After the battle, about 10,000 Romans were taken as prisoners, and they were settled in a Central Asian city named Turfan. Some of them became mercenaries for Hun chieftains who were fighting the Chinese (Farrokh 2007: 136–40). But the Roman infantry technique had no appreciable effect in war making along the northern frontiers of China or in Central Eurasia. This was because the legionaries with their *pilums* and swords were sitting ducks against the mounted archery of steppe nomads in the wide open plain. And even in garrison duties or for guarding narrow passes, Chinese infantry equipped with crossbows were more effective than *pilum* and short-sword-equipped legionaries. We must not overemphasise the military effectiveness of the Western armies in general and Roman infantry in particular across all terrains and against different force structures.

Just before his assassination, Caesar was preparing to invade Parthia. We could speculate that if Caesar survived the assassination and invaded Parthia, his famed legions would have faced the same fate that waited Antony's distinguished legions. In March 36 BCE, Antony, with 10,000 Iberian cavalry and 100,000 infantry moved against Parthia. The Parthians mustered 50,000 cavalry (Farokh 2007: 144). Antony soon retreated in disgrace. For Antony's mishap, the maxim that those who do not learn from history are bound to repeat it is applicable. Antony should have learnt from Crassus's sad case. In fact, Antony was lucky. Internal troubles within the Parthian domain enabled his defeated force to retreat. The Parthian Kingdom was a feudo-federal political entity. It was divided into 18 provinces (each was under a governor) and several semi-autonomous principalities. The nobility owned landed estates and relied upon the lesser nobility to provide troops when the kingdom faced an external invasion. Artabanus III, King of Parthia from 10/11–38 CE, was challenged by one pretender, Tiridates in his own country and Mithridates (not to be confused with the king of same name who fought Sulla) in Armenia. Armenia was in a sort of vassal relationship with the Parthian Kingdom. Trajan's (r. 98–117 CE) campaign in Parthia in 114–16 CE was not very successful. Tussles between two Parthian commanders, Sanatruces and his cousin, Parthamaspates, allowed Trajan to retreat with some safety. Further, the war between the Kushanas and the Parthians in Hyrcania and constant conflict with the Sakas at Khorasan prevented the Arsacids from turning their full fury against the Roman legions stationed in Syria (Ash 1999: 114, 117; Bivar 2006: 91; Benjamin 2018: 156, 159).

We can say that the operational effectiveness of the heavy infantry ended on the west side of the Euphrates and Tigris rivers. In the end, the Parthians were defeated not by the Romans but by the Sasanians who established their rule in Iran. The Kushana Empire, Parthian Kingdom and even the Roman Empire to a great extent were weakened by the epidemic of smallpox which spread along the Silk Road. The Black and Pale horsemen of Apocalypse started their journey from Southern Arabia and then travelled along the trade routes to Iran, Punjab, China and in the west across the Mediterranean lands. During 165–66 CE, due to the smallpox epidemic, both the Parthians and the Romans had to stop their military operations along the Euphrates. Almost one-fourth of the Roman Empire's population perished due to this epidemic (Bivar 2006: 93–4). Smallpox, by weakening the Parthians, facilitated the rise of the Sasanians. The Sasanians, as the next chapter shows, proved to be tough foes for the Eastern Roman Empire until the advent of the Arab Caliphate.

South Asia

The Maurya Empire collapsed due to an amalgam of internal (faction fights within the royal family, breakdown of civil–military relations) and external factors (invasion by Antiochus and the attacks of the Greek Kingdom in Bactria). After the collapse of the pan-Indian Maurya Empire, the various successor polities failed to retain control over the north-west frontier. In 186 BCE, Pushyamitra Shunga, a general of the Maurya Empire, by executing the last Maurya ruler, Brihadratha, established the short-lived Shunga Empire. His tenuous authority extended over North and Central India but not in North-West India. Taking advantage of this fact, the external invaders launched repeated raids into the interior of India. First came the Greeks, who were displaced from Bactria, followed by the Parthians, Sakas and, finally, the Kushanas. The Greeks of Bactria created an Indo-Bactrian Kingdom along the north-west frontier of India, which in turn was destroyed by the Parthians and the Sakas. By 150 BCE, Punjab came under the control of the Greek rulers of Bactria (i.e. Indo-Bactrian rulers). The Parthians established a short-lived Indo-Parthian Kingdom in the Punjab-Indus region after the collapse of the Indo-Bactrian Kingdom. The Indo-Parthian Kingdom was destroyed by the Sakas, who in turn established an Indo-Saka Kingdom in Punjab, Baluchistan, Sind and Kashmir in around 130 BCE (Grainger 2013: 61–74).

Against the cavalry-heavy armies of the Central Asian conquerors, the Indian rulers perceived that their light infantry-elephant-centric armies had no chance in open battles in the field. Hence, the Indian rulers resorted to the construction of defensive architecture. Indian military theory focused on siege warfare as well as on the techniques of pacification in order to avert internal rebellions. And these fortified cities were able to withstand the repeated raids launched by the Indo-Greeks, Indo-Parthians and the Indo-Sakas. In fact, the post-Maurya age could be categorised as the 'Golden Age' of Indian military architecture due to proliferation of techniques and ideas on siege warfare, and military fortifications in ancient India diverged from the military engineering trajectory of the Mediterranean landscape.

Dieter Schlingloff asserts that, to a great extent, Kautilyan theory influenced the military architecture of India. We can reconstruct the history of ancient India's fortifications from literary sources and archaeological findings. The Buddhist text *Milindapanha* notes that the city of Sagala was fortified with a wall and had moats and gates with towers. Jain literature tells us that cities were protected with ramparts and walls with battlements. Arrow- and stone-throwing machines were placed on the battlements. The *Arthashastra* speaks of a city with either four, eight or 12 gates surrounded with moats, ramparts and outer walls. Kautilya speaks of three parallel ramparts of earth and walls made of brick and stone. Further, the *Arthashastra* notes that the defensive walls of the cities should never be circular but should be square or shaped like a quadrilateral (Schlingloff 2014: 12–16, 46–7). In fact, the *Arthashastra* at one point notes that geometric plans should be adopted while building the walls, but autonomy should be given to the city builders to take advantage of the irregular terrain instead of rigidly adhering to textbook instructions. The military engineers of ancient India generally followed the *Arthashastra*'s principles regarding fortifications.

Generally, the fortified cities of India were situated on the plain, especially near the river valleys. Before 500 BCE the walls were constructed with wood. However, wood was vulnerable to rotting in the humid climate of North India and was also susceptible to fire. By about 300 BCE, wood was replaced with earth. In fact, Mo Zi, the Chinese military theoretician discerning the vulnerability of wood to fire, proposed covering the wooden fortifications with layers of earth. Probably the lack of availability of stones in Central and South China encouraged Mo Zi to go for wooden fortifications. The earthen walls of the Indian forts were covered and strengthened with bricks and stones. The river supplied water for the moat, which constituted the first line of defence. The earth excavated while digging the moat was used for building a wall with gates and bastions (Mo Zi 2013: 349–52; Schlingloff 2014: 44–5). This wall, with its defensive paraphernalia (flanking towers and gates with guardrooms), constituted the second line of defence. This sort of complex military architecture was absent in the cases of the Hellenistic, Roman, Achaemenid and Parthian empires. However, the Indians adopted certain techniques from the Central Asian nomadic invaders. For instance, the Kushanas in Punjab during the second century CE introduced curtain walls pierced with loopholes and semi-circular towers (Deloche 2007: 37). From these loopholes, the garrison shot arrows at the besiegers. This sort of curtain wall became common in later Indian fortifications.

Some ideas about the size of fortifications of ancient India's cities and continuous evolution of the defensive fortifications can be gleaned from archaeological records. The city of Rajgir in Magadha between 500 BCE and 100 CE had a 173-foot-wide wall of earth. Such massive earthen walls which protected Chinese and Indian cities were not vulnerable to the stone-throwing machines used by the Greek and Roman armies. Ahicchatra in North India was fortified during 300 BCE. The rampart, which was in the shape of an irregular triangle, enclosed an urban area of 1.5 sq km. In around 100 BCE, the mud-baked brick rampart was further protected with several rectangular towers. Between 300–200 BCE, Sisupalgarh, located on the right bank of the Bhargaui River, was heavily fortified. The city covered some 1.36 sq km of area. It was surrounded by a clay rampart about 26 feet high and 10 feet wide. Bastions were constructed at the four corners, and there were eight gates opening towards the outside, two on each flank. In the first century CE, two brick walls further surrounded the rampart. The city of Ujjayini had about 60,000 inhabitants. Fortifications of Ujjayini were started in 600 BCE, and this trend was accelerated during 300 BCE. A rampart 40 feet high and pentagonal in shape surrounded the city. The pentagonal rampart first came in Western Europe during the early modern era – the age of Parker's so-called military revolution. There were eight gates. Further, the rampart was surrounded by a water-filled moat which was 148 feet wide and 21 feet deep.

On the left bank of the Jamuna River was the city of Kausambi, which was an important transit point in the trade route which connected Punjab with North India and Tamralipti harbour in South Bengal, the gateway to the Bay of Bengal. The almost square-shaped city's outline was protected with an earthen rampart 42.5 feet high. The rampart, which enclosed a city area of 2.3 sq km, was encased in several layers of bricks. There were 11 gates and several towers, and the latter's height rose to about 72 feet. Outside the rampart was a big water-filled moat which was 475 feet wide. The moat was supplied with water from the Jamuna by a canal. All these fortifications were completed by the first century BCE, when the population of this city came to about 180,000 persons. As a point of comparison, Athens's wall enclosed an urban space of 2.5 sq km, and the population of this Greek metropolis came to about 100,000 souls (Deloche 2007: 39–44; Schlingloff 2014: 18, 20–2, 32).

By all accounts, the biggest city in the ancient world was the Maurya capital Pataliputra in Bihar at the confluence of the Ganga and Son rivers. Besides the account of the Greek (Seleucid) ambassador to the Maurya Court, Megasthenes, archaeological excavations provide us with data on this city as it existed during 300 BCE. The city was shaped like a parallelogram. The circumference of the city came to about 33.8 km, with a surface area of 25.5 sq km, and the whole area was enclosed by a wall. Alexandria was half its size. Rome under Emperor Marcus Aurelius (r. 161–80 CE) had a wall 18.8 km long which enclosed an area of 13.7 sq km. Pataliputra's protective wall had 570 towers and 64 gates. Interestingly, the wall of Pataliputra had towers at every 175 foot interval. Kautilya's *Arthashastra* notes that the walls should have towers with 177 foot intervals between them in order to provide interlocking fire for the benefit of the defenders. It is to be noted that the effective range for arrows came to about roughly 180 feet (Schlingloff 2014: 32–5). From 300 CE onwards, the fortification system became more complex, when walls were crenellated with rectangular or serrated merlons (Deloche 2007: 49). A brief comparison with Roman fortifications is in order.

The Roman Empire until *c.* 100 CE was on an expanding spree. After that date, when it went on a strategic defensive, Rome started giving attention to building fortifications. In Western Europe, the Romans faced mainly tribes who lacked the skill and elaborate paraphernalia required to conduct large-scale sieges. Against them, the Romans constructed mud brick forts and thin rectangular stone forts surrounded by ditches. For example, at Newstead in Scotland, the Roman ditch was 9 feet deep (Webster 1969: 175). The Roman forts in Algeria and Tunisia were plain square or rectangular buildings with no corner turrets, and each had only one gateway. The so called Danubian Limes which was breached by Attila in 446 CE had a series of single towers along the right bank of the Danube River. The curtain walls of the big forts varied between 7 and a maximum 32 feet (Toy 1985: 46–8). In the Near East, with the rise of the Sasanian Empire (224–651 CE), the strategic scenario changed. The Sasanians not only had a large cavalry-centric field army but also had the capability to conduct lengthy and complex sieges. So Roman cities in West Asia registered elaborate and complex fortifications. The Roman fortification in West Asia actually took off under the Byzantine Empire, when it went on a strategic defensive mode.

The Mauryan emperors also understood the danger posed by armed peasantry. Asoka (r. 268–232 BCE), after the conquest of Kalinga (Orissa), tried to demilitarise armed peasantry. But he failed. Even the Qin Empire in the third century BCE attempted the demilitarisation of the countryside, with limited success (Bodde 1986: 56; Thapar 1989: 62, 158–9). Probably the vast demographic resources of China and India made such a task impossible. Asoka's rock edicts mention the presence of prisons (Singh 2017: 51). It was an administrative innovation under Mauryan rule.

Manusamhita (Manu's Code of Law), composed around the beginning of the Common Era, has a section on counterinsurgency (COIN). This work is probably in reaction to the problems faced by the Maurya Empire in controlling the vast heartland of India, full of armed peasants and tribesmen. Like most of ancient India's works, we are not sure about the authorship of this work. We are not sure of even a historical Manu. The *Manusamhita* lays down that the principal duty the ruler is to protect his subjects. In case the king fails to provide rule of law, then uprising by the people occurs. This work enjoins the king to use both kinetic and non-kinetic measures to prevent rebellion. Good administration is necessary. Punishment should be equitable to offence. The king is warned not to impose unjust or heavy taxation. Honest officials organised in a hierarchical manner constitute, according to Manu, a facet of good governance. Intelligence and surveillance systems are essential components of the royal arsenal to guard against evil political entrepreneurs conspiring against the ruler. All these techniques fall under the rubric of non-kinetic components. Ultimately it is the army (the kinetic element) which constitutes, according to *Manusamhita*, the last and ultimate line of defence (*Manava-Dharmasastra* 2006: 154–66). As a verse notes: 'it is with military force, therefore, that he should subdue all creatures' (quoted from *Manava -Dharmasastra* 2006: 159).

Eastern Eurasia

The Han Empire in China existed from 202 BCE to 220 CE. The Western (Former) Han lasted from 202 BCE to 9 CE and the Eastern (Later) Han from 25 to 220 CE. In 157 CE, the population of the Western Han Chinese Empire was 56 million, and the population of the Roman Empire at that time was 46 million (Higham and Graff 2012: 4). The Han army comprised cavalry supported by infantry equipped with crossbows, spears and halberds (Ven 2000: 12). The Western Han Empire, like the preceding Qin Empire (221–206 BCE), conscripted peasants for manning the armies. In contrast, military service remained voluntary in Indian history. In the Western Han Empire, all adult males aged 23 to 56 were liable for military service. The men were drafted for two years, and the conscripts served their time in their native regions under the local chieftains who became their commanders during war (Loewe 1986: 151; Dreyer 2012: 30). The

Jin army, from a population base similar to the Han Empire, had 700,000 men under arms in 280 CE. In comparison, the Maurya Empire, from a population of 50 million, maintained an army that numbered 690,000 men, and the combined strength of the armies of both the Eastern and Western Roman Empires came to about 645,000 men in 400 CE (Smith 1914: 124–6; Liebeschuetz 1995: 265; Graff 2002: 43; Guha 2010: 30–1).

An important task of the Qin and Han bureaucracies was to register the population and oversee military exercises (Ven 2000: 12). The population under the Qin Empire was divided into household groups of five and 10 and then on up into larger units, and each group was legally responsible for the behaviour of its members under the mutual responsibility system. In the Qin army, squads of five men each and platoons of 10 each were drawn from the household units, so that one man from each of a group of five households served in a particular squad (a five-man unit) of the army. This ensured that men who served in a tactical formation knew each other intimately and were probably related to each other by blood/marriage, so they were prepared to fight to the death to save the other members of their unit (Yates 2009: 31). Thus, the Qin military system was geared to create primary group solidarity among the military formations. The very fact that the Qin and Han Empires registered their population and conducted census surveys to tax and mobilise them for military purpose shows that the tentacles of the state reached the bottom levels of the society. The state bureaucracy amassed information about the lower levels of the social strata. In 5 BCE, the civil service of the Han Empire had 120,285 personnel (Loewe 1986: 122). By ancient standards, the Han Empire had a big bureaucracy. Compared to the Han Empire, the Roman Empire appears as an under-administered state. Hence, the Qin and the Han polities were strong states.

In the Western Han Empire, some 2 per cent of the population was conscripted for military purpose (Loewe 1974: 81). Military conscription was, however, unpopular with the peasants. In 209 BCE, a widespread revolt started against the Qin dynasty against the heavy burden of forced labour and military conscription (Graff 2012: 44), and peasant rebellion due to natural disasters brought down the Western Han Empire. If peasant rebellion and external invasion by the steppe nomads occurred simultaneously, then the reigning Chinese monarch was sure to lose the 'mandate of heaven'.

The combat effectiveness of the conscripted peasants was limited. The conscripted infantry was useful only in police actions (Lewis 2000: 37). The Hsung-nu/Xiongnu inhabiting the region along the northern frontier of China were herdsmen and hunters. Like the Scythians, they had no permanent place of abode and were extremely mobile. Hence, they were difficult to bring under subjection by the sedentary states. From the second century BCE, Xiongnu started attacking Northern China regularly. They inflicted widespread damage on livestock and civilian property (Loewe 2009: 70, 74). In 117 BCE, a specialist cavalry force was raised by the Han Empire which could fight the Xiongnu effectively (Dreyer 2012: 30). Neither China nor India, with hot climates and paddy fields, were suited for breeding good quality horses. Both the Chinese and Indian rulers acquired warhorses from Central Eurasia. When the sedentary Chinese polities were strong, the emperors projected power into the steppe region. And when the empires became weak, the nomads punched holes into the defensive systems of the agrarian bureaucratic empires. In 119 BCE, the Western Han Emperor Wu ti (r. 141–87 BCE) launched a massive punitive expedition into the Xiongnu homeland. In 133 BCE and in 10–11 CE, Emperors Wu ti and Wang Mang mustered 300,000 soldiers to deter Xiongnu attacks (Bielenstein 1986: 237; Wright 2012: 61). The Roman Empire was never able to mobilise such a large number of soldiers at any one time. The late Roman Empire, for a particular expedition, could muster only about 20,000 troops (Goldsworthy 2000: 193). In the case of China, such massive military mobilisations were possible due to a huge population, generation of surplus agrarian wealth and the presence of an elaborate bureaucracy.

To an extent, both the Roman and Han empires attempted to reach an accommodation with the non-state enemies beyond their frontiers across the Rhine and the Ordos. Neither Wu ti's successors nor the emperors who came after Octavian attempted permanent annexation of Mongolia or Germania. Neither the forest-covered, marshy Germania nor the treeless arid Mongolia was economically productive. Further, the Chinese sedentary armies could not maintain themselves in the arid Mongolian steppes against the highly mobile nomadic cavalry forces. One Han official noted that the Xiongnu had an army that arrived like gale winds and departed like retracting lightning. Their occupation was raising animals, and they did not stay anywhere for long (Olberding 2013: 91–2). It was horribly expensive to maintain a large number of imperial troops amidst the forests of Germania or in the Ordos Loop. Trade concessions, and in the case of China the present of imperial princesses as brides, along with a show of military deterrence across the borders, kept uneasy peace for some time.

Though Sun Tzu is against conducting lengthy sieges, from the end of the Warring States era, socio-economic changes resulted in the rise of cities. The urban centres became nodes of commercial and military power. And to protect the cities against peasant uprisings, nomadic invaders and other competing states, the sedentary polities started constructing elaborate walls. Elaborate defensive constructions which involved moving massive amount of earth were done by corvee labour imposed on the peasants, vagabonds etc. Hence lengthy sieges became a feature of the Chinese military landscape (Bodde 1986: 61; Sawyer 2009: 45–64). More or less the same factors operated in case of Mauryan India. The rise of cities as economic and political nodal centres made siege warfare a horrible reality in the subcontinent. And Kautilya's *Arthashastra* points out the various siege techniques (like the use of armoured elephants as battering rams and *sataghnis*, i.e. stone-throwing machines, etc.) to be used for taking a city (Roy 2014: 488–9).

Mo Zi or Master Mo (b. 479–d. 381 BCE?) was the most famous military theoretician of ancient China who focused on conducting siege warfare and urban combat. Unlike Sun Tzu, who focuses on politics and grand strategy, in Mo Zi's

The Book of Master Mo, the bedrock for the foundation of the Mohist School of thought provides technical details concerned with construction of field works and fortifications. This book was composed between the fifth and third centuries BCE but influenced siege warfare during the Later Han Empire and afterwards. As regards some of the technical details provided by Mo Zi, his book notes:

> in general, with regard to the methods of defending a city, city gates are prepared that are 'hanging gates' (portcullis gates), with a mechanism for raising and lowering them. . . . A trench should be dug . . . with a width comparable to that of the gate leaves. . . . At the head of the trench there should be a sluice gate.

The focus on technical details encouraged some scholars to argue that Mo Zi was from the artisan class. Master Mo noted that not fate but rational calculations enable human beings to overcome problems. Master Mo, like the Enlightenment philosophers of France, searched for universal timeless principles for defending a city. Mo Zi laid down draconian punishment for someone refusing proper duty in the battlefield and also rewards for those who performed well at the forward edge of the battle area (Mo Zi 2013: vii–xxi; the quotation is from p. 350). As regards the necessity of awarding draconian punishment for underperforming when the state faced a crisis, one could trace in Mo Zi's philosophy the influence of the Legalist School which developed in Qin China.

The types of Chinese artillery used for conducting sieges was impressive. A huge crossbow mounted on a framework or carriage was known as arcuballista (*lithobolon*). The Greek and Roman *lithobolons* shot arrows, javelins and stones. The Hellenistic militaries used tripod mountings for such machines. The Chinese used rectangular mountings which were often wheeled, and this in turn raised the mobility of such machines. The Chinese engineers used to couple two to three bowstaves together in order to increase the energy of tension and hence the initial velocity of the bolt shot from such a machine. Greek and Roman artillery, as we have seen in an earlier chapter, also used the torsion of sinew or hair. The range of a Greek torsion catapult was about 1,230 feet, and that of a Chinese arcuballista was 1,500 feet. One of the Mohist engineers noted that Chinese crossbow catapult machines were mounted on ramparts 30 feet above the ground and relied on a trigger mechanism. Each such machine was about 20 feet in length. The defenders were deployed on the ramparts of the fort with multiple bolt-shooting arcuballistas, and some of machines were also placed on carriages which had two axles. The length of the bolt shot by such a machine was 10 feet long (Needham and Yates 2002: 187–91).

Massive peasant rebellions occurred towards the end of the reign of Wang Mang (r. 9–25 CE) because the Yellow River changed its course. This in turn ruined the cultivation, and the state, requiring money, in desperation increased taxation on the hapless peasants. Red Eyebrows became the principal peasant insurgent groups. And the local gentry, to protect their landholdings, also supported the peasant insurgents against the central government. Without the aid of the local gentry who commanded the provincial forces, the royal armies were unable to crush the widespread peasant uprisings (Bielenstein 1986: 240–51). The Western Han Empire came to an end after the overthrow of Wang Mang. And after a lengthy civil war, Guangwu (r. 25–57 CE) ascended the throne as the first emperor of the Eastern Han.

The Eastern Han Empire anticipated the danger posed by an armed peasantry. Further, civil war had reduced the taxable population by one-third. So the Eastern Han Empire in 30 CE replaced universal military service with long-term volunteer recruits. The personnel were recruited from the 'barbarians' (steppe nomads like Southern Xiongnu), criminals, vagabonds and volunteers. The Xiongnu, Xianbei and Qiang people who settled in Northern China were partly Sinicised but retained their skill of horse archery, and they constituted the crack force of the Eastern Han army. For paying the troops, a special tax was introduced. By the first century CE, the Han Empire possessed a professional army. So it would be wrong to argue that only Western Eurasia possessed professional military forces from the era of Classical Greece onwards. The Eastern Han Empire went through the process which the Roman Empire would face during the fourth century CE. Guang wu decentralised the royal monopolies and reduced the revenues from salt and iron. Agriculture became the main item of state taxation. The small and middle peasantry sought protection from taxation by allying themselves with the local landlords, who were also in charge of the lower levels of military command in the provinces. These local strongmen would later lead armed rebellions against the state whenever the central government faced a crisis. The local landlords, with the support of native peasantry, also hampered military recruitment of the peasants in the Later Roman Empire. After the collapse of the Eastern Han, several kingdoms emerged in China which fought amongst themselves. Among these, foremost was the Jin Empire (266–420 CE). Military households with a hereditary obligation to provide soldiers were the principal source of recruits for these polities (Lewis 2000: 39–75; Crespigny 2009: 93, 110–11; Dreyer 2009: 119, 141; Yates 2009: 23).

The Han and Jin empires maintained a cavalry-centric force which differed from the heavy infantry-centric force of the Romans. However, the Jin, unlike the Han but like the Sasanians, deployed heavy cavalry for shock combat along with light cavalry (mounted bowmen). The Western Han Empire and its enemy, the Xiongnu, deployed light unarmoured cavalry who fought with bows and arrows. After 300 CE, the Jin Empire deployed heavy armoured cavalry. Like the Sasanian cavalry, both the mount and the cavalryman of Jin Empire were covered with lamellar armour made of either iron or leather. The introduction of the stirrup from Southern Manchuria made this shift possible. The evolution of edged weapons raised heavy cavalry's shock value. The Western Han cavalrymen used straight, slender, double edged swords which were used for slashing as well as thrusting. After 200 CE, the heavy curved single-edged sabre suitable for chopping emerged, and this became the basic weapon of the Jin heavy cavalry (Graff 2002: 41–2).

Unlike Greece and Rome, China and India did not face any serious threat from the sea. The principal threats for the Chinese and Indian rulers were the steppe nomads who attacked along the northern frontier of China and north-west frontier of India, respectively. However, China's rulers constructed riverine navies for controlling the rivers. Rivers were important for trade and commerce. Further, control over the rivers was necessary for transporting troops and military supplies for conquest, consolidation and pacification of the interior of China. Oared boats were probably used in Neolithic China from 5000 BCE. The first evidence of the use of ships for military operations in China occurred in *c.* 1045 BCE, when King Wu of Zhou ferried 300 chariots and 48,000 infantry across the Yellow River in Mengjin in 47 ships to attack the Shang capital. These were not specialised warships but cargo boats commandeered for military operation (Lorge 2012: 82–3).

The Paleolithic Period in Japan spanned from 150,000 to 10,500 BCE. It was followed by the Jomon era (10,500–300 BCE). A 3 foot-long bow emerged during this period, and such bows were probably used for both hunting and fighting. The Yayoi Period in Japan, from 300 BCE to 300 CE, witnessed the rise of wet rice agriculture. During this period, bronze and iron metallurgy came into Japan from Korea. Yayoi immigrants from Korea introduced improved bronze and iron weapons (tanged swords, axes and spears). Rice cultivation and development in metal production resulted in expansion of the population of this island to about 4 million during the third century CE. Most of the population was concentrated in Western Japan. Large settlements came up which were surrounded with ditches and protected by watchtowers. In Northern Kyushu, artisans manufactured bronze spears and halberd blades (Farris 1999: 49).

According to one theory, the horse riders from North Asia established their sway in Korea during the fourth and fifth centuries CE, and then from Korea the cavalry influence went to Japan (Edwards 1983: 265–95). From 250 CE onwards, immigrants from North-East Asia introduced cavalry to Japan. Men on horseback, equipped with iron weapons, established the Yamato Kingdom in the Nara Plain. Horse riders and massive irrigational projects which extended rice cultivation constituted the backbone of the Yamato Kingdom. Gradually, this polity subjected the whole of Japan under its sway and also projected power into Korea. The presence of iron ore in Korea was one of the principal reasons behind Japanese power projection in this region (Brown 2006: 108–21; Kidder 2006: 87–107).

In the middle of the third century CE, Emperor Sujin of Japan ordered the provinces to build ships. These ships were used for coastal communications and also for ferrying troops. The absence of good roads, due to mostly mountainous terrain, made sea communication for internal trade and administrative purposes very important. The chronology of Japan's ancient history is problematic. Either in 200 CE or in 362 CE, rebels aided by the Korean Kingdom of Shiragi (Silla) killed the Japanese Emperor Chuai. In retaliation, Empress Jingo decided to invade Korea. The armada set out from North-West Kyushu crossed the Korea Strait and reached Kyongju (Keishu) on the south-east coast of Korea (Marder 1945: 2–4). In 529 CE, the Yamato Kingdom sent a force of 60,000 soldiers to shore up its position in Southern Korea (Brown 2006: 146). The transport capacity of the Yamato fleet was certainly greater than the Roman fleet at its heyday. Further, the Yamato fleet was technically superior to its Mediterranean counterparts because the former had to cross the treacherous Sea of Japan and not an inland sea like the Mediterranean.

Conclusion

Several times in history, the destruction of the state in China was caused by both interstate and intrastate warfare. In fact, the Western Han collapsed due to a large-scale peasant uprising. The Massive peasant uprisings resulting in dynastic changes occurred repeatedly in China's history. However, this was absent in the case of Indian history. This is partly because the monarchs of India never introduced military conscription and because of the low level of agricultural taxation and the Hindu *karma* doctrine (one's position in society was the product of actions in previous life). Why did no large-scale peasant uprising occur in the Roman Empire? This was probably because the Roman Empire was able to separate its army from society through an elaborate mechanism of training, discipline and introduction of a bureaucratic organisational framework. The period under review saw the continuous evolution of the Roman force within the paradigm of a heavy infantry force. In the paradigm of Roman warfare, light infantry and cavalry, which were supplied by their allies, played a secondary role in the battlefield. The core of the Roman army, both during the republic and the early empire, remained the heavy infantry. When Rome faced external opponents who used a different paradigm of war, like Carthage, the Romans innovated by adopting non-Western elements like elephants, etc., and used these engines of war with great effect against the Macedonian Kingdom. But after the conquest of the Mediterranean lands, the Roman military machine reverted to the traditional Greco-Roman paradigm of heavy infantry-centric force.

In contrast, in Eastern Eurasia, the rise of the horse nomads of Central Eurasia forced the sedentary polities of India and China to change their infantry-centric force structure to a cavalry-centric force structure. Ancient India and China, in response to external invasions by the mounted steppe nomads, went on a fort-building spree. For guarding such forts, infantry was required. The Chinese infantry equipped with crossbows was more effective than Roman-style infantry equipped with javelins and swords. This was because a crossbow had greater range than a javelin and the rate of fire of the Chinese trigger-equipped crossbow was greater than a soldier throwing a javelin. And Roman legionary armour was useless against the arrow shot by a compound bow used by the steppe nomads. So the Roman heavy infantry was useful only in a particular context. Hence, the

Indians, Chinese and the West Asians saw no particular reason to evolve such infantry. Towards the Later Roman Empire, the rise of external enemies (Huns, Avars, etc.) with a cavalry-centric force accelerated the transition of the Roman military machine from a heavy infantry-centric force to one based mainly on cavalry. And this military transformation in Mediterranean and Western Europe gave birth to a convergence in military practices all over Eurasia, which is the subject of the next chapter.

References

Alston, Richard, 'Warfare and the State-The Military and Politics: The Late Republic and the Principate', in Philip Sabin, Hans Van Wees and Michael Whitby (eds.), *The Cambridge History of Greek and Roman Warfare*, vol. 2, *Rome from the Late Republic to the Late Empire* (Cambridge: Cambridge University Press, 2007), pp. 176–97.

Anon, *Manu's Code of Law: A Critical Edition of the Manava-Dharmasastra*, Patrick Olivelle with the editorial assistance of Suman Olivelle (2005, reprint, New Delhi: Oxford University Press, 2006).

Ash, Rhiannon, 'An Exemplary Conflict: Tacitus' Parthian Battle Narrative (Annals 6.34–35)', *Phoenix*, vol. 53, no. ½ (Spring–Summer 1999), pp. 114–35.

Bagnall, Nigel, *The Punic Wars: 264–146 BCE* (2002, reprint, New York/London: Routledge, 2003).

Benjamin, Craig, *Empires of Ancient Eurasia: The First Silk Roads Era, 100 BCE-250 CE* (Cambridge: Cambridge University Press, 2018).

Bielenstein, Hans, 'Wang Mang, the Restoration of the Han Dynasty, and Later Han', in Denis Twitchett and John K. Fairbank (eds.), *The Cambridge History of China*, vol. 1, *The Chin and Han Empires, 221 BC-AD 220* (Cambridge: Cambridge University Press, 1986), pp. 223–90.

Bivar, A.D.H., 'The Political History of Iran Under the Arsacids', in Ehsan Yarshater (ed.), *The Cambridge History of Iran*, vol. 3(1), *The Seleucid, Parthian and Sasanian Periods* (1983, reprint, Cambridge: Cambridge University Press, 2006), pp. 21–99.

Bodde, Derk, 'The State and Empire of Chin', in Denis Twitchett and John K. Fairbank (eds.), *The Cambridge History of China*, vol. 1, *The Chin and Han Empires, 221 BC-AD 220* (Cambridge: Cambridge University Press, 1986), pp. 20–102.

Brice, Lee L., 'Disciplining Octavian: A Case Study of Roman Military Culture, 44–30 BCE', in Wayne E. Lee (ed.), *Warfare and Culture in World History* (New York/London: New York University Press, 2011), pp. 35–59.

Brown, Delmer M., 'The Yamato Kingdom', in John Whitney Hall, Marius B. Jansen, Madoka Kanail, and Denis Twitchett (eds.), *The Cambridge History of Japan*, vol. 1, *Ancient Japan* (1993, reprint, Cambridge: Cambridge University Press, 2006), pp. 108–62.

Caesar, Julius. *The Civil War Together with the Alexandrian War, The African War, and The Spanish War by Other Hands*, translated with an Introduction by Jane F. Gardner (1967, reprint, Harmondsworth: Penguin, 1984).

Campbell, R. Brian, 'The Roman Empire', in Kurt Raaflaub and Nathan Rosenstein (eds.), *War and Society in the Ancient and Medieval Worlds: Asia, The Mediterranean, Europe, and Mesoamerica* (Washington, DC: Center for Hellenic Studies, distributed by Harvard University Press, 1999), pp. 217–40.

Charles, Michael B. and Rhodan, Peter, 'Magister Elephantorvm: A Reappraisal of Hannibal's Use of Elephants', *Classical World*, vol. 100, no. 4 (2007), pp. 363–89.

Crespigny, Rafe De, 'The Military Culture of Later Han', in Nicola Di Cosmo (ed.), *Military Culture in Imperial China* (Cambridge, MA: Harvard University Press, 2009), pp. 90–111.

Daly, Gregory, *Cannae: The Experience of Battle in the Second Punic War* (London/New York: Routledge, 2002).

Deloche, Jean, *Studies on Fortifications in India* (Pondicherry: Institutfrançais de Pondichery, 2007).

Dreyer, Edward L., 'Military Aspects of the War of the Eight Princes, 300–307', in Nicola Di Cosmo (ed.), *Military Culture in Imperial China* (Cambridge, MA: Harvard University Press, 2009), pp. 112–42.

Dreyer, Edward L., 'Continuity and Change', in David A. Graff and Robin Higham (eds.), *A Military History of China* (Kentucky: University Press of Kentucky, 2012), pp. 19–38.

Edwards, Walter, 'Event and Process in the Founding ofJapan: The Horserider Theory in Archaeological Perspective', *Journal of Japanese Studies*, vol. 9, no. 2 (1983), pp. 265–95.

Fagan, Garrett G. and Trundle, Matthew, 'Introduction', in Garrett G. Fagan and Matthew Trundle (eds.), *New Perspectives on Ancient Warfare* (Leiden/Boston: Brill, 2010), pp. 1–19.

Farris, W. Wayne, 'Japan to 1300', in Kurt Raaflaub and Nathan Rosenstein (eds.), *War and Society in the Ancient and Medieval Worlds: Asia, The Mediterranean, Europe, and Mesoamerica* (Washington, DC: Center for Hellenic Studies, distributed by Harvard University Press, 1999), pp. 47–70.

Farrokh, Kaveh, *Shadows in the Desert: Ancient Persia at War* (Oxford: Osprey, 2007).

Gilliver, Catherine M., 'Battle: The Late Republic and the Principate', in Philip Sabin, Hans Van Wees and Michael Whitby (eds.), *The Cambridge History of Greek and Roman Warfare*, vol. 2, *Rome from the Late Republic to the Late Empire* (Cambridge: Cambridge University Press, 2007), pp. 122–57.

Goldsworthy, Adrian, *Roman Warfare* (London: Cassell, 2000).

Goldsworthy, Adrian, *Caesar's Civil War: 49–44 BCE* (2002, reprint, New York/London: Routledge, 2003).

Goldsworthy, Adrian, *In the Name of Rome: The Men Who Won the Roman Empire* (2003, reprint, London: Phoenix, 2004).

Goldsworthy, Adrian, 'War: The Late Republic and the Principate', in Philip Sabin, Hans Van Wees and Michael Whitby (eds.), *The Cambridge History of Greek and Roman Warfare*, vol. 2, *Rome from the Late Republic to the Late Empire* (Cambridge: Cambridge University Press, 2007), pp. 76–121.

Graff, David A., *Medieval Chinese Warfare: 300–900* (London/New York: Routledge, 2002).

Graff, David A., 'State Making and State Breaking', in David A. Graff and Robin Higham (eds.), *A Military History of China* (Kentucky: University Press of Kentucky, 2012), pp. 39–56.

Grainger, John D., *Rome, Parthia, India: The Violent Emergence of a New World Order 150–140 BC* (Barnsley: Praetorian Press, 2013).

Guha, Sumit, *Health and Population in South Asia: From Earliest Times to the Present* (2001, reprint, Ranikhet: Permanent Black, 2010).

Higham, Robin and Graff, David A., 'Introduction', in David A. Graff and Robin Higham (eds.), *A Military History of China* (Kentucky: University Press of Kentucky, 2012), pp. 1–18.

How, W.W., 'Arms, Tactics and Strategy in the Persian War', *Journal of Hellenic Studies*, vol. 43, Part 2 (1923), pp. 117–32.

Hoyos, Dexter, *Hannibal Rome's Greatest Enemy* (Exeter: Bristol Phoenix Press, 2008).

Hoyos, Dexter, *Mastering the West: Rome and Carthage at War* (New York: Oxford University Press, 2015).

Kidder, Jr., J. Edward, 'The Earliest Societies of Japan', in John Whitney Hall, Marius B. Jansen, Madoka Kanai and Denis Twitchett (eds.), *The Cambridge History of Japan*, vol. 1, *Ancient Japan* (1993, reprint, Cambridge: Cambridge University Press, 2006), pp. 48–107.

Lendon, J.E., *Soldiers and Ghosts: A History of Battle in Classical Antiquity* (New Haven/London: Yale University Press, 2005).

Lewis, Mark Edward, 'The Han Abolition of Universal Military Service', in Hans van De Ven (ed.), *Warfare in Chinese History* (Leiden: Brill, 2000), pp. 33–75.

Liebeschuetz, Wolfgang, 'The End of the Roman Army in the Western Empire', in John Rich and Graham Shipley (eds.), *War and Society in the Roman World* (1993, reprint, London/New York: Routledge, 1995), pp. 265–76.

Loewe, Michael, 'The Campaigns of Han Wu-ti', in Frank A. Kierman, Jr. and John K. Fairbank (eds.), *Chinese Ways in Warfare* (Cambridge, MA: Harvard University Press, 1974), pp. 67–120.

Loewe, Michael, 'The Former Han Dynasty', in Denis Twitchett and John K. Fairbank (eds.), *The Cambridge History of China*, vol. 1, *The Chin and Han Empires, 221 BC-AD 220* (Cambridge: Cambridge University Press, 1986), pp. 103–222.

Loewe, Michael, 'The Western Han Army: Organization, Leadership, and Operation', in Nicola Di Cosmo (ed.), *Military Culture in Imperial China* (Cambridge, MA: Harvard University Press, 2009), pp. 65–89.

Lorge, Peter, 'Water Forces and Naval Operations', in David A. Graff and Robin Higham (eds.), *A Military History of China* (Kentucky: University Press of Kentucky, 2012), pp. 81–96.

Marder, Arthur J., 'From Jimmu Tenno to Perry: Sea Power in Early Japanese History', *American Historical Review*, vol. 51, no. 1 (1945), pp. 1–34.

Miles, Richard, *Carthage Must Be Destroyed: The Rise and Fall of an Ancient Civilization* (2010, reprint, London: Penguin, 2011).

Mo Zi, *The Book of Master Mo*, translated and edited with Notes by Ian Johnston (2010, reprint, London: Penguin, 2013).

Needham, Joseph and Yates, Robin D.S., *Science and Civilisation in China*, vol. 5, *Chemistry and Chemical Technology*, Part 6, *Military Technology: Missiles and Sieges* (1994, reprint, Cambridge: Cambridge University Press, 2002).

Oakley, Stephen, 'The Roman Conquest of Italy', in John Rich and Graham Shipley (eds.), *War and Society in the Roman World* (1993, reprint, London/New York: Routledge, 1995), pp. 9–37.

Olberding, Garret, 'The Debate between Wang Hui and Han Anguo: A Case Study of Early Han Military Addresses', in Peter A. Lorge (ed.), *Debating War in Chinese History* (Leiden/Boston: Brill, 2013), pp. 79–106.

Rankov, Boris, 'Military Forces: The Late Republic and the Principate', in Philip Sabin, Hans Van Wees and Michael Whitby (eds.), *The Cambridge History of Greek and Roman Warfare*, vol. 2, *Rome from the Late Republic to the Late Empire* (Cambridge: Cambridge University Press, 2007), pp. 30–75.

Rathbone, Dominic, 'Warfare and the State-Military Finance and Supply: The Late Republic and the Principate', in Philip Sabin, Hans Van Wees and Michael Whitby (eds.), *The Cambridge History of Greek and Roman Warfare*, vol. II, *Rome from the Late Republic to the Late Empire* (Cambridge: Cambridge University Press, 2007), pp. 158–76.

Rawlings, Louis, 'The Carthaginian Navy: Questions and Assumptions', in Garrett G. Fagan and Matthew Trundle (eds.), *New Perspectives on Ancient Warfare* (Leiden/Boston: Brill, 2010), pp. 253–87.

Rich, John, 'Introduction', in John Rich and Graham Shipley (eds.), *War and Society in the Roman World* (1993, reprint, London/New York: Routledge, 1995), pp. 1–8.

Rosenstein, Nathan, 'Republican Rome', in Kurt Raaflaub and Nathan Rosenstein (eds.), *War and Society in the Ancient and Medieval Worlds: Asia, The Mediterranean, Europe, and Mesoamerica* (Washington, DC: Center for Hellenic Studies, distributed by Harvard University Press, 1999), pp. 193–216.

Rosenstein, Nathan, 'Phalanges in Rome?', in Garrett G. Fagan and Matthew Trundle (eds.), *New Perspectives on Ancient Warfare* (Leiden/Boston: Brill, 2010), pp. 289–303.

Roy, Kaushik, 'Hinduism', in Gregory M. Reichberg, Henrik Syse with the assistance of Nicole M. Hartwell (eds.), *Religion, War, and Ethics: A Sourcebook of Textual Traditions* (New York/Cambridge: Cambridge University Press, 2014), pp. 471–543.

Sawyer, Ralph D., 'Martial Prognostication', in Nicola Di Cosmo (ed.), *Military Culture in Imperial China* (Cambridge, MA: Harvard University Press, 2009), pp. 45–64.

Schlingloff, Dieter, *Fortified Cities of Ancient India: A Comparative Study* (2013, reprint, London/Delhi: Anthem, 2014).

Singh, Upinder, *Political Violence in Ancient India* (Cambridge, MA/London: Harvard University Press, 2017).

Smith, Vincent A. *The Early History of India: From 600 BC to the Muhammadan Conquest Including the Invasion of Alexander the Great* (Oxford: Clarendon Press, 1914).

Tarn, W.W., *The Greeks in Bactria and India* (1951, reprint, New Delhi: Munshiram Manoharlal Publishers, 1980).

Thapar, Romila, *Asoka and the Decline of the Mauryas* (1963, reprint, Delhi: Oxford University Press, 1989).

Toy, Sidney, *Castles: Their Construction and History* (1939, reprint, New York: Dover, 1985).

Ven, Hans van De, 'Introduction', in Hans van De Ven (ed.), *Warfare in Chinese History* (Leiden: Brill, 2000), pp. 1–32.

Webster, Graham, *The Roman Imperial Army of the First and Second Centuries AD* (London: Adam & Charles Black, 1969).

Wileman, Julie, *Warfare in Northern Europe before the Romans: Evidence from Archaeology* (Barnsley: Pen & Sword, 2014).

Wright, David C., 'The Northern Frontier', in David A. Graff and Robin Higham (eds.), *A Military History of China* (Kentucky: University Press of Kentucky, 2012), pp. 57–79.

Yates, Robin D.S., 'Law and the Military in Early China', in Nicola Di Cosmo (ed.), *Military Culture in Imperial China* (Cambridge, MA: Harvard University Press, 2009), pp. 23–44.

6

MILITARY CONVERGENCE IN MEDIEVAL EURASIA

Introduction

Traditional historiography highlights discontinuity in military matters between Roman and post-Roman Western Europe (Oman 1885; Ferrill 1988). However, revisionist historiography argues that the medieval European armies continued the Roman heavy infantry and siege warfare tradition (Bachrach 1995: 64–91, 2001, 2002: 119–33; Petersen 2013). The assertions of revisionist historiography are rather overdrawn, however. Bernard S. Bachrach and Leif Inge Ree Petersen overemphasise the importance of siege warfare in the Mediterranean lands and Middle East during the medieval era. They, like Speros Vryonis Jr., assume that the Huns, Arabs, Slavs and German tribes learnt the approaches of siege warfare from the Romans (Vryonis Jr. 2008: 105–17). We need to move away from the Eurocentric narrative that Rome was the only source of military knowledge in Western Eurasia during the medieval era. That narrative forgets that the use of battering rams, arrow-, bolt- and stone-throwing machines, wooden towers, tunnelling and counter-tunnelling techniques, etc., first emerged in Assyria long before Classical Greek civilisation. The Byzantine Empire probably learnt these siege techniques from West Asia. Further, the Huns may have learnt the style of siege warfare from the Chinese.

The picture that post-Roman Mediterranean powers practised infantry-oriented battles and siege warfare while Asia carried on cavalry-centric campaigns focusing on skirmishing and ambushing is tainted. Instead of military divergence, we have convergence as regards matters military in medieval Eurasia – cavalry and not infantry dominated medieval Eurasian warfare. In fact, different forms of cavalry warfare came into existence. While the steppe nomads conducted mounted archery with light cavalry, to counter them the sedentary states used heavy cavalry equipped with lances, swords and battle axes. And the

DOI: 10.4324/9780429284069-7

Mongols, despite being nomads, by integrating various forms of cavalry war with techniques of siege warfare as practised by the sedentary states created a hybrid army which gave them 'global' (read Eurasian) supremacy.

Rather than Rome, the principal impetus for the shift towards cavalry warfare in Eurasia came from the steppe nomads. In fact, Charles Oman's old hypothesis that the Battle of Adrianople (378 CE) marked the beginning of the 'Age of Cavalry' holds water (Oman 1885). From the time of Later Roman Empire, the role of cavalry was increasing in Eurasia, and the age of cavalry continued till the rise of the Western way of warfare in the post-1500 CE period. Some fringe areas of Eurasia, such as Scandinavia, Japan until the ninth century CE and South-East Asia, remained outside the cavalry system of warfare, primarily due to ecological reasons. The period roughly between *c.* 400 CE and *c.* 1500 CE saw the rise and fall of great pastoral and agrarian bureaucratic empires. But overall, we see the rise of cavalry elites and decentralisation of state power leading to fragmentation of central authority of the polities. And this process could be termed as feudalisation.

We can broadly agree with R.J. Barendse's formulation that the sedentary societies of medieval Eurasia experienced feudalism. He rightly notes that in the eighth century, the Tang Empire became decentralised and the horse nomadic dynasties reigned supreme. During the tenth and eleventh century, the Baghdad Caliphate also had to go for devolution of power to the horse nomads. Eleventh-century Byzantium saw the decline of central government's power and the rise of provincial landlords with their private armies. Further, Western Europe, India and Japan saw the rise of mounted aristocracy: knights, Rajputs and samurai. Barendse notes that feudalism could be defined as the subjugation of the peasantry by the mounted aristocracy, who proved more powerful than any central institution (Barendse 2003: 503–29). To a great extent, the rise of feudalism in Eurasia was possible due to increasing military effectiveness of the cavalry. Three great waves of steppe nomadic invaders – Huns, Arabs and the Mongols – accelerated the transition towards the rise of cavalry in general and horse archers in particular during this period. The breaking out of the Vikings from Scandinavia roughly between the eighth and eleventh century further sped up the process of the feudalisation of Western Eurasia.

The Huns and the sedentary polities of Eurasia

From the third century CE, the Western Roman Empire went on a strategic defensive mainly due to the attack by the German tribes who were known as Goths (they were divided into Visigoths/Western Goths and Ostrogoths/Eastern Goths), Franks, Alemanni, Vandals etc. In Gaul, to thwart the German raiders who lacked access to siege technology, the localities were surrounded with walls made of stones. Semi-circular towers were constructed along the walls, which provided platforms to the archers and crossbowmen for defence against the raiders. By the end of the fourth century, the Roman army numbered a maximum of 500,000 men, and the legions had substantial poliorcetic capabilities. During the first half of the fifth century CE, the Western Roman Empire mobilised rural militias under the command of the great landowners for local defence (Bachrach 1999: 277, 284; Petersen 2013: 45–9).

However, 'barbarian' pressure increased due to the intrusion of the Huns into Europe. The Roman Empire, like the Chinese empires, provided annual subsidies and trading privileges to the favoured chieftains as part of their 'divide and rule' policy. Again, the sons and relatives of the chieftains were taken as hostages and brought to the Roman Court in order to impress the power of the empire and instil Roman values in them. The objective was that these persons would become rulers of the tribes, remain loyal to Rome and remember the power and prestige of the empire. However, in the long run, such policies proved detrimental to the security of the Roman Empire. Subsidies, trading privileges, Roman political and military support to the selected chieftains resulted in the formation of bigger political units among the Germans. And they could bring to the field a greater number of warriors against Rome. In addition, a stronger hereditary element intruded at the top end of the German political formations. The result was generation of more robust and politically stable German polities along the frontiers of the Roman Empire (Heather 2010: 233, 235).

Historian Peter Heather writes that the Huns exacerbated the politico-military crisis of the Roman Empire by pushing the German tribes inside the Roman boundaries. The Huns did not enter Europe in one single stream. There was no single Hun invasion of Europe. They came in groups of various sizes under different clan leaders. Moreover, they were breathing down the neck of the Germans, who were forced to attack the Western Roman Empire. Most of the German invaders came to the Western Roman Empire between 376 and 410 CE. However, the German raiders were willing to accept the political suzerainty of the Roman Empire, which was able to incorporate the 'barbarians' within the imperial framework. Heather continues that the direct attacks of the Huns on the Western Roman Empire was less dangerous than the implosion of Hun power when their great leader Attila died in 453 CE. The political vacuum caused by the disappearance of Hun power resulted in a maelstrom, and the various Germanic groups now released from their control entered the Western Roman Empire with greater vigour. As the German invaders took over the lands of the Western Roman Empire, the latter's financial power base, based on agrarian production, started shrinking. And the local magnates in the countryside, realising that Rome could not defend them, made separate political arrangements with the Germanic invaders. This development accelerated the dissolution of the Western Roman polity (Heather 1995: 4–41).

A powerful and interesting argument no doubt! Nevertheless, it has certain loopholes. It does not explain why the Western Roman Empire collapsed but the Eastern Roman Empire/Byzantine Empire continued to function until the mid-fifteenth century. The Ostrogoths entered into the Balkan Peninsula. Still, the Byzantine Empire weathered the crisis. Again, we need to see why the Huns were able to overwhelm the Germanic invaders that were a terror to both the Western and Eastern Roman empires. What were the reasons behind Hunnic military superiority vis-à-vis both the German tribes and the Roman armies? Further, why were the Huns entering Europe in fifth century CE? Let us have a flashback.

The Eurasian steppe extended from Korea in the east till Hungary in the west and from Manchurian, Siberian and Russian forests in the north to the Caucasus, Pamirs and Yellow River in the south. The steppe is a treeless pasture, a grassy plain between the mountains with sparse rainfall, unsuited to agriculture unless expensively irrigated. The nomads lacked the technology and capital to irrigate their land. Abundant grazing ground allowed the breeding of cattle, sheep, goats and especially horses. The vegetation consisted mostly of grass. The climate was extremely cold and dry in winter. Most of the nomads followed pastoral nomadism. The flesh, skins and faeces of their flocks provided them with food, drink and clothing plus materials for their weapons and fuel. Since the steppe was a deficit economic zone, the only option for the nomads was to negotiate with the sedentary polities for favourable trade exchange. The nomads exchanged dairy products, meat, hides and horses for grain and iron. Iron was required for manufacturing both hunting tools and weapons. In the absence of trade, the nomads had to opt for mercenary military service with the agrarian polities or attacking the sedentary agriculturists for plundering. One-third of the length of the steppes bordered on North China. So throughout recorded history, China was the principal target of the steppe nomadic incursions. Different groups of nomads fought with each other for the best grazing land (Saunders 1971: 11). So political disunity debilitated their military power. However, when leaders like Attila and Genghis Khan rose who were able to establish unity among the nomads, then the strategic scenario became precarious for the sedentary polities. This was because of the superior military capabilities of the steppe nomads over the agrarian armies until the advent of gunpowder weapons.

The steppe nomadic tribes were noted for their horsemanship and marksmanship. In Mongolia, 3-year-old children were tied on top of horses. Moreover, the steppe horses were hardy. They could fend for themselves, even digging through the snow with their hooves to find edible roots. During winter, the steppe ponies were able to find shrubs beneath the snow. The traditional weapon of the steppe nomads was the composite bow. The bow was made of wood, horn/bone and sinew, and all these materials were available in the steppe land. This curved bow, which could be folded, was short enough to be used from horseback. The effective range of the Hun bow varied from 150 to 320 feet. From their childhood, Huns started practising with the composite bows for hunting fast-footed animals and birds. They became adept in shooting downwards and upwards as well as backwards even when their horses were galloping. The steppe nomadic horse archers could shoot one arrow every five seconds, 11 arrows per minute (on average) and 600 per hour. This was impressive firepower. Along the arid, waterless, unfriendly steppes, the nomadic horse archers could maintain themselves by drinking *kumiz* (fermented mare's milk). When it was not available, then they would bleed their horses and drink the blood. It would not be an overstatement to say that the steppe nomads ate and slept on their horses (Kennedy 2002: 17–31; Man 2005: 34, 86). So nomadic armies with a minimal logistical paraphernalia had a large operational range. Their mobility and wide operating radius provided them with immense strategic reach. The ecology of Outer Mongolia with its rich pastures was suited for large-scale horse breeding. Life was generally hard in the steppe land. This in turn made the nomads capable of undertaking the hardships of war. In Mongolia, from June to August, the steppe is full of grass and flowers due to rainfall. The cold is severe during September, and snowstorms occur during October. By November, the water courses are frozen. Frequent snowfalls occur until May, and the wind blows ferociously through the steppe. Worse, due to lack of oxygen, the nomads are not always able to light a fire (Saunders 1971: 44). So the harsh environment made the nomads hardy.

The Han dynasty was a near contemporary of the Roman Empire. The Huns were located north of the Ordos Loop of the Yellow River. They were probably a Turkish tribe. West of them was the Yueh-chih (Kushanas), an Indo-European tribe. On the eastern side of the Huns were the Tung Hu. They were also pastoral nomads who fought on horseback. In 89 CE, some Manchurian clans defeated the Huns, and the latter started migrating westwards. The long westward journey of the Huns started a chain reaction in Eurasia. In about 100 CE, they reached Southern Kazakhstan and then Syr Darya (the Jaxartes River) in around 120 CE. Around 160 CE, the Huns were north of the Sea of Azov. In 350 CE, the Huns had crossed the Volga River. The Sarmatians wore conical caps and mailed coats and fought with lances. One of their subgroups was known as the Alans. These groups were pushed west in Europe by the westward migrations of the Huns. After their defeat, the Alans were accepted within the Hun confederation. Later, the Mongols also would follow this policy vis-à-vis the defeated tribes. The next target was the Ostrogoths, who lived across the Dnieper River. In 376 CE, the Hun and Alan cavalry smashed the Ostrogothic force north of the Black Sea (Lattimore 1963: 1; Man 2005: 32, 48, 50, 72–5). The latter, as a consequence of this defeat, moved into the Balkans. In desperation, the Ostrogoths were forced to enter the boundary of the Eastern Roman Empire and in the First Battle of Adrianople fought on 9 August 378 CE (we will call it the First Battle of Adrianople to differentiate it from the second one fought on 14 April 1205) destroyed the field army of the Byzantine Empire (Kennedy 2002: 25). The arguments of historians such as Thomas S. Burns and others that the First Battle of Adrianople was an infantry encounter are not convincing (Burns 2008: 165–74). The German infantry definitely played a part, but the charge of the Ostrogothic cavalry decided the issue. However, the extensive fortifications of Constantinople saved the Byzantine Empire from the mounted Ostrogothic warriors, who lacked expertise in conducting siege warfare.

In Central Asia, the Huns were divided into two groups. One group of Huns went westwards and came to be known as the Western Huns; they subsequently attacked the Eastern and Western Roman empires. The other group, known as the Eastern Huns (White Huns), moved south of Oxus and Jaxartes, occupied Bactria and attacked Persia and India. After the collapse of the Kushana Empire, the Gupta Empire (319–605 CE) came to control North India. The Gupta Empire was not as centralised as the Maurya Empire was but was a feudo-federal polity. The decline of long-distance trade resulted in a shortage of money, so the empire was forced to pay its officials with land grants. And the army, comprising elephants, horses and infantry, was mostly supplied by the feudatories. During a march, the army was provided by the villagers with food and transport. In addition, forced labour was also imposed on the peasants. Emperor Skandagupta (r. 455–67) had to face the wrath of the Huns. From Bactria, the Eastern Huns attacked along the Hindu Kush Mountains. After his death, the Huns entered North and Central India through Punjab. The Guptas tried to adapt their forces by adopting the mounted archers equipped with curved bows. From the Kushanas, the Guptas borrowed the use of leather boots, long swords, belts and coats. The Chinese cavalry also borrowed all these tools and the techniques associated with riding horses from their enemy and neighbour, the steppe nomads. However, India imported warhorses from Central Asia and Afghanistan, and these two sources of supply were cut by the Huns. Further, Gupta cavalrymen's skill in mounted archery was inferior to the *mlechchas* (Huns). Skill in horse archery, better quality of horses and use of superior stirrups were responsible for repeated Hun victories against the Gupta armies. By the mid-sixth century, Hun power in India disintegrated (Mookerji 1997: Plates 4–5; Jha 2004: 158–67). The hot and humid climate of India along with the absence of grasslands were responsible for the Huns' failure to establish a stable polity within India. These two factors also prevented permanent conquest of India later by the Mongols.

To an extent, climate change also played a role in exacerbating the crisis of the Western Roman Empire in particular and the sedentary empires of Eurasia in general. The Warm Classical Optimum lasted from 200 BCE to 300 CE. The 500-year Roman Warm Period, particularly the temporary northward extension of the Mediterranean climate zone in Europe, somewhat facilitated Roman expansion by boosting agricultural productivity. During this period, the Mediterranean region experienced a prolonged period of warm and stable climate. A positive North Atlantic Oscillation (NAO) pushed warm winds west towards Scandinavia, glaciers retreated and the Mediterranean littoral experienced dry summers and winter rainfall. This in turn facilitated the extension of farming (the Roman open field farming practice) and rising fertility rates, especially in Gaul. The Roman grain (wheat and millet) and wine agriculture displaced the Celtic mixed farming based on a mixed crop and cattle combination, which was more resilient in the erratic colder northern climate. The global population reached an estimated 200 million. Cities became larger, trade routes extended and the iron weapons-equipped heavy infantry armies of Rome were ranging further afield. The period from 500 CE to 800 CE was much cooler. Ice melts cooled North Europe, and the NAO reversed, bringing deep cold. The strong El Niño influence resulted in the cooling of the North Atlantic sea surface. Temperature dropped in Gaul by 1 degree Centigrade in 400 CE compared to the previous centuries. As climate started becoming cooler, Roman agriculture in Gaul and Britain started declining. Moreover, the Germanic tribes, who in the preceding warmer period had adopted the Roman farming in North-Western Europe, had experienced demographic growth due to increasing agricultural surplus. Now, the cooler climate from fifth century onwards resulted in the collapse of their Roman-style grain-based cultivation. In desperation, the Germans tried to push southwards into the empire. So food shortages gave rise to political and military upheavals. Climate change functioned as a contributor or amplifier of historical change. In addition to a cooler and wetter climate, Germanic inroads plus heavy taxation of the farmers further reduced the agricultural output of the Roman Empire. Lower food yields resulting in nutritional deficiency further reduced the productivity of the agricultural workers. Egypt was the breadbasket of the Roman Empire, and by this time, erratic rainfall in the Ethiopian Highlands and the resultant Nile floods disrupted grain-based agriculture in the province. Further, the Antonine and Cyprian plagues during the second and third century CE further weakened the demographic base of the Roman Empire. A severe and prolonged half-century-long drought in Central Eurasia during the mid-fourth century CE also pushed the Huns towards the sedentary empires along the rim of Eurasia (McMichael et al. 2017: 5, 75, 141–9). And in this crisis scenario emerged Attila, the 'scourge of God'.

Besides the advent of the Huns in Europe, the rise of Sasanian/Sassanid power in Persia (254 to 651 CE) also accelerated the transformation of the Later Roman army from a legionary-centric force into a cavalry-centric organisation. The Sasanian Empire maintained both light cavalry equipped with javelins as well as heavy cavalry. However, the Sasanians, unlike the Parthians, did not maintain an all-cavalry force. In order to capture the Roman strongholds along the Euphrates, the Sasanian force used light and heavy infantry, engineers and elephants. In the course of campaigning in Punjab, the Sasanians adopted war elephants from the Indians. While the Sasanian light infantry branch had archers and slingers, the heavy infantry was equipped with spears. The Sasanian infantry were experts in mining, countermining and sapping. In response, the Roman fortifications became more elaborate with the construction of bastions, which provided interlocking fields of fire. The Sasanian army was a well-tuned force, with battle groups comprising 1,000 men each. Each Sasanian division had about 12,000 warriors. As a point of comparison, each German infantry division during the Second World War had about 12,000 men. There were special officers in charge of military police for maintaining discipline in the Sasanian units. The Sasanian army, which included both regular soldiers paid by the government as well as noblemen and their retainers (feudal levies), at best had 150,000 men. However, the core element of the Sasanian field armies mostly comprised cavalry. This was because they faced threats along multiple frontiers. The Romans were on their western frontier, the Arabs threatened the south-west frontier, and the Western

Turks posed a threat along their north-east frontier. Finally, along the north on the Caucasus Front, the Khazars were a standing threat. In order to shift force in accordance with the changing strategic scenario, only mobile cavalry could be deployed quickly across the roadless and riverless (except the Euphrates and Tigris running north to south in Mesopotamia/Iraq, there were no navigable rivers running east to west) arid Iranian Plateau (Petersen 2013: 35–6; Farrokh et al. 2018: 11, 13–20, 25–6).

The rising military effectiveness of cavalry in Eurasia and the force structures of the enemies of the Roman Empire compelled the transition from infantry- to cavalry-oriented Roman military machine. The Hun cavalrymen were armed with two types of swords: double-edged long swords and single-edged swords plus composite bows. The use of these swords spread through the steppe nomads among the Avars, Sarmatians and the Sasanians (Nickel 1973: 131–42). The Huns did not use spurs but urged their horses with whips. They had wooden saddles, but poorer Huns used padded cloth or skin saddles (Kennedy 2002: 30). During the early fifth century CE, the Chinese cavalrymen used wooden saddles covered with bronze plates fastened with nails (Dien 2005b: 186). The Mongols in the thirteenth century would use iron stirrups, which in turn further raised the military effectiveness of their horse armies.

Around 420 CE, the Western Huns were led by Rua and his brother Octar. Joint rule by two kings seemed to be the practice among the Western Huns. In 434, Attila ruled with his brother Bleda. Later, Attila murdered him and became the sole ruler of the Western Huns. Attila emerged as the 'strongman' of Eurasia between 433 and 453 CE. Attila maintained mobile headquarters. When Attila was campaigning against the Eastern Roman Empire, his headquarters was in a Pannonian village. Priscus, the historian cum diplomat of the Byzantine Empire, visited Attila's Court in the summer of 449. At Attila's camp, Priscus met Romulus, the envoy of the Western Roman Empire. Attila's palace was built of wood and lit by torches. The gloom of the interior was somewhat lightened by gleams of looted silver and gold. It was believed that Attila was planning an attack on the Persian Empire through the Caucasus in the near future. At that time, Attila's eldest son, Ellac, was in charge of the region near the Sea of Azov. Overall, the Roman diplomats were not impressed by the simplicity of Attila's headquarters (Thompson 1945: 112–15; Browning 1953: 143–5, Kennedy 2002: 38, 45; Man 2005: 53, 65). However, they were terrified of Attila's military power.

In 441, Attila had attacked the Byzantine Empire. The Huns occupied the northern cities of Illyricum and defeated the Eastern Roman army led by Aspar at the Battle of Chersonese. Attila demanded the annual tribute to be tripled to 952 kg of gold and an immediate down payment of 2,721 kg of gold. The annual payment to Attila amounted to 2.5 per cent of the total budget of the Eastern Roman Empire. The first great attack by the Huns had actually occurred in 395 when they crossed the frozen Danube and the Caucasus Mountains and then pillaged and plundered Thrace and Anatolia (Bayless 1976: 176–7). The repeated attacks by the Huns obstructed both the Eastern and Western Roman Empires' attempts to check the German tribes. In 431, the Eastern Roman Empire sent an expedition against the Vandals but recalled it when the Huns threatened the Balkans in 434. Similarly, a Hun attack in 442 forced the cancellation of another expedition against the Vandals. After the Vandals gained control of Carthage in 439, the grain shipment from North Africa to Rome declined and the city suffered continuously from grain shortages (Clover 1973: 108, 116). Without the pressure exerted by the Huns, the German tribes could not have entered the Roman frontiers in such large numbers. And supposing they had entered the Roman territory, then the Roman army, without any distraction posed by the Huns, would have been able to checkmate the intruders.

The breakdown of civil-military relations within the Roman Empire also debilitated the capacity of the empire to deal with the external invaders. For instance, in the fourth century, the Roman garrisons in Britain and Gaul, under ambitious generals, repeatedly challenged the authority of the Roman Emperor. In fact, Walter Goffart asserts that the emperors at Rome deliberately weakened the Roman garrisons in the outlying provinces of Britain and Gaul to prevent the rise of pretenders (Goffart 2008: 1–32).

This breakdown of civil-military relations was not a uniquely Roman problem. In fact, several Chinese imperial dynasties repeatedly faced the problem of threats posed by successful generals commanding large forces along the frontier. The Tang dynasty (618–907) suffered due to the delegation of large amount of civil-military power to the generals deployed along the northern frontier of China. Later Chinese imperial dynasties tried to contain the dangers posed by overmighty generals by inculcating a Confucian ethos of civility and projecting martiality as somewhat inferior, and through the deployment of a powerful body of mandarins who were civil bureaucrats recruited through a pan-Chinese examination system (Lorge 2008: 1–20). The Hindu rulers contained the *senapatis* (generals) by utilising the *varna* (caste) system (the warrior caste Kshatriya's position was lower in the *varna* system than the Brahmins, who were civilian advisors to the *rajas*) and the ideology of non-violence propagated through the medium of various religions such as Vaishnavism. Kautilya's *Arthashastra* notes that the position of the *senapatis* was much lower than the *mantris* (civilian ministers). Not only did a civilian minister control the military department, but the Kautilyan handbook also lays down that the salary of a *mantri* should always be higher than that of a *senapati*. As a result, after the coup of *Senapati* Pushyamitra Shunga, India, until the advent of Islamic Turks, did not exhibit any revolt by the military against their civilian superiors. What Goffart does not explain is how the Byzantine Empire, unlike its western counterpart, was able to control the generals.

In 451, Attila crossed the Rhine River and invaded Gaul. Metz fell to him on 7 April. Aetius, with a Roman–Visigoth force, was able to check Attila's force. In accordance with the steppe nomadic tradition, Attila constructed a defensive wagon laager, which formed his base during the campaign. From the base, Attila launched waves after waves of his mounted archers, who carried out pillage and plunder deep into the enemy countryside. Aetius and his Visigoth allies were well acquainted with the Hunnic art of war. In fact, Aetius had previously employed Hun mercenaries to keep the Visigoths in check. In the Battle of Chalons (20 June 451 CE) Attila's force was defeated but not destroyed. The Huns, carrying a lot of loot, withdrew from Gaul (Bachrach 1993: 57; Barnes 2015: 18–19). Attila's invasion of Italy was not very successful. The Roman Court relocated to Ravenna, which was surrounded by marshes and hence could not be crossed by Hun cavalry. Moreover, the narrow mountainous terrain of Italy was not exactly horse country. Worse, an outbreak of plague forced Attila to withdraw.

In the end, Hun military power in Western and Central Europe was limited by the same ecological factors that debilitated Mongol military power in these regions 800 years later. In Europe, the pastureland ends in Hungary, and this limited the number of mounted archers that the Huns could maintain in Europe. The Hungarian Plain, according to one calculation, had some 42,400 sq km of pastures, which is 4 per cent of the pasture available in modern Mongolia. The Hungarian pasture could support only a limited number of mounts, as a horse living by grazing alone requires 60–250 acres of pasture per year. The nomads allowed their horses to graze rather than stall feeding them like the sedentary polities, and a nomadic mounted archer required at least 3–5 horses while campaigning. The limited number of mounted Hun archers which Attila could maintain was adequate for raiding and pillaging the Eastern and Western Roman empires but not enough to destroy these two political entities. While campaigning in Gaul and Italy, Attila could maintain only a small number of mounted archers, and even that for only a short time. Further, the Mongols would adopt counterweight trebuchets manned by Chinese and Middle Eastern Islamic crews. Attila's force lacked such sophisticated siege machines in significant numbers. So Attila was almost impotent against the fortified cities of the Romans. For conducting *Stellungskrieg* (positional war), Attila tried to adapt his military machine by adopting infantry, which in turn reduced the speed, mobility and operational reach of his army. For instance, the Huns used infantry archers protected by screens of hide and wheeled siege towers, battering rams suspended from chains and also rams mounted on wheels for taking fortified cities (Sinor 1972: 171–83; Kennedy 2002: 34–5; Petersen 2013: 46–7). The use of such siege machines reduced the mobility of the Hunnic army. The shock and awe generated by the mobile horse archers constituted the principal strength of the Huns. In fact, Hunnic victories occurred because Attila, by using his superior mobility, increased the 'speed of battle' and achieved surprise over his enemies. In addition, his infantry was less effective than the heavy disciplined Roman infantry in positional war involving sieges, and he could not maintain a large cavalry force west of the Carpathians permanently. Therefore, at times Attila's army received a bloody nose from the Roman forces. The Carpathians marked the boundary of nomadic pastoralism (Lindner 1981: 3–19) and hence effective Hun and Mongol military power.

Arther Ferrill asserts that the Roman infantry during the Later Roman Empire was not drilled and disciplined because the Roman generals adopted the tactics of fighting in open order. Hence, Roman infantry failed against the Huns and the Germans (Ferrill 1988). He does not explain why the Roman generals adopted the open order tactics. The Roman generals probably adopted the open order technique to deal with the onrush of the German tribes, who initially fought on foot and later resorted to cavalry combat. Before the sixth century CE, the German tribes fought mostly as infantry equipped with spears, swords and battle axes. From *c.* 600 CE onwards, the Germans besides infantry also deployed cavalry equipped with swords. Mounted archery was unknown among them. Historian E.A. Thompson writes that the secret of Roman defeat against the Germans was their inferior morale and bad (fractious) politics (Thompson 2008: 119–46). Why and how these two factors operated is of course not spelled out in his article.

We should not eliminate the purely military tactical factors from the equation. Closely packed legionaries would have been easy targets for the Germanic infantry onrush. Anyway, if the Roman legions of the fifth century CE had been drilled and disciplined as in the days of the Roman Republic and deployed in close order like closely packed sardines in a tin can, then they would have been static targets for the Hunnic mounted archers like Crassus's legionaries against the Parthian horse archers. Rather, an open order infantry formation offered less easy targets for the Hunnic mounted bowmen. What Ferrill forgets is that the nature of warfare was changing in fifth century CE and the cavalry's military effectiveness was increasing. In the end, ecology saved post-Roman Western Europe from the steppe nomads.

Post-Roman Western Europe and the Byzantine Empire

By the fifth century CE, Chinese armoured cavalry had come into vogue. The horse was covered with plate armour and a plate neck guard. The Persians also came up with similar heavy cavalry around that time. In the next century, the Chinese heavy cavalry used lamellar armour and the cavalrymen used lances (Dien 2005a: 115, 117). A somewhat similar war machine also came into existence in Western Europe around roughly this time. But unlike Western cavalry, the Tang emperors (especially Taizong, r. 626–49) stressed training on mounted archery to counter the threat posed by the steppe nomads like the Eastern Turks (Graff 2002: 38).

Among the German raiders who swept into Western Europe in the fifth century CE, the Franks were most successful. In the later seventh century, however, the Frankish Kingdom broke up. Charles Martel, the Mayor of the Palace to the Merovingian Kings, reasserted Frankish rule with the establishment of the Carolingian dynasty (752–911). He was able to turn back an Arab raiding force at the Battle of Poitiers (732). His son Pippin III campaigned extensively in Southern Gaul and across the Rhine. Pippin III's son, Charles the Great, or Charlemagne

(r. 768–814), was the greatest Carolingian Emperor. He controlled France, West Germany, North Italy and the Pyrenees (Hooper and Bennett 1996: 12).

The rise of Carolingian power in Gaul to an extent was facilitated by the fact that the Justinian plague which devastated the Byzantine Empire between the sixth and eighth centuries did not have any adverse demographic effect in Gaul. Gaul was an agrarian society, and 90 per cent of the people were engaged in agriculture. Not only were more and more lands brought under cultivation, but productivity also increased with the replacement of Roman latifundia estates (large estates of the lords and the agricultural work was done either by the slaves or peasants) with bipartite estates. Gaul during eighth and ninth centuries experienced a demographic growth rate of 1 per cent annually. The population of Charlemagne's Empire came to about 20 million (Bachrach 2013: 9–17).

Feudalism slowly rose during the Merovingian times and continued under the Carolingians. It was an ancient Frankish custom for a lord to provide his retainers with arms and horses, and these mounted retainers were under obligation to protect the lord, his family and his property. In Gaul, the Merovingian kings employed the personal followers/armed retainers of the large landholders for policing and military duties. The loyalty of these magnates was bought by the bestowal of church lands (Bachrach 1993: 3–4, 712). The Merovingian kings appointed their own vassals as dukes and counts in the official positions and paid them with benefices. Public office became associated with landholdings and vassalage. In about 730 CE, in the Frankish state under Charles Martel, military service was required of the benefice holders. A feudal warrior, in return for a grant of land (a benefice, which later came to be known as a fief), was under the obligation to serve in the army of his lord for a particular period. With the passage of time, the fief became heritable. The Carolingian state under Charlemagne's successors was not as bureaucratic as the Roman Empire. The Carolingian monarchy to a great extent was a German tribal polity. This state unlike the Roman Empire in its heyday could neither pay its troops in cash nor provide equipment. So the Carolingians could not maintain a professional standing bureaucratic army. Further, Western Europe was threatened by the Muslims and the Vikings. The traditional infantry levy was too slow to meet the Islamic cavalry and the fast-moving Viking raiders. So the only answer was military feudalism: mounted knights who provided military service to the state in return for grants of land known as fiefs (Beeler 1972: xvi–29).

While sixth-century land grants only allowed the collection of royal dues from specified territories, seventh-century grants of land for rewarding the aristocratic followers involved granting of permanent ownership of land. Such land grants facilitated the emergence of territorial lords with their private armies. Then why did the royal court issue such types of land grants? Guy Halsall provides an answer. The military forces of the Merovingian Kingdom comprised landowners' private militias in the south and Franks in the north commanded by royal officials. However, by the end of the seventh century, *beneficia* and vassalage became common. The rise of the Carolingians further accelerated this trend. The Carolingians were one of the most successful of a number of powerful aristocratic Frankish factions. They faced a number of regional and local elites with a strong landed power base. The Carolingians negotiated with the local landed elites. In return for oaths of vassalage, the local landed elites received royal ranks and offices. This was the political price the Carolingians had to pay for maintaining their royalty. The Guptas in North India followed a similar strategy vis-à-vis the high caste landed elites known as *thakurs* (also known as *samantas*). The latter in return for an oath to the emperors received royal ranks and permanent land grants and claimed Rajput (*rajaputra* – sons of the *raja*/king) status. Facing threats of confiscation of landed property by the royalty and to prevent the ancestral land from slipping away from their family's hands, the Frankish landlords gave parts of their land to the abbeys, only to receive them back as royal grants for maintaining royal authority. Similarly, the Rajputs gave part of their lands to the temples and later received them back as royal grants for maintaining a military force for the *raja*. At times, the Frankish lords gave parts of their ancestral land to the temples and abbeys, and one of their sons entered the church and administered the land on behalf of their families (Halsall 2003: 73–5; Chattopadhyaya 2005).

In 847, the three grandsons of Charlemagne declared that every free man must have a lord. In 877, Charles the Bald, King of the West Franks (r. 840–77) gave official sanction to the reality that benefices were hereditary. So every man had to have a lord, every official was a landholder and every holder of a benefice had to serve as a soldier, and further, offices, benefices and military obligations became hereditary. Feudalism had attained its full form, at least in Western Europe. The result was decentralisation of the polity and parcelisation of royal authority. Though the counts were formally king's officials, they were hereditary officials. Therefore, the monarchs could neither control them nor remove them from their offices for disloyalty or inefficiency. Further, the fighting men of the kingdom were vassals of the feudal lords to whom they were obligated for performing military service in return for grants of fiefs. The feudal warriors were loyal to their lords rather than to the distant central government. Frequently, the vassals, despite the prevalence of the ethos of chivalry, betrayed their lords. They, like the samurai and the Rajput chiefs, changed sides for opportunistic reasons. The landholders started building castles surrounded by ditches. The construction of motte and bailey castles further undermined the 'absolute' authority of the rulers. Worse, the vassals and the lords carried on private wars with their private armies without any reference to the monarchs (Beeler 1972: 28, 39–40; France 1999: 5, 40).

The traditional view is that Charlemagne's army mainly comprised feudal cavalry (Beeler 1972: 24). The recent argument put forward mainly by Bernard Bachrach asserts that infantry armed with spears, swords, bows and arrows played an important role in the Carolingian army. This was because the principal operational task of the Carolingian army was to capture enemy fortifications, which involved digging, tunnelling, climbing and operating siege machines (Bachrach 1993: 3, 28, 32). Bachrach writes: 'The decisive arm of the military forces of

Charles Martel, and his sons, was not cavalry; infantry and artillery were their most useful units' (quoted from Bachrach 1993: 75). Bachrach rhetorically asks: 'Where during the early Middle Ages did the charge of "heavy cavalry" win the day against a prepared enemy?' (quoted from Bachrach 1993: 20). Somewhat following Bachrach, Charles R. Bowlus asserts that cavalry was not the dominant arm in the Carolingian army and horsemen were used for side duties like reconnaissance and skirmishing. In main battle, the horsemen played a secondary role. Rather, infantry and river flotillas constituted the core of Carolingian force (Bowlus 2008: 291–320).

From the opposite side, one may add the very fact of cavalry inferiority resulted in the collapse of the Carolingian Ostmark in the early tenth century due to the invasion of the Magyars from the steppes of South Russia. We could speculate the defeat of the Bavarian army at Pressburg near Bratislava in 907 encouraged the Carolingians to come up with heavy cavalry geared for mounted shock combat.

The traditional argument is partly convincing. The socially prestigious component and the principal striking force of the Carolingian army remained feudal heavy cavalry. The first part of Bachrach's argument (supported by a host of other historians like John France and Guy Halsall), that couched lance technique by heavy cavalry did not emerge under Charles Martel (France 2008: 322), is acceptable. Couched lance technique of the heavy cavalry came into Western Europe during the tenth and eleventh century. But the second part of Bachrach's argument, that cavalry was not important for military operations in post-Roman Europe, is unacceptable. For battles and campaigns, cavalry remained the queen in medieval Europe as in other parts of Eurasia. And yes, for siege operations infantry was required. For instance, in 737 CE during the Siege of Avignon, Charles Martel used infantry equipped with battering rams and rope ladders. But for long-distance campaigns at the operational level, cavalry was necessary. Even before the Carolingians, the Merovingian kings maintained horse-breeding estates under the control of the constable, who was an important member of the royal household. Each estate was managed by a bailiff supported by several wranglers and grooms. Charles's vassals could muster about 35,000 horsemen. The wealthy vassals served as armoured cavalrymen, who constituted the core of his force. And the poor free men who were armed with spears and bows constituted an undisciplined infantry. From the late eighth century, the Frankish cavalrymen started using stirrups. At that time, the Franks fought on horsebacks with swords and lances. The lances were used as throwing weapons rather than for launching a charge. This development occurred later. The Byzantine cavalry initially used mail armour. But Charlemagne's heavy cavalry used scale armour (Bachrach 1993: 7, 53, 440). Scale armour or lamellar cuirass first emerged in Chinese Turkestan, and the Eurasian nomads transmitted this piece of defensive technology to the western part of Eurasia (Nicolle 2002: 232).

By around *c.* 900, the heavily armoured Frankish knights, like the Rajputs of early medieval India, believed in frontal charges and considered tactical retreat as disgraceful. The knight wore a short-sleeved and knee-length mail shirt and carried a large kite shaped shield. The knight was armed with a long lance and a long cutting sword. This equipment more or less remained unchanged until the twelfth century. The development of the couched lance technique became possible in the eleventh century with the adoption of high wrap-around saddles. In addition, a development in harness paraphernalia (chest and cupper straps) which fixed the saddle more securely to the horse also made possible the rise of the couched lance technique. And this improved harness technique actually first emerged in Central Asia (Nicolle 2002: 12). Further, the Carolingian cavalry-men, instead of the straight swords, adopted the *tulwars* (curved swords) known as *spata* India (swords from India). The *tulwar* was more effective for slashing in close-quarter combat and was introduced in India by the Muslims. Compared to the straight sword used by the Rajputs, Romans and Carolingians, *tulwar* was lighter and had a longer reach, and its sharpened blade could be used with slight pressure to cut the veins on the hands of the enemy rider. In contrast, a heavy straight sword was used for stabbing and, once stuck in the abdomen of an enemy rider, could rarely be used again. Further, just holding the heavy straight sword was tiring for the mounted man (Hooper and Bennett 1996: 12–13, 154; Coupland 2008: 249–70; Upadhyaya 2008: 180). The case of Carolingian cavalry showed a classic case of the West borrowing certain combat techniques and military technologies from the East.

This work is not trying to revive the traditional thesis that the introduction of the couched lance technique by the Carolingian heavy cavalry resulted in feudalism in Western Europe. Probably before the ninth century, the Carolingians lacked heavy cavalry but certainly had medium and light cavalry. Even before the emergence of couched lance technique in the tenth century, cavalry was important in the Carolingian army. And feudalism for me means providing land grants to the nobles for maintaining cavalry contingents. This trend was present in Eurasia from the fifth century CE onwards. Provision of land grants for mounted service was not linked with the couched lance technique. The couched lance tactic by heavy cavalry was absent in the region east of an imaginary line running from Stettin–Prague–Trieste. But the medieval societies east of this line experienced feudalism.

Horses for conducting warfare were also at a premium in England. The Saxon kings prohibited the export of horses. Royal policy laid down that in each diocese, the bishop was to provide food for the king's horses raised in that area and for the men who cared for them. The landholders lent their horses to the king and also performed military services for the monarch (Bachrach 1993: 711).

An equine military culture among the sedentary states of medieval Eurasia gave birth to a homogeneous military ethos among the cavalrymen. This culture is known as chivalry in Western Europe, *Kshatradharma* in India and *bushido* in Japan. Chivalry refers to the code of values of the knights, which originated around 1100 CE. It refers to deeds of horsemanship and arms. Chivalry appreciated loyalty, generosity, courage and military skill regarding horsemanship and arms. In accordance with the tenets of chivalry, the vanquished knights, instead of being killed, were ransomed (Hooper and Bennett 1996: 156). A code of conduct known as *dharmayuddha* (just war) also came into play among the Rajputs. Rajput ethics demanded that a retreating enemy should not be attacked; ambushes and nocturnal attacks were considered as unjust, etc. Those Rajputs who implemented the 'unjust' techniques of war (commando attacks, surprise raids, etc.) were considered as *pariahs* (fallen) for not following *Kshatradharma*. For the Rajputs, war became a sort of sport. And they paid a high price for this when the Islamic Turks, who were masters of deception, treachery, ambuscades and sudden nocturnal attacks, confronted them from the ninth century onwards (Roy 2004: 67–9).

Mail-clad knights on horseback dominated warfare in Western Europe between 1000 and 1300. With the passage of time, the weight of armour carried by the knight rose. Further, the horse was also covered with iron armour. The knights started wearing leather jerkins (it came from the Islamic lands) over their mail armour. In the eleventh century, a knight carried 32 kg of weight, which by the thirteenth century had increased to 45 kg. Therefore, a knight carried a greater load than a hoplite soldier did. In order to carry the increasing load, the size of the horses (which were bred) rose from 10 hands to 16 hands. The mount of heavy cavalry of the modern British army, in comparison, measured about 15.3 hands. William the Conqueror's horses, as depicted in the Bayeux Tapestry, have mounts as large and heavy and certainly bulkier and stronger than the Arab steeds. William's horse is portrayed with a well-developed and thickly muscled neck, shoulders and chest. The croup is markedly arched, and the haunches and thighs are heavy. The average weight of such a horse is calculated to be 680 kg. A feudal force comprised of heavily armoured knights mounted on horses covered with plate armour (heavy cavalry) was supported by not so well-clad chevals (they could be categorised as medium cavalry) and pedites (infantry). Unlike the traditional Greco-Roman military system, the infantry enjoyed less social prestige and was marginally effective in the battlefields. At the Battle of Hastings (14 October 1066), the charges of the Norman heavy cavalry decimated Harold's Anglo-Saxon infantry armed with swords and battle axes (Bachrach 2002: 174–9; Davis 2006: 85–100; Pierce 2006: 63–83).

Gradually, the infantry, especially those armed with bows, became an important factor in the European battlefields. Composite bows were never popular in Western Europe, mainly for ecological reasons. In the damp conditions of Western Europe, the glue dissolved and the composite bow fell apart. Further, the knights, like the Rajputs of medieval India, were sedentary horseman. They were not pastoral nomads, and so they were not skilled enough to use the composite bows from horseback. After all, the composite bow was a steppe nomadic horseman's weapon and hunting tool from their childhood. Rather, the Western European infantry, like the Chinese foot archers, used crossbows. It is to be noted that neither the Genoese crossbowmen nor the English infantry armed with longbows would have stood any chance against the high-speed swarming and flanking attacks by masses of horse archers. However, longbows and crossbows were quite effective at times and in certain terrains against the clumsy frontal charges of mounted knights armed with the unwieldy long lances, swords and axes. The best swords were manufactured at Damascus and Toledo by the Muslims with crucible steel imported from India. Overall, Western Europe in regard to land war at that time did not enjoy any technological superiority over the Islamic world.

The sizes of the feudal forces gathered by the warlords of Western Europe in comparison with the Chinese and Indian polities were generally small due to weak logistical infrastructure. For instance, the Norman army which conquered England in 1066 had 14,000 personnel, including 7,000 effectives. Of the latter, 3,000 were cavalry (France 1999: 13–31). As a point of comparison, in December 964, the Song field army numbered 60,000 infantry and cavalry (Lorge 2013: 123). To give an example from India, King Harsha of Kanauj, the ruler of North India (r. 606–48 CE), maintained 100,000 cavalry besides numerous war elephants and infantry. The size of his infantry exceeded the number of cavalry under his control (Roy 2015: 70–1). The vast demographic resources of China and India and extensive paddy cultivation along the fertile river valleys in these two countries enabled the rulers to maintain large armies. In India, land was double cropped and India's population came to about 100 million (Habib 1991: 163–4)

From the sixth century onwards, the Byzantine Empire gave priority to the introduction of stirrups and horse archery in order to confront the challenge posed by the Sasanians in Persia, nomads in South-East Europe and Arab military power in the Middle East. The Byzantine field force comprised 50 per cent infantry and 50 per cent cavalry. This force structure was significantly different from the Roman army under Julius Caesar. This was due to the cavalry-centric force structure of the newly emerging threats and rising combat effectiveness of the horsemen. Both the Byzantine Empire and the Carolingian Empire were threatened by the Avars, who were mounted archers equipped with composite/compound recurve bows (Bachrach 1993: 13, 438; Alofs 2014: 425–44).

Justinian I (r. 527–65 CE), the Byzantine emperor, attempted to reconquer the Western Roman Empire. The mounted archers (nomadic mercenaries) in the forces led by the Byzantine generals Belisarius and Narses repeatedly defeated the Ostrogothic cavalry equipped with swords and spears. Before the latter could make contact, the mobile mounted archers in the Byzantine army wiped them out (Kaegi, Jr. 2007: 239–40). However, the bubonic plague that came in 542 CE devastated the economic and demographic potential of Justinian I's domain, and this in turn had a deleterious effect on the Eastern Roman Empire's military power. During the next two centuries, this deadly plague caused the death of almost 100 million people in the Eastern Mediterranean region, West Asia, the Arabian Peninsula and Europe. The bubonic plague would return centuries later in Europe as the Black Death. It is still debated whether the plague originated in India or in East Africa. This disease was spread by blood-feeding fleas from the infected black rats. The rat population acquired this infection via occasional contact with the natural hosts of the plague bacterium, the ground-dwelling rodents that live in the underground burrows located in Central Asia, Southern China, Mongolia, North-West India, Arabian Peninsula and the Great Lakes Region in Eastern Africa. When the rat population declined, the fleas jumped to human bodies and infected the latter. Then the infection spread from person to person. From the Great Lakes region, the merchants in Egypt spread this disease. And from Alexandria, this disease made a hop to Constantinople (McMichael et al. 2017: 150–5). This city, which was the terminus of eastern and western trade routes, enabled the disease to spread in both Asia and Europe.

In the sixth century, most of the Byzantine army comprised contingents provided by the big politically powerful landholders (Petersen 2013: 61). By the middle of the eleventh century, the bulk of the Byzantine army was comprised of indigenous mercenaries supplemented by Norman, Russian, Turkish and Frankish mercenaries. Most of the army comprised cavalry, except the Varangian Guards (Scandinavians), which constituted the elite infantry corps. A large number of the military commanders were landowners (Haldon 1999: 246). The Byzantine Empire, unlike the Western Roman Empire, survived because of the former's greater adaptability to cavalry force structure and possession of a more extensive defensive architecture. The nucleus of the Byzantine army comprised cavalry. While the Byzantine Empire raised armoured heavy cavalry from its subjects, for horse archers, it depended on the mercenary nomads. The Byzantine heavy cavalry, armed with lances, sabres and iron maces in tight formation and in total silence, trotted towards the enemy and hit the hostile force en masse and not as isolated individuals like the Western knights (Dennis 2007: 234). One historian rightly notes the similarities in the theory and practice of warfare as conducted by Tang China and the Byzantine Empire. Both mainly relied on cavalry force to ward of their nomadic enemies. Besides raising cavalry from the sedentary societies, these two empires also hired 'barbarian' mounted archers (Graff 2016). And the Byzantine emperors paid the nomadic mercenaries with land and not money (Magdalino 2007: 182). In fact, the Delhi Sultanate in India (1206–1526) also followed the same model.

Arabs, Turks and Crusades

Petersen's assertion that, in the ninth century CE, the Roman culture of siege warfare held sway from the Middle East to Scandinavia (Petersen 2013: 429) is erroneous. Rather, cavalry culture dominated the vast stretches of Eurasia. The Arabs emerged from the desert of Arabia and defeated the two superpowers of that time: the Sasanian Empire and the Byzantine Empire. While the Byzantine Empire was almost extinguished in Asia, the Sasanian polity was completely destroyed. Scholar Eduard Alofs asserts that continuity characterised the nature of warfare in Persia between the sixth and fourteenth century CE. The Persian army was largely composed of cavalry and lacked heavy armoured infantry.

Light infantry was equipped with sickle swords, light lances and bows. Mounted archery constituted the basic Iranian tactics. However, the Iranian horse archers were heavily armoured, unlike the non-armoured Hunnic and Parthian horse archers. Both the mount and the rider were covered with lamellar armour. In addition to the composite reflex bow, which had replaced the cupid reflex bow, each cavalryman also carried a light lance. Unlike the medieval feudal knights, they did not hold the lance with two hands but used it only with one hand. Hence, Persian lances like the Achaemenid cavaliers' lances were light throwing spears/javelins and not the heavy lances of the medieval feudal knights of Western Europe. Alofs continues that battles constituted a minor part of the Persian military doctrine. Rather than waging decisive battles, the Persian military tradition focused on skirmishes, raids, manoeuvres and nocturnal attacks punctuated by negotiations. The chief objective was not destruction of the hostile force but absorption of the enemy troops into its own ranks (Alofs 2015a: 4–27, 2015b: 132–54).

To an extent, Alofs' argument is similar to the colonial British soldier-scholars' depiction of later Mughal warfare in eighteenth-century South Asia. More recently, Jos Gommans in his study of the Mughal army has reiterated this view. Avoidance of battle, focusing on manoeuvring, skirmishing and negotiations, incorporation rather than elimination of enemy troops, etc. are stressed by Gommans, John Keegan and V.D. Hanson as characteristics of the so-called Eastern way of warfare. In contrast, the Western way of warfare, according to them, was battle centric and focused on waging decisive battles with heavy infantry geared for close-quarter combat aiming at the destruction of the enemy forces (Keegan 1994; Hanson 2001: 21–2; Gommans 2002). We have seen earlier that the Sasanian army was a multi-branch force with a strong emphasis on cavalry. And the Sasanian army was capable of conducting sieges and waging decisive battles. So Alofs' view about the Sasanian military machine is just wrong.

Let us now analyse the Arab military culture during the course of its expansion in the Middle East. Besides incessant raiding, the Arab military doctrine focused on fighting battles. The Battle of Yarmuk in Syria and the Battle of al-Qadisiya in Iraq, both fought in 636, smashed the power of the Byzantine and Sasanian empires. Iraq fell like a ripe plum into the folds of the Arabs. At Yarmuk, the size of the Muslim force varied between 20,000 and 40,000 men. At al-Qadisiya, the size of the Muslim army was between 2,000 and 4,000 men. The Muslim armies were numerically inferior and less equipped than their opponents. Again, the Muslim order of battle was traditional. The Muslim armies generally deployed in the traditional formation: centre, right wing and left wing (Kennedy 2005: 2–5). Then how does one explain their victories?

Most of the men who joined the Arab force initially were of Bedouin origin. Raiding, fighting and defending themselves against the raids of enemy tribes were an integral part of their lives. The adult males had some skill in riding, using spears, swords and archery, etc. So they were ready-made soldiers. Most of the Muslims had camels when they embarked on the career of foreign conquests.

Nomadic life in Arabia centred on the camel as transport and source of food. The Arab forces used camels for transport in Arabia, Syria, Iraq and Mesopotamia. Camel was an excellent means of transport, especially in the desert region. The camel-mounted Arab columns moved out of the desert and launched raids in Western Iran and then escaped back to their desert bases on their dromedaries. The camels, by providing milk and meat, also solved the logistical problem of the Arab forces. So the Arab armies, unlike their opponents, could maintain themselves in the waterless desert. The heavy Sasanian cavalry, due to lack of water and pastures, could not pursue the Arabs inside their desert hideouts. The Arabs used the camels as a means of transport and especially for carrying their infantry and baggage. The Arab infantry was equipped with lances and bows. A riding camel was able to cover 50 miles per day for a fortnight, so the camel-mounted Arab force had a huge operating radius in inhospitable desert terrain. But the Sasanian army, with its heavy cavalry, infantry and elephants geared for decisive battle and sieges against the Eastern Roman army, failed against incessant large-scale deep raids conducted by the marauding Arab raiders. During the second half of the seventeenth century, the Mughal army, like the Sasanian army, also failed to evolve an effective response against large-scale deep raids conducted by the Maratha light horse in Deccan. After capturing Syria (634–38), the Arabs were able to acquire horses. They would travel on camels to the battlefield and then would change to horses before battle and charge the enemy on horseback during the clash on the battlefield. Besides superior mobility, morale was an important factor for the success of the Muslim armies. The men were not paid salaries. They fought for their religion, and the prospect of booty and because their friends and fellow tribesmen were fighting (Hill 1975: 32–42; Kennedy 2002: 20, 2005: 1, 6).

At Yarmuk, the Islamic army implemented an innovative technique. The Islamic force had light Bedouin cavalry suited for skirmishing and raiding but not for fighting a pitched battle. The core element in this battle was Islamic infantry. The Islamic infantry took a defensive position surrounded by camels. The Byzantine cavalry failed to pierce the defensive position of the Islamic infantry (Jandora 1986: 103). Probably the Byzantine horses became unsteady due to the smell of the camels.

At al-Qadisiya, the Arab infantry took up a well-defended position. In its front was the western channel of the Euphrates. The Lake of Najaf and the canal leading from it to the river protected the left flank of the army. The right flank was covered by swamps. In the rear was the ditch, and beyond it was the desert. The Sasanian commander, enjoying 3:1 numerical superiority, decided to smash the Arab infantry's defensive position by launching a frontal charge. The 30,000-strong Persian cavalry was deployed in 13-deep ranks. However, the narrow ground in front of the excellent Arab defensive position (only 2 km wide) allowed only a front of 2,000 horsemen, so the Persian cavalry was unable to take advantage of its numerical superiority and mobility. The Persian heavy cavalry lacked space to manoeuvre. The battle lasted for three days. On the third day, the

Muslims launched a nocturnal attack. The Persian army started to retreat, which turned into a rout (Jandora 1986: 104–8). Yarmuk and al-Qadisiya did not represent the superiority of infantry over cavalry. In both these two battles the Byzantine and Sasanian cavalry failed due to their defective tactical deployments. In fact, the Arabs during the course of their conquest of foreign territories absorbed the use of cavalry in their military system. And cavalry became the main strike force of the Islamic war machine.

As Islamic power expanded into Persia, many non-Muslims, like the men from Daylam in Northern Iran, converted and joined the Muslim force. They proved to be noted infantry soldiers. It was Caliph Umar (r. 634–44) who introduced the principle that the Muslim soldiers must be settled in the garrison cities rather than being dispersed in the countryside. This mechanism, he hoped, would enable them to maintain their military control and prevent them from becoming assimilated to the surrounding society, thus losing their religious and ethnic identity. During this time, the soldiers and their families were paid pensions and salaries. The defeat of the Umayyad Caliph Marwan II by the Abbasid Abd Allah Ali in 750 at the Battle of Tell Kushaf marked the end of Umayyad power (Kennedy 2005: xiii, 5, 7; Peacock 2005: 208–9).

By the 680s, wooden stirrups were replaced with iron stirrups in the Umayyad cavalry, and heavy cavalry instead of light cavalry became the principal striking force in the Arab armies. Alofs asserts that from the sixth century CE onwards the Arabs became adept in mounted archery. Only in the eighth century did the Syrian and Khorasani Arabs's mounted archery skill decline. Alofs does not provide the reasons for the decline in the horse archery skill of the Arabs. Nor does he explain how the Central Asian Turks were able to supplant the Arabs in the Abbasid Caliphate army. The Abbasids depended on the Persian bureaucrats and Central Asian nomadic Turks rather than the Arab tribes (who were supporters of the Umayyads) for filling the military ranks. From the 770s, the Turks inducted in the Caliphate army as mamluks (slave soldiers also known as *ghulams*) were known for their mounted archery skill. The Turks, unlike the Arabs, were skilled horse archers (Crone 1999: 320–1; Alofs 2014: 424–30). Nor were the Arab horses as good as the Central Asian ones. So the Central Asian Turkish nomads were militarily more effective than the Arabs. Gradually, the mamluk institution spread throughout the Islamic world from Egypt in the west to India in the east. The mamluks were paid with *iqtas* (*jagirs*, i.e. hereditary land grants).

Meanwhile, the *Dar-al Islam* was in the throes of expansion both in the eastern and in the western direction. Umayyad Caliph Abd al Malik (r. 684–705) and his son Walid continued the conquest of the North African shoreline. Musa ibn Nusayr was sent as Governor of North Africa in 697. During 707–08, Musa captured Tangiers. Yulian, Governor of Ceuta, deserted to the Islamic side due to differences with Rudhriq, the Visigoth ruler of Spain. In 710, Musa sent a reconnaissance expedition across the Strait of Gibraltar under Tariq. In 711, Tariq confronted the Visigoth army. Treachery in the high command resulted in the defeat of the Visigoths. During 711–12, Tariq captured the Visigoths' capital,

Toledo. The Muslim force swarming into Spain consisted mainly of cavalry. During July–August 712, Musa captured the fortified town of Carmona through treachery. A nocturnal commando attack did the trick (James 2012: 45–54).

In 739–40, a Syrian army under Kulthun was sent to North Africa to crush the Berber rebellion. Kulthun marched into Egypt with 27,000 men and summoned another 3,000 regular soldiers from Egypt. Then many irregular Arabs also joined his side. On reaching Tangiers, Kulthun gave Mughith command of the infantry and Harun command over the cavalry. In total, he commanded 30,000 combatants, besides a host of non-combatants. Maysara, the Berber leader, collected a huge force and confronted Kulthun at Nafdura. Viewing the approach of the huge hostile force, Kulthun constructed a defensive ditch. Harun and Mugalith advised him to remain on the defensive behind the ditch. Harun and Mugalith wanted to conduct harassing attacks against the hostile force. However, Kulthun's second in command, Balj, wanted an immediate decisive battle. Kulthun agreed with Balj's proposal. Balj charged with 7,000 Syrian cavalry. The battle was decided by the cavalry encounter. The Berber cavalry charged and pushed back Balj's Syrian cavalry, and in the ensuing battle, Kulthun was killed. One-third of the Muslim personnel were killed and another one-third captured. The defeated Muslim army took refuge in Ceuta, where it was besieged by the Berbers. Caliph Abd al Malik (r. 724–43) sent another army under Hanzala ibn Safwan that defeated the Berbers in 741–42. When the last Umayyad Caliph, Marwan (r. 744–50), was defeated by the Abbasids, the Umayyad survivors fled to North Africa (James 2012: 63–7).

In July 751 at the Battle of Talas River, an Arab force of the Abbasid Caliphate defeated a Chinese force under the Korean General Go Seonji. This battle stopped the southward advance of the Tang Empire (618–907 CE) and established Islamic supremacy in Central Asia. In 712 CE, the Arabs occupied Sind. *Manjaniqs* (mangonels), a camel siege train and heavy armoured Syrian cavalry defeated the elephant–infantry force of the Hindu ruler Rai Dahir. Towards the end of tenth century, the amirs (nobles) of the Samanid rulers of Bokhara and Balkh defeated the Hindu Sahi rulers of Afghanistan and extended Islam's sway up to the doorsteps of India. The superiority in mounted archery of the Islamised Central Asian Turks (who were originally steppe nomads) gave victories to the Islamic force against the elephant-centric army of the Hindu Sahi rulers (Nizami 1981: 1–45). In the late twelfth century, Muhammad Ghori (b. 1149–d. 1206), from Ghur in Afghanistan, after defeating the Rajput rulers of Punjab and North India, established the Delhi Sultanate. The Rajputs had inferior Indian horses, which in terms of stamina and speed were no match against the Turks' Central Asian mounts. Moreover, the heavy Indian simple bamboo bows in terms of speed and accuracy were inferior compared to the composite bows and crossbows in the hands of the Islamic invaders. The Rajputs' stirrups made of ropes and wood were inferior compared to the iron stirrups of the Turks, which enabled the mounted archers to stand and shoot even when their mounts were galloping. Further, the use of horseshoes raised the operational radius of the Turkish cavalry force. Lastly, for close-quarter mounted combat, the Rajput heavy sword, suited for stabbing (it was sharpened only at the tip), was less effective than the long *tulwar*. Furthermore, one dangerous Turkish tactic was to shoot at the Hindu *rajas* sitting on their *howdahs*. With the death of the ruler, the Hindu host melted away. Picking off the ruler sitting on the back of the elephant was a tactic first used by the mounted archers in Alexander's payroll during the Battle of Hydaspes (Roy 2004: 62–6).

It is difficult to see what other possible options the Hindu rulers had. In open battle, they had no counter to Muslim horse archery. India had to import horses in exchange for gold and silver from Afghanistan and Arabia, and by this time these two sources were cut off from the Hindu monarchs' reach. Disciplined heavy infantry in the style of the Macedonians would have been made mincemeat of by the horse archers in the vast open plains of India. In desperation, the Hindu rulers took refuge in mountain forts. Even these were destroyed by the mangonels of the Islamic rulers. The Greco-Roman military culture came up with the torsion-tension trebuchets. And the Chinese in the fourth century BCE constructed the traction trebuchet. During the twelfth century in West Asia, counterweight trebuchets came into existence. And the Mongols used such trebuchets against the Southern Song (Allsen 2002: 267). The Islamic armies brought the traction trebuchets in India. The Delhi Sultanate's army was able to destroy the Hindu mountain forts with the aid of counterweight trebuchets.

The Turks were at the forefront of expanding Islamic power in Asia, Africa and Europe. The Seljuq Turks from Central Asia conquered most of the Middle East in the eleventh century. Many more Turks migrated to this region during the Mongol period (Peacock 2005: 205). The rise of Arab and Turkish power in Anatolia forced the weakening Byzantine Empire to ask for succour from Western Europe, and the military interventions by the Western European knights further weakened the legitimacy of the Byzantine state and also its material base. Crusading warfare involved clashes between two types of cavalry: heavy lance-wielding armoured cavalry of the West against the light unarmoured horse archers of Islam. In 1081, Guiscard, with 15,000 soldiers, attacked the Byzantine Empire. While war between the Christians was governed by a certain code of conduct, this ethical value system did not apply when war broke out between the Christians and the pagan heretics, Saracens, etc. (France 1999: 11–13). The First Crusade lasted from 1096–99. The Western force, this work argues, enjoyed no technological advantage over the Islamic ground force. The soldiers of both Islam and Christendom were motivated by religious feelings. The Crusaders were lucky that the caliphate was breaking up during this period. John France asserts that a combined infantry–cavalry tactical combination was responsible for the defeat of the Islamic forces (France 1994). This was not a new tactic. In fact, it was a throwback to Julius Caesar's tactical formula. However, the Western infantry–cavalry combined arms tactics was successful because the Islamic horse archers were few in number. When later the mounted archers intervened in large numbers, then things went wrong for the Crusaders.

In 1189–90, during the Third Crusade, the Crusaders under the leadership of the German Emperor Frederick I, marched through the Bulgarian and Byzantine territories. Meanwhile, the Cumans and the Bulgars threatened the Balkan territories of the Byzantine Empire. The Cumans were originally from South-East Siberia. They migrated westwards and inhabited the land between the Volga and Don rivers. During the twelfth and thirteenth centuries, the Cuman–Kipchak Khanate spilled into the Danube region. The capture of Constantinople by the Crusaders on 12 April 1204 put an end to the Byzantine polity for more than half a century (Vasary 2009: 6–49).

On 14 April 1205 was fought the Second Battle of Adrianople. The Cumans, under their leader Qoja, implemented the time-tested steppe nomadic tactic of feigned retreat. As the Latins pursued them blindly, the Cumans turned around and attacked them. Many Latins were felled by the Cuman lassos and then were killed in close-quarter combat with daggers. The commander-in-chief of the Latin force, Baldwin, ruler of the Latin Empire in the Near East, was captured and taken to the Bulgarian capital, Tarnovo. The Western Europeans had no counter to the nomadic light cavalry archers during the twelfth and thirteenth centuries. The slow-moving heavily armoured feudal knights mounted on armour-clad horses were unable to catch up with the fast-moving light nomadic cavalry, which fought from a distance with their composite bows. The Cumans' battle tactics involved retreating in face of the enemy and then implementing Parthian shots. Once the enemy force was disordered, then the Cumans turned back and engaged in close-quarter combat with spears and swords. The Cumans were interested in pillage and plunder of the sedentary societies. Since they did not have any political ambition of establishing a polity, they functioned as mercenaries for both the Bulgar Empire and the Byzantine Empire. On 9 March 1230, the Bulgars, with the aid of Cuman mercenaries, were able to defeat the Greek Byzantine force of Theodoros at the Battle of Klokotnica. As a result of this battle, the Bulgars annexed Thessaly (Vasary 2009: 50–63).

To sum up, continuous innovations and a balanced hybridisation enabled the Islamic armies to establish Islamic rule from India in the east to Spain in the west. Initially, the camel-borne Arab Islamic soldiers while breaking out of Arabia adopted lance-sword wielding cavalry from the Sasanians and trebuchets from the Byzantine Empire. Next, while campaigning in Central Asia, the Islamic military adopted mounted archers and from South Asia the war elephants. The military effectiveness of the hybrid Islamic military establishment was seriously reduced due to the break-up of the caliphate and establishment of several competing Islamic polities. Internal bickering among the Islamic rulers allowed the Latins to intrude into West Asia. Meanwhile, a storm was brewing in the east. And from the east would come the greatest threat for Islam.

The rise and fall of the Mongol 'world' empire

The Mongols' date with history started with the rise of Temujin/Genghis Khan (b. 1165/67–d. 18 August 1227), the greatest warlord in history. Under Genghis

Khan, the Mongols created the largest land empire in world history; it was still expanding when he died. His military exploits were greater than that of Alexander, Hannibal, Julius Caesar, Attila, Napoleon and Hitler. Temujin first united the Mongolian tribes, then turned his attention to China, and finally further west. In 1197, Temujin attacked the Merkit tribe. Subedei (b. 1175–d. 1248) had joined Temujin's force at the age of 14. During the Merkit Campaign, Subedei Bahadur, a junior commander, was in charge of 100 soldiers. Later, he would become Genghis's foremost general. In 1206, Temujin became the leader of all the tribes of Mongolia. He took the name Genghis/Jenghiz Khan (Gabriel 2006: 3–9).

The combat effectiveness of the Mongol army was greater than the Huns, because the former in addition to light cavalry archers also included heavy cavalry. They also learnt to conduct sieges. The steppe bred small horses (equivalent to ponies) which were hardy but stunted in size because of harsh winters in the years when they were still growing. The mounted archers used such horses as light cavalry. Between the seventh and tenth century, the Turks who were in Sinkiang (Chinese Turkestan) practised agriculture in the irrigated strips around their oases. The Turkish nobles who controlled patches of the agricultural land were able to develop bigger horses. They selected good-looking colts, stall-fed them and were able to produce bigger and stronger horses. The nobles wearing armour rode on them. Thus emerged the cataphracts (heavy armoured cavalry), which were used for charging at the hostile force (Lattimore 1963: 2). The Mongol cavalrymen wore armour made of the thick hides of buffaloes dried by fire and hence were extremely hard and strong but lighter compared to the plate armour of the Western knights. This along with better horses gave the Mongol heavy cavalry more speed compared to the lumbering Western European knights. Not only were the Mongol heavy cavalrymen more agile than the Western knights, but Mongol light cavalry also had a longer reach, as these cavalrymen were equipped with composite bows (Hildinger 2001: 121). The Mongol horse archer's composite bow had a range from 600 to 900 feet. The heavy cavalrymen were covered with armour and were equipped with swords and lances, and the light cavalrymen wore no armour. The latter carried composite bows and javelins. Each horse archer carried two to three composite bows, three quivers and files for sharpening the arrows' heads. When dipped in red-hot brine, the arrows were able to pierce armour. In winter, the Mongol soldier wore a fur cap, coat and leather boots. His food comprised yoghurt (curdled milk), or *kumiz*. In addition, he carried a bag of millet meal which sustained him for some days and a waterproof pouch with a change of clothing for crossing swamps and rivers. The Mongol battle formation was organised into heavy cavalry, forming the first two lines, and light horse archers, constituting the third, fourth and fifth lines. The horse archers moved through the first two lines and discharged volleys of arrows, which caused chaos and confusion within the enemy formations Then, the heavy cavalry charged (Saunders 1971: 63–4).

The Mongol army was arranged in accordance with the decimal system, like the Parthian and Sasanian armies. Later, the Mughals of India also copied this aspect. The lowest unit was a squad of 10 men, with above them a company of 100 men, and until 1197 the largest unit was 1,000 men regiment (*mingan*). Genghis introduced the 10,000 strong unit (*tumen*) for conducting large-scale deep operations. A *tumen* occasionally functioned as an independent task force capable of sustained operation over a long distance. When several *tumens* were grouped together for constituting a field army/force, an *orlok* (somewhat equivalent to a field-marshal) was appointed. Each tribe in the Mongol army was led by a khan. Khan was a military title, and despite popular notion was not an Islamic title. Below the khan came the *noyan*, and below that was the *bagatur* (Gabriel 2006: 8–27). To reduce the power of the tribes, Genghis Khan gradually broke up the tribal units and established new formations comprising different ethnic communities and tribes mixed indiscriminately within each unit. His genius was to introduce impersonal organisational loyalty in place of tribal loyalty in the Mongol army. Ata-Malik Juvaini, who served the Mongol rulers of Persia, in his history of Genghis Khan noted the high motivation of the Mongol soldiers, their obedience to the commanders above them and their capability to undergo hardship. The Mongol army was held by iron discipline. During inspection, the troops' kits were rigorously inspected, and those whose equipment was found deficient were severely punished. Further, for maintaining both civil and military communications throughout his rapidly expanding empire, Genghis Khan established a series of *yams* (post houses) at regular intervals, each of which was provided with food, drink, horses and other supplies. Messengers arriving at a *yam* showed their passes and were provided with food, drink and fresh horses to continue with their journey (Saunders 1971: 63–5; Juvaini 1997: 31–2). The Suri Sultanate (1486–1545) established by the Afghans introduced this postal system in North India.

However, Genghis Khan's greatest contribution lay in the fact that he politically united the tribes of Mongolia. Juvaini, the thirteenth-century Persian historian, wrote about the conditions of the Mongols before the advent of Genghis Khan in the following words:

> Before the appearance of Chingiz Khan they had no chief or ruler. Each tribe or two tribes lived separately; they were not united with one another, and there was constant fighting and hostility between them. Some of them regarded robbery and violence, immorality and debauchery as deeds of manliness and excellence.
>
> *(quoted from Juvaini 1997: 21)*

Genghis was a nation builder. He gave the Mongols *yasa*: a code of laws, ordinances etc. His Mongol nation was an agglomeration of steppe nomadic tribes of Mongolia such as Merkits, Kereits, Tartars, Naimans etc., held together largely by his charisma and force (Gabriel 2006: 26). Once they were united, like the Huns, their military power was virtually unstoppable against the armies of the sedentary quasi-bureaucratic agrarian empires of Eurasia.

In 1206, war with the Jurchen-Jin Empire (1115–1234) of North China (not to be confused with the Jin Empire that ruled China after the collapse of the Han dynasty) and Manchuria started. In response to the Jurchen-Jin force structure, the Mongol army also underwent transformation. The Mongols adopted iron helmets. Later, while fighting against the Muslim polities of Central and West Asia and also in South Russia, the Mongols took care to capture the artisans and blacksmiths and transport them back to Mongolia in order to keep their logistical base running (Gabriel 2006: 16). In 1211, Genghis deployed 110,000 men against the Jurchen-Jin, whose army was six times numerically superior to the Mongol force (Martin 1943: 46). Nevertheless, the Mongols defeated them. The figures give one an idea about the scope and intensity of warfare which occurred in the medieval China–Mongolia theatre. Europe did not mobilise such a huge force before the advent of Napoleon. Still, historians (except a handful of China specialists) display a blissful ignorance about the range and variety of warfare beyond Europe. Most of the time, the Mongol forces, like the *Wehrmacht* (German armed forces during the Second World War), did not enjoy numerical superiority over their enemies. The Mongols did enjoy substantial hardware superiority (especially in the cavalry branch) over their opponents. Further Mongol advantages, like the *Wehrmacht*, principally lay in the sphere of software: command, doctrine and training. In 1218, Jebe, commanding 20,000 soldiers, killed the ruler Guchulug and absorbed the Kara Khitai polity, which encompassed Kazakhstan within the expanding Mongol dominion. In the same year, the Muslim governor of the city of Otrar (which belonged to the Khwarazm Empire) on the Syr Darya River massacred a Mongol caravan on charges of espionage. Sultan Muhammad II, the ruler of Khwarazm Empire (1200–1220), rightly considered the Mongols as a dire threat (Saunders 1971: 55; May 2007: 15–16).

In 1219, Genghis marshalled 150,000 combat troops against the Khwarazm Shah, whose army numbered more than 300,000 soldiers (Martin 1943: 46). During the Khwarazm campaign, the Mongol army was also able to conduct sieges. They learnt siege techniques during their military operations in China. Like China under the Southern Song (1127–1279), Sultan Muhammad's realm was studded with fortified cities. And Muhammad, realising his army's incapability to meet the Mongols in open battle, decided to follow a positional strategy based on these fortified cities. Thanks to the Mongols' siege techniques, the fortified centres of Muhammad fell one by one. Juvaini tells us that on 21 April 1219, the Mongol army came up before Jand and filled up the moat. Then they deployed battering rams, catapults and scaling ladders. Under cover of the stone-throwing catapults, the Mongol storming parties scaled the walls and threw open the gates of the town (Juvaini 1997: 89). Next came the town of Bokhara. One man was able to escape from the Mongol sack of Bokhara and came to Khorasan. When he was asked about the attack, he said: 'They came, they sapped, they burnt, they slew, they plundered and they departed' (quoted from Juvaini 1997: 107). Sultan Muhammad stationed a force of 110,000 men (60,000 Turkish lancers and 50,000 Tajiks) along with 20 elephants for defending Samarkhand. Against the Khwarazm field force, the standard Mongol tactic was to cut off their line of communications and then attack the isolated force. First, the Mongols sent horse archers (light cavalry), whose barrages dislocated Muhammad's troops, and then the demoralised soldiers were charged by the Mongol heavy cavalry armed with lances and swords (Juvaini 1997: 116–25).

Genghis Khan's greatest commander, Subedei Bahadur, crossed the Caucasus in 1221–1222 and defeated the Georgian army. The Georgians were planning to join the Fifth Crusade when the Mongols demolished them. In 1223, Subedei defeated the combined forces of the Kipchak Turks and the Russian princes at the Battle of Khalkha River in 1223. The Cumans, who had been a terror of the Byzantine force and the Frankish knights, were allied with the Kipchaks. This battle completely destroyed Cuman power in South-East Europe and gave the Byzantine Empire a new lease on life. In 1224, the Tanguts of Xi-Xia rebelled against Mongol authority. In 1225, the might of the Mongol military machine was turned against Xi-Xia, and by the next year, the Tangut Kingdom was on the verge of collapse. As the Mongol forces were closing towards the Tangut capital, Genghis, while hunting, fell from his horse and was seriously wounded. He died of his injuries on 18 August 1227 (May 2007: 17; Vasary 2009: 63).

Ogedei was ruler of the Mongol Empire from 1229 to 1241 CE. In 1233, the Mongols had captured the Jurchen-Jin's new capital, Kaifeng. The Jurchen-Jin Emperor Aizong fled to Caizhou, and the Mongols laid siege to this place in October 1233. The city was starving, and the Mongol soldiers diverted a river to flood the dying city. This shows that the Mongols had acquired substantial engineering capability. Caizhou surrendered in February 1234. A great *kuriltai* (grand assembly) was held in 1235 under Ogedei Khaghan (a title superior to khan), and it was decided to pursue the conquest of Korea, Song China and the lands west of the Volga River. The Mongols were capable of conducting multifront wars simultaneously. In 1236, 150,000 Mongol soldiers under the redoubtable Subedei and Batu (the son of Jochi, who had died in 1225) invaded the realms of the Kipchak Turks and the Bulgars around the Volga. For logistical purpose, Subdei used wagons with wooden wheels which had iron rims. Attila also used such wagons during his invasion of Italy. Kipchak resistance collapsed, and some were incorporated into the Mongol army and the rest entered Hungary as refugees. The Mongol menace was advancing towards Europe. In 1239, Armenia and Azerbaijan fell to the Mongols. In 1240, the Mongols sent a reconnaissance expedition to Tibet in order to assess the probability of attacking the Song Empire from west through the country of the lamas (Wylie 1977: 108–10; Gabriel 2006: 17; May 2007: 18).

Tibet, surrounded by Kunlun and Karakorum mountains on its north and north-west and the Himalayas in the south, adopted the cavalry as the premier arm of the ground force when the Tang Empire was ruling China. The infantry, who played a marginal role, was armed with swords, spears and wicker shields and wore lamellar armour. Tibetan heavy cavalry protected with chain mail armour carried lances whose length varied from 6 to 12 feet. Armoured cavalrymen at times also carried round shields made of hide with iron bosses. Such shields were common in India, Nepal and Bhutan. The iron helmets of the cavalrymen were adopted from Han Chinese and the Mongols. The saddles were adopted from China. From the seventh century CE onwards, Tibet had an efficient iron-working industry. Tibet had acquired the techniques of manufacturing wrought iron from Han China (LaRocca 2006: 3–27).

The principal theatre for the Mongol military machine was China. In 1127, the population of China was 140 million, or twice that of Tang Empire. The Song army's size rose from 378,000 men in 960 to 900,000 men in 1000 and to 1,259,000 men in 1041. About 80 per cent of the empire's revenue went to sustain this gargantuan military establishment. Genghis Khan had a plan to raze North China and eliminate 10 million people for the sake of providing pastures to his horses. It was somewhat similar to Adolf Hitler's intention of killing the Slavs and transforming Soviet Russia into an open space for exploitation by the German settlers. Luckily, neither plan was implemented. Nevertheless, Mongol depredations took their toll on China's demographic assets. By 1393, as a consequence of Mongol occupation, the population of China had declined substantially (Waterson 2013: 16, 59).

The core of the Song army comprised infantry, and the cavalry arm constituted a subsidiary branch. This was because the sedentary South Chinese were not good horsemen and China always imported horses from Mongolia, a source of supply now cut off. The Song soldiers used two types of swords: the *jian*, which was a straight, slender double-edged sword with tapering blade, and the *dao* for the mounted rider, which was a sturdy single-edged sabre with a slight curvature. The basic weapon of the Chinese infantry was the handheld crossbow. It was made of mulberry wood and brass, and its effective range was 328 feet. The Song crossbowmen were deployed in three lines and maintained a continuous volley fire. This manner of deployment was flexible enough to accommodate handguns for volley firing by the infantry. However, the Song, despite their knowledge of fire lances and handguns, did not adopt them wholesale, as those two gunpowder weapons were less effective compared to crossbows and composite bows. In 1259, the Song came up with repeating crossbows, each of which was operated with a lever mechanism and had a magazine box attached to it. The repeating crossbow was the medieval light machine-gun. In fact, the Qin dynasty used heavier crossbows mounted on wooden frames (221–206 BCE). It was the pre-modern equivalent of a tripod-mounted heavy machine-gun. The Song military engineers increased the effective range of such crossbows to about 3,000 feet. The Song army was well organised. The lowest unit named section comprised 10 men. A section comprising two subsections and five sections made a *dui* of 50 men (Waterson 2013: 53–6).

Despite some criticism in recent times, the ecological limitation of Mongol military power is a convincing explanation for the limits of its reach. We can agree with Denis Sinor's claim that the grasslands end in Europe at Hungary, and the Hungarian pastures were inadequate to maintain a sizeable number of Mongol *tumens* for conducting sustained operations for a long time in Central and Western Europe. This factor also debilitated Attila's military capacity to conquer the Western Roman Empire. For instance, the Mongols, like the Huns, did not feed their horses with fodder; they were always pastured. So grass and water were essential prerequisites for conducting steppe nomadic warfare in its classic form. Each horse required a large area of pasture land annually. The Hungarian pastureland could support from 205,000 to 320,000 horses. Each steppe nomadic warrior had three to six horses. So a maximum 35,000 to 68,000 horse archers could be maintained (Rogers 1996: 10–11), and this number was adequate for conducting raids but not for the permanent annexation of Western and Central Europe. True, the Mongols attempted to campaign in areas (like Burma/Myanmar, Vietnam, Java, etc.) without pastures. But then they had to replace their paradigm of cavalry-centric warfare with other military techniques, and this in turn reduced the military effectiveness of these deadly steppe nomads. Overall, the Mongol campaigns in South-East Asia were not spectacularly successful, which is discussed in the next chapter. So we can conclude that logistical limitations (lack of pastures) stopped the Mongol *Blitzkrieg* (lightning attack), just as logistical breakdown (lack of adequate railroads and absence of good roads) to a great extent was responsible for stopping the *panzers* west of Moscow in December 1941.

Horses were the currency of power in the Mongol world, and pastures (somewhat equivalent to oil during the industrial era) were an essential component for maintenance and expansion of the power base. Struggle for pastures to an extent accelerated the breakup of the Mongol Empire. Since no more pasturelands existed in Eurasia for further conquest, the attention of the great Mongol clan leaders turned inwards. Faction politics, to an extent, was the product of this struggle. In 1241, Ogedei died. In 1246, Ogedei's son Guyuk ascended the throne but was challenged by Batu Jochid Khan (r. 1227–1255), the ruler of South Russia and head of the Golden Horde Mongols. In 1251, Mongke ascended the throne of Genghis Khan and ruled the Mongol Empire till 1259 CE. Under him, the Mongol army numbered 1 million men. The core of the force comprised horse archers and spear- and sword-equipped armoured cavalry. Besides them, the army also included engineers, artillery crew and infantry (May 2007: 21; Waterson 2013: 65–6).

Meanwhile, the Mongol advance in China was continuing but slowing down. This was because in South China, the Mongol cavalry was obstructed by a large number of rivers and canals (which were guarded by a large number of ships and infantry garrisons equipped with deadly crossbows and various types of trebuchets) and frequently bogged down in the extensive paddy fields. Mongke's brother Kublai Khan (founder of the Yuan dynasty and ruler of China from 1260–1294) was granted North China as his appanage. He was charged with destroying the Dali Kingdom (modern Yunnan) in the south-west of Song Empire. The aim was to outflank the Song Empire by attacking it through Yunnan. So the invasion of Dali was a prelude to the final conquest of Central and South China. Kublai directly controlled about 20 per cent of the military assets of the Mongol Empire. In 1258, the Mongol invasion of Central China started.

About 90 to 100 *tumens* were on the march against what can be called the Mongols' 'Eastern Front'. Each *tumen* had two-thirds of its theoretical strength. Even then, some 600,000 soldiers were on the move. In Europe, only in 1812 was Napoleon able to gather such numbers during his fatal march to Moscow. Mongke himself took the field. In order to overcome their numerical inferiority against the huge Song army, the Mongols also conscripted Han Chinese soldiers and artisans. This trend would reach its culmination point under Kublai's Yuan dynasty. About 5 per cent of the total invasion force comprised Chinese troops. The Mongol onrush in China was temporarily stopped due to civil war. Between 1260 and 1263, Kublai and his brother Ariq Boke fought against each other (Waterson 2013: 68–72).

The Mongols adapted their military machine by adopting the paraphernalia of siege warfare, such as trebuchets, from China and the Khwarazm Empire supplemented by disciplined Chinese, Georgian and Iranian infantry. And for operating in the Middle East, the Mongols adopted the Islamic force's camel-borne logistical system (Amitai 2002: 229–30). Therefore, the Mongols not only won battles in the open field but were also able to capture fortified cities which had previously baffled the Goths and Huns in Europe. Thus, the Mongols were able to create an empire bigger than that of Attila. But they could not maintain a large cavalry force outside the grasslands, and this hobbled their operating radius in Central Europe, South India, and South-East Asia, respectively.

In 1241, Bela, King of Hungary, took position in the plain of Mohi and chained his wagons for protection of his camp. It was a typical steppe nomadic practice, and Zahir ud-din Babur, the founder of the Mughal Empire, would also use this tactic against Ibrahim Lodhi, the Sultan of Delhi, during the First Battle of Panipat (21 April 1526). The Mongols used catapults as field artillery to throw burning tar and naphtha against Bela's wagon laager (Hildinger 2001: 145–6). Bela's Hungarian army was outmanoeuvred, outfought and destroyed by the Mongols. However, Batu Khan, the leader of the Mongol force in East Europe, retreated and established the Golden Horde Khanate in South Russia. Lack of adequate pastures west of the Carpathian Mountains was the factor that prevented the Mongols, like the Huns before them, to subjugate Europe. Europe was thus saved from Mongol occupation, which in turn enabled this region to experiment with gunpowder technology from *c.* 1400 onwards, which would give the West its supremacy vis-à-vis the 'rest' in the early modern era. The point to be noted is that the Song used a lot of gunpowder weapons (grenades, fire lances, etc.) against the Mongols. However, these weapons did not prove to be effective. In fact, the Mongols during their campaign along the Euphrates River transmitted gunpowder weapons to the Middle East and then to Europe. Gunpowder weapons, due to slow experimentation and evolution, only became effective (even then marginally against the steppe nomads) in the sixteenth century.

The Chagatai Khanate of Central Asia launched several expeditions against the Delhi Sultanate. The size of the Mongol invading force under their commanders,

Qutlugh Khwaja and Targhi, varied between 20,000 and 50,000 cavalry. This time, India was ready. Sultan Alauddin Khalji (r. 1296–1316) ruler of the Delhi Sultanate commanded the vast demographic and economic resources of Hindustan. He raised an army of 450,000 men, including 150,000 horsemen. Further, the Turks settled in India were well acquainted with the Mongol art of war. And luckily for Alauddin, the Chagatai Khanate was at loggerheads with the Ilkhans of Persia (Nizami 1981: 149–269). The Mongols were repulsed, and they would not make any further attempt to conquer India. The cases of Egypt under the Mamluk Sultanate and India under the Delhi Sultanate show that a Turkish cavalry force backed by the resources of sedentary society could block a Mongol force. Had India and Egypt been attacked by the Mongols when they were united, then things might have turned out differently.

Amir Timur (also known as Tamerlane, b. 1336–d. 1405) attempted to resurrect Genghis Khan's empire in Central and West Asia. Timur, with some modifications, continued the Mongol military system. All the nomadic males were under the obligation to perform military duty. For increasing the military manpower pool, one soldier was conscripted from every nine households of the settled population. Further, mercenaries from Khorasan and Afghanistan were used. Timur, like Genghis Khan, did not allow the tribes to function as autonomous entities in his polity. The tribal confederations were broken up and subsumed within his politico-military format (Manz 1999: 96–100). Cavalry remained the principal shock weapon of his force, as panzers were for the *Wehrmacht*. Just like Hitler, who took a personal interest in evaluating tanks and had a deep technological knowledge about them, Timur had a thorough understanding of the various breeds of horses and took special interest in the types of mounts used by his army (*Tamerlane* 2018: 3). The Timurid army consisted mostly of armoured mounted archers. The mounted riders wore lamellar or plate armour. Timur practiced the hunting operation of the Mongols for training his army. Later the Mughals of India (1526–1707) also adopted this training regimen. Timur's army had a sophisticated logistical system. Timur used pioneers, whose duty was to dig wells to provide water supply points along the routes through which his army marched (Hildinger 2001: 173–90). His command system was quite sophisticated. With the aid of various types of kettledrums and cymbals, in the battlefield, Timur was able to redeploy his *qushuns* (cavalry regiments) in accordance with the changing tactical situations (*Zafarnamah* 2008: 222–3). Timur's military operations stretched from Delhi in the east to Ankara in the west. Warfare among the Eurasian steppe nomads further debilitated them, which in turn allowed Europe a fresh lease on life and gave them the opportunity to continue their experimentation with gunpowder warfare. For instance, Timur's devastating attack on the Golden Horde in 1395 seriously weakened the latter. This in turn enabled the rise of Muscovy and saved the Balkans from further nomadic depredations. In February 1405, Timur died while preparing for launching an invasion of China. After his death, his quasi-nomadic empire broke up.

Due to the Mongol influence, Muscovy also adopted the format of cavalry warfare. The Rus and the early Muscovites were sedentary people, but they adopted the model of steppe nomadic cavalry warfare because it was the most effective military method. This way of war enabled Muscovy not only to survive but also to gradually expand its control throughout Russia. Ivan III especially used the steppe nomadic style to fight the Golden Horde in South Russia (Poe and Lohr 2002: 3). Like the steppe nomads, the Muscovite archers used the reflex composite bows and deployed in battle like Timur's army in five divisions: right, left, centre, front and rear. In 1408, in a battle, the Lithuanian armoured lancers were defeated by the mounted archers of Muscovy. The Mongol style of war's superiority over the armoured European lancers, which was evident from the Battle of Leignitz (9 April 1241) and Battle of Mohi (11 April 1241), continued in the fifteenth century (Ostrowski 2002: 23). Mounted archers remained the principal arm of the Muscovite military organisation until the 1550s, when infantry arquebusiers started emerging as the core component (Hellie 2002: 41). Both Ivan III (d. 1505) and Vasily III (d. 1533) granted the newly conquered land to their nobles. In return for the grant of landed estates, the nobles had to provide fully equipped cavalry during the expeditions launched by the Grand Prince. This land grant system, known as *pomeste*, was similar to the *jagir* system of the Islamic polities. The *pomeste* system operated till the 1570s, and after that date, gradually, state-run bureaucratic infantry forces armed with arquebuses started appearing in the Russian military landscape (Martin 2012: 20–1).

Conclusion

From Korea in the east to the coastline of Atlantic in Western Europe, men on horseback held sway between *c.* 400 CE to 1500 CE. This was the 'Age of Horses'. When the steppe nomads were able to conquer sedentary polities, they adapted their political and military infrastructures. The result was that men on horseback settled down in the cities as rulers, and their armies, besides horse archers, also included the paraphernalia of the sedentary military polities. Several sedentary military polities, in order to check the supremacy of the horse riders, also raised cavalry forces. However, sedentary cavalrymen could not be as effective as the nomadic horse archers, since they were not born and brought up in pastoral nomadic settings of the steppes. The sedentary polities in turn came up with heavily armoured lancers and swordsmen as the second-best solution. However, they were no match against superb horse archery and the lance–sword equipped heavy cavalry of the steppe nomads. Credit is due to Genghis and his successors for constructing a hybrid army by blending the sedentary and steppe nomadic military traditions. Further, we have seen that the armies of the quasi-nomadic polities were well-organised and disciplined entities. The argument of some of the Western scholars that military discipline and a rational bureaucratic organisation were unique features of the West is just untenable.

Then again, the most dreaded steppe nomads, even when backed by resources of the conquered sedentary states, failed to adopt the techniques of naval power projection efficiently. A horse ethos just cannot go with seamanship. However, we should not give overt importance to cultural factors. At least one quasi-nomadic society, that is, Yuan dynasty, tried to experiment with naval power. It failed, as the next chapter shows, due to geopolitical and geo-economic factors. The gunpowder revolution and the rise of oceangoing ships, which in turn facilitated maritime commerce in the course of time, would displace the horse warriors from their dominant position. And this military transformation occurred in the Western part of Eurasia. The next chapter deals with the evolution of warfare in the maritime space of medieval Eurasia.

References

Ahmad Ibn Arabshah, *Tamerlane, The Life of the Great Amir*, Ahmad Ibn Arabshah, translated by J.H. Sanders, Introduction by R.D. McChesney (1936, reprint, London/New York: I.B. Tauris, 2018).

Allsen, Thomas T., 'The Circulation of Military Technology in the Mongolian Empire', in Nicola Di Cosmo (ed.), *Warfare in Inner Asian History: 500–1800* (Leiden/Boston: Brill, 2002), pp. 265–93.

Alofs, Eduard, 'Studies on Mounted Warfare in Asia I: Continuity and Change in Middle Eastern Warfare, c. CE 550–1350-What Happened to the Horse Archer?', *War in History*, vol. 21, no. 4 (2014), pp. 423–44.

Alofs, Eduard, 'Studies on Mounted Warfare in Asia II: The Iranian Tradition-The Armoured Horse Archer in the Middle East, c. CE 550–1350', *War in History*, vol. 22, no. 1 (2015a), pp. 4–27.

Alofs, Eduard, 'Studies on Mounted Warfare in Asia III: The Iranian Tradition-Cavalry Equipment, Infantry, and Servants, c. CE 550–1350', *War in History*, vol. 22, no. 2 (2015b), pp. 132–54.

Amitai, Reuven, 'Whither the Ilkhanid Army? Ghazan's First Campaign into Syria (1299–1300)', in Nicola Di Cosmo (ed.), *Warfare in Inner Asian History: 500–1800* (Leiden/Boston: Brill, 2002), pp. 221–64.

Ata-Malik Juvaini, *Genghis Khan: The History of the World Conqueror*, by Ata-Malik Juvaini, translated and edited by J.A. Boyle with an Introduction by David O. Morgan (1958, reprint, Manchester: Manchester University Press, 1997).

Bachrach, Bernard S., *Armies and Politics in the Early Medieval West* (Aldershot: Variorum, 1993).

Bachrach, Bernard S., 'On Roman Ramparts', in Geoffrey Parker (ed.), *The Cambridge Illustrated History of Warfare: The Triumph of the West* (Cambridge: Cambridge University Press, 1995), pp. 64–91.

Bachrach, Bernard S., 'Early Medieval Europe', in Kurt Raaflaub and Nathan Rosenstein (eds.), *War and Society in the Ancient and Medieval Worlds: Asia, The Mediterranean, Europe, and Mesoamerica* (Washington, DC: Center for Hellenic Studies, distributed by Harvard University Press, 1999), pp. 271–307.

Bachrach, Bernard S., *Early Carolingian Warfare: Prelude to Empire* (Philadelphia: University of Pennsylvania Press, 2001).

Bachrach, Bernard S., *Warfare and Military Organization in Pre-Crusade Europe* (Aldershot: Ashgate, 2002).

Bachrach, Bernard S., *Charlemagne's Early Campaigns (768–777): A Diplomatic and Military Analysis* (Leiden/Boston: Brill, 2013).

Barendse, R.J., 'The Feudal Mutation: Military and Economic Transformations of the Ethnosphere in the Tenth to Thirteenth Centuries', *Journal of World History*, vol. 14, no. 4 (2003), pp. 503–29.

Barnes, Cameron, 'Rehorsing the Huns', *War & Society*, vol. 34, no. 1 (2015), pp. 1–22.

Bayless, William N., 'The Treaty with the Huns of 443', *American Journal of Philology*, vol. 97, no. 2 (1976), pp. 176–9.

Beeler, John, *Warfare in Feudal Europe: 730–1200* (1971, reprint, Ithaca/London: Cornell University Press, 1972).

Bowlus, Charles R., 'Warfare and Society in the Carolingian Ostmark', in John France and Kelly DeVries (eds.), *Warfare in the Dark Ages* (Aldershot: Ashgate, 2008), pp. 321–39.

Browning, Robert, 'Where Was Attila's Camp?', *Journal of Hellenic Studies*, vol. 73 (1953), pp. 143–5.

Burns, Thomas S., 'The Battle of Adrianople: A Reconsideration', in John France and Kelly DeVries (eds.), *Warfare in the Dark Ages* (Aldershot: Ashgate, 2008), pp. 165–74.

Chattopadhyaya, Brajadulal, *The Making of Early Medieval India* (1994, reprint, New Delhi: Oxford University Press, 2005).

Clover, Frank M., 'Geiseric and Attila', *Historia*, vol. 22, no. 1 (1973), pp. 104–17.

Coupland, Simon, 'Carolingian Arms and Armour in the Ninth Century', in John France and Kelly DeVries (eds.), *Warfare in the Dark Ages* (Aldershot: Ashgate, 2008), pp. 249–70.

Crone, Patricia, 'The Early Islamic World', in Kurt Raaflaub and Nathan Rosenstein (eds.), *War and Society in the Ancient and Medieval Worlds: Asia, The Mediterranean, Europe, and Mesoamerica* (Washington, DC: Center for Hellenic Studies, distributed by Harvard University Press, 1999), pp. 309–32.

Davis, R.H.C., 'The Warhorses of the Normans', in John France (ed.), *Medieval Warfare: 1000–1300* (Aldershot: Ashgate, 2006), pp. 85–100.

Dennis, George T., 'The Byzantines in Battle', in John Haldon (ed.), *Byzantine Warfare* (Aldershot: Ashgate, 2007), pp. 223–36.

Dien, Albert E., 'A Study of Early Chinese Armor', in Peter Lorge (ed.), *Warfare in China to 1600* (Aldershot: Ashgate, 2005a), pp. 83–144.

Dien, Albert E., 'The Stirrup and Its Effect on Chinese Military History', in Peter Lorge (ed.), *Warfare in China to 1600* (Aldershot: Ashgate, 2005b), pp. 185–208.

Farrokh, Kaveh, Karamian, Gholamreza and Maksymiuk, Katarzyna, *A Synopsis of Sasanian Military Organization and Combat Units* (Siedlce/Tehran: Siedlce University of Natural Sciences and Humanities, 2018).

Ferrill, Arther, *The Fall of the Roman Empire: The Military Explanation* (1986, reprint, London/New York: Thames and Hudson, 1988).

France, John, *Victory in the East: A Military History of the First Crusade* (Cambridge: Cambridge University Press, 1994).

France, John, *Western Warfare in the Age of Crusades: 1000–1300* (Ithaca/New York: Cornell University Press, 1999).

France, John, 'The Military History of the Carolingian Period', in John France and Kelly De Vries (eds.), *Warfare in the Dark Ages* (Aldershot: Ashgate, 2008), pp. 321–39.

Gabriel, Richard A., *Genghis Khan's Greatest General: Subotai the Valiant* (2004, reprint, Norman: University of Oklahoma Press, 2006).

Goffart, Walter, 'Rome, Constantinople, and the Barbarians', in John France and Kelly DeVries (eds.), *Warfare in the Dark Ages* (Aldershot: Ashgate, 2008), pp. 1–32.

Gommans, Jos, *Indian Frontiers and High Roads to Empire: 1500–1700* (London/New York: Routledge, 2002).

Graff, David A., 'Strategy and Contingency in the Tang Defeat of the Eastern Turks: 629–30', in Nicola Di Cosmo (ed.), *Warfare in Inner Asian History: 500–1800* (Leiden/Boston: Brill, 2002), pp. 33–71.

Graff, David A., *The Eurasian Way of War: Military Practice in Seventeenth-Century China and Byzantium* (London/New York: Routledge, 2016).

Habib, Irfan, 'Population', in Tapan Raychaudhuri and Irfan Habib (eds.), *The Cambridge Economic History of India*, vol. 1, *c. 1200-c. 1750* (1982, reprint, Hyderabad: Orient Longman, 1991), pp. 163–71.

Haldon, John, 'The Byzantine World', in Kurt Raaflaub and Nathan Rosenstein (eds.), *War and Society in the Ancient and Medieval Worlds: Asia, The Mediterranean, Europe, and Mesoamerica* (Washington, DC: Center for Hellenic Studies, distributed by Harvard University Press, 1999), pp. 241–70.

Halsall, Guy, *Warfare and Society in the Barbarian West: 450–900* (Oxon: Routledge, 2003).

Hanson, V.D., *Carnage and Culture: Landmark Battles in the Rise of Western Power* (New York: Doubleday, 2001).

Heather, Peter, 'The Huns and the End of the Roman Empire in Western Europe', *English Historical Review*, vol. CX, no. 435 (Feb. 1995), pp. 4–41.

Heather, Peter, 'Holding the Line: Frontier Defense and the Later Roman Empire', in Victor Davis Hanson (ed.), *Makers of Ancient Strategy: From the Persian Wars to the Fall of Rome* (Princeton/Oxford: Princeton University Press, 2010), pp. 227–46.

Hellie, Richard, 'The Costs of Muscovite Military Defense and Expansion', in Eric Lohr and Marshall Poe (eds.), *The Military and Society in Russia: 1450–1917* (Leiden/Boston: Brill, 2002), pp. 41–66.

Hildinger, Erik, *Warriors of the Steppe: A Military History of Central Asia, 500 BC to 1700 AD* (Cambridge, MA: Da Capo, 2001).

Hill, R., 'The Role of the Camel and the Horse in the Early Arab Conquests', in V.J. Parry and M.E. Yapp (eds.), *War, Technology and Society in the Middle East* (London: Oxford University Press, 1975), pp. 32–43.

Hooper, Nicholas and Bennett, Matthew, *Cambridge Illustrated Atlas: Warfare, The Middle Ages, 768–1487* (Cambridge: Cambridge University Press, 1996).

James, David, *A History of Early Al-Andalus: The Akhbar Majmua, A Study of the Unique Arabic Manuscript in the Bibliotheque Nationale de France, Paris*, with a translation, notes and comments (London/New York: Routledge, 2012).

Jandora, J.W., 'Developments in Islamic Warfare: The Early Conquests', *Studia Islamica*, no. 64 (1986), pp. 101–13.

Jha, D.N., *Early India: A Concise History* (New Delhi: Manohar, 2004).

Kaegi, Jr., Walter Emil, 'The Contribution of Archery to the Turkish Conquest of Anatolia', in John Haldon (ed.), *Byzantine Warfare* (Aldershot: Ashgate, 2007), pp. 237–49.

Keegan, John, *A History of Warfare* (1993, reprint, New York: Vintage, 1994).

Kennedy, Hugh, *Mongols, Huns and Vikings: Nomads at War* (London: Cassell, 2002).

Kennedy, Hugh, *The Armies of the Caliphs: Military and Society in the Early Islamic State* (2001, reprint, London/New York: Routledge, 2005).

LaRocca, Donald J., Clarke, John, Heller, Amy and Jamspal, Lozang, *Warriors of the Himalayas: Rediscovering the Arms and Armor of Tibet* (New Haven/London: Yale University Press in association with the Metropolitan Museum of Art, New York, 2006).

Lattimore, Owen, 'The Geography of Chingis Khan', *Geographical Journal*, vol. 129, no. 1 (1963), pp. 1–7.

Lindner, Rudi Paul, 'Nomadism, Horses and Huns', *Past and Present*, no. 92 (Aug. 1981), pp. 3–19.

Lorge, Peter, *War, Politics and Society in Early Modern China: 900–1795* (2005, reprint, London/New York: Routledge, 2008).

Lorge, Peter, 'Fighting Against Empire: Resistance to the Later Zhou and Song Conquest of China', in Peter Lorge (ed.), *Debating War in Chinese History* (Boston/Leiden: Brill, 2013), pp. 107–39.

Magdalino, Paul, 'The Byzantine Army and the Land: From *Stratiotikon Ktema* to Military *Pronoia*', in John Haldon (ed.), *Byzantine Warfare* (Aldershot: Ashgate, 2007), pp. 167–88.

Man, John, *Attila: A Barbarian King and the Fall of Rome* (London: Bantam, 2005).

Manz, Beatrice Forbes, *The Rise and Rule of Tamerlane* (1989, reprint, Cambridge: Canto, 1999).

Martin, H. Desmond, 'The Mongol Army', *Journal of the Royal Asiatic Society of Great Britain and Ireland*, no. 1 (April 1943), pp. 46–85.

Martin, Janet, 'Economic Effectiveness of the Muscovite *Pomeste* System: An Examination of Estate Incomes and Military Expenses in the Mid-16th Century', in Brian J. Davies (ed.), *Warfare in Eastern Europe: 1500–1800* (Leiden/Boston: Brill, 2012), pp. 19–34.

May, Timothy, *The Mongol Art of War: Chinggis Khan and the Mongol Military System* (Barnsley: Pen & Sword, 2007).

McMichael, Anthony J. with Woodward, Alistair and Muir, Cameron, *Climate Change and the Health of Nations: Famines, Fevers, and the Fate of Populations* (Oxford/New York: Oxford University Press, 2017).

Molana Sharf-ud-din Yezdi, *Amir Timur Beg, 1336–1396, English Rendering of Molana Sharf-ud-din Yezdi's Zafarnamah*, vol. 1, edited by Khan M. Atif and H.A. Qureshi (Lucknow: New Royal Book Company, 2008).

Mookerji, Radhakumud, *The Gupta Empire* (1973, reprint, Delhi: Motilal Banarasidass, 1997).

Nickel, Helmut, 'About the Sword of the Huns and the "Urepos" of the Steppes', *Metropolitan Museum Journal*, vol. 7 (1973), pp. 131–42.

Nicolle, David, *Warriors and Their Weapons Around the Time of the Crusades* (Aldershot: Ashgate, 2002).

Nizami, K.A. (ed.), *Politics and Society During the Early Medieval Period: Collected Works of Professor Mohammad Habib*, vol. 2 (New Delhi: People's Publishing House, 1981).

Oman, Charles, *The Art of War in the Middle Ages: AD 378–1515* (London: Oxford, 1885).

Ostrowski, Donald, 'Troop Mobilization by the Muscovite Grand Princes: 1313–1513', in Eric Lohr and Marshall Poe (eds.), *The Military and Society in Russia: 1450–1917* (Leiden/Boston: Brill, 2002), pp. 19–40.

Peacock, Andrew C.S., 'Nomadic Society and the Seljuq Campaigns in Caucasia', *Iran & the Caucasus*, vol. 9, no. 2 (2005), pp. 205–30.

Petersen, Leif Inge Ree, *Siege Warfare and Military Organization in the Successor States (400–800 AD): Byzantium, the West and Islam* (Leiden/Boston: Brill, 2013).

Pierce, Ian, 'Arms, Armour and Warfare in the Eleventh Century', in John France (ed.), *Medieval Warfare: 1000–1300* (Aldershot: Ashgate, 2006), pp. 63–83.

Poe, Marshall and Lohr, Eric, 'Introduction: The Role of War in Russian History', in Eric Lohr and Marshall Poe (eds.), *The Military and Society in Russia: 1450–1917* (Leiden/Boston: Brill, 2002), pp. 1–15.

Rogers, Greg S., 'An Examination of Historians' Explanations for the Mongol Withdrawal from East Central Europe', *East European Quarterly*, vol. 30, no. 1 (Spring 1996), pp. 3–26.

Roy, Kaushik, *From Hydaspes to Kargil: A History of Warfare in India from 326 BC to AD 1999* (New Delhi: Manohar, 2004).

Roy, Kaushik, *Warfare in Pre-British India: 1500 BCE to 1740 CE* (London/New York: Routledge, 2015).

Saunders, J.J., *The History of the Mongol Conquests* (London: Routledge, 1971).

Sinor, Denis, 'Horse and Pasture in Inner Asian History', *Oriens Extremus*, vol. 21 (1972), pp. 171–83.

Thompson, E.A., 'The Camp of Attila', *Journal of Hellenic Studies*, vol. 65 (1945), pp. 112–15.

Thompson, E.A., 'Early Germanic Warfare', in John France and Kelly DeVries (eds.), *Warfare in the Dark Ages* (Aldershot: Ashgate, 2008), pp. 119–46.

Upadhyaya, Rashmi, 'Military Strategy of the Rajputs During the Early Medieval Period', in Shahabuddin Iraqi (ed.), *Medieval India 2: Essays in Medieval Indian History and Culture* (New Delhi: Manohar, 2008), pp. 173–90.

Vasary, Istvan, *Cumans and Tatars: Oriental Military in the Pre-Ottoman Balkans, 1185–1365* (2005, reprint, Cambridge: Cambridge University Press, 2009).

Vryonis Jr., Speros, 'The Evolution of Slavic Society and the Slavic Invasions in Greece: The First Major Slavic Attack on Thessaloniki, AD 597', in John France and Kelly DeVries (eds.), *Warfare in the Dark Ages* (Aldershot: Ashgate, 2008), pp. 105–17.

Waterson, James, *Defending Heaven: China's Mongol Wars, 1209–1370* (London: Frontline Books, 2013).

Wylie, Turrell V., 'The First Mongol Conquest of Tibet Reinterpreted', *Harvard Journal of Asiatic Studies*, vol. 37, no. 1 (1977), pp. 103–33.

7

CONVERGENCE OR DIVERGENCE?

Naval warfare in medieval Eurasia

Introduction

The dominant view among historians that Asia lacked a maritime cum naval tradition before the rise of European naval supremacy in the early modern age is questioned in this chapter through evaluation of the military effectiveness of the pre-gunpowder Asian navies as compared to their European counterparts. This chapter analyses whether there are any traces of superiority in the technical aspects or in the range of operations of Western European warships vis-à-vis Asian vessels. Whether the evolution of naval warfare in Asia followed a similar or divergent trajectory as compared to the development of navies in the Mediterranean and the Atlantic is discussed. This chapter also examines why the big Asian empires lacked ocean-going navies in the fifteenth century. It argues that the evolution of the maritime tradition of a particular region cannot be separated from the dynamics of land warfare. Hence, the linkages between geopolitics, geo-economics and the maritime tradition are highlighted here. For the reason mentioned earlier, littoral warfare and amphibious warfare, along with battles in the sea, are also included within the ambit of naval operations. We will not delve into naval conflicts which occurred in the American waters or in the Pacific before *c.* 1500 CE because of the paucity of sources.

This chapter is organised along the various regions of Eurasia. Broadly, we can discern four distinct maritime spaces with their distinctive maritime traditions. However, there were interactions and commingling among these four maritime traditions. China, Korea and Japan constituted the Far Eastern maritime tradition. The Indian Ocean maritime tradition comprised South-East Asia, India and Arabia. The Maghreb, the Near East, the Balkans and Italy formed the Mediterranean maritime tradition. And finally, Scandinavia, North-Western Europe and Iberia gave birth to the Atlantic maritime tradition.

DOI: 10.4324/9780429284069-8

China–Korea–Japan

A coastal navy developed in pre-modern China because sea communication was quicker than land transportation. For instance, until *c.* 1800, it took government couriers 32 days to travel by horse and 56 days by foot to cover the distance between Guangzhou (Canton) and Beijing, despite being provided with the best travel facilities. The distance between these two cities was approximately 1,900 miles. Pre-modern ships could cover this distance in 22 days (Jung-pang 2012: 25).

China developed a navy capable of projecting offensive power across the seas under the Liu Song dynasty (420–79 CE). In 446 CE, this dynasty sent a fleet to teach Champa a lesson. Champa was a loosely formed alliance of ancient kingdoms in Central Vietnam. In 446 CE, a naval detachment with marines sent by Liu Song pillaged and burnt the capital of Champa. The technology used at that time for shipbuilding was quite impressive. During the mid-fifth century CE, the ships plying between China and India were capable of sailing in the open water without landfall for 80 days at a stretch (Jung-pang 2012: 5–6; Lien 2017: 1). No contemporary polities bordering the East Atlantic and the Mediterranean possessed ships with such long-range capacity. In fact, the Mediterranean galleys of Classical Greece had to be beached every night so that the crew could cook their food, wash themselves and take rest.

The rise of the Sui Empire (581–618 CE) in the late sixth century, to an extent, was possible due to the suppression of the rival polities in the south with the aid of naval power. At that time, some of the Chinese ships were five-deckers (rising 100 feet above water) with masts that were 49 feet high, and each vessel had a carrying capacity of 800 personnel. Such big ships were absent in the Mediterranean as well as in the Baltic and in the English Channel. In 589 CE, Emperor Wen of the Sui dynasty decided to attack the Chen state (557–89 CE) in South China. The Sui state sent a river fleet and a high sea fleet. The high sea fleet sailed down from Shandong's northern coast to strike at the Chen capital, Nanjing, from the east. Simultaneously, the river fleet was to attack the Chen capital from the western side. The river fleet compared to the high sea fleet had bigger ships with 50-foot-tall masts (Jung-pang 2012: 44–5). We could speculate that the bigger, top-heavy ships could not be used in the turbulent water of the East and South China seas, respectively.

After subduing the land south of Yangzi, the Sui emperor in 605 CE launched a naval attack against Champa and destroyed its capital, located north of modern Hue. The overseas invasion of Champa was launched due to several factors. Champa was rich in gold, which its rulers had amassed through commerce and piratical activities. Fan Yangmai, the ruler of Champa, had a strong navy. In 421 CE, he had sent 100 ships to threaten the Chinese control of the port of Hanoi (Jung-pang 2012: 6, 43). To prevent recurrence of such a threat, Champa's naval power needed to be destroyed.

During the period of Five Dynasties (907–60 CE), naval engagements between the Wu-Yue and Nan Han states along the southern coast of China saw the use of a new weapon – flaming oil fired from canisters. It was used to set fire to the sails and riggings of hostile ships (Jung-pang 2012: 6, 14). The Byzantine navy also used a similar weapon, known as Greek fire. Probably, the use of naphtha (the source of Greek fire) as a weapon for use in naval conflicts had developed independently in China and in the Eastern Mediterranean area.

By the end of the tenth century CE, the economic, cultural and demographic centres of Chinese civilisation shifted from the southern bend of the Yellow River to the Yangtze Delta. In 1086, the population of China (960–1276 CE) came to about 130 million. From the eleventh century onwards, coastal parts of South-East China housed more than 50 per cent of the country's population. Further, this region also generated almost two-thirds of the total revenue of China. Obstructions along the overland caravan route to the west and the military forays of the steppe nomadic tribes along the northern frontier of China generated a push towards overseas commerce in South-East China under the Song dynasty (Jung-pang 1969: 59).

The Song navy came into existence in 1132 CE. The Song monarchs used to offer prizes for new naval inventions. Randall J. Sasaki writes that this led to the invention of the compass (Sasaki 2015: 23). The Indians were also using the compass from *c.* 900 CE onwards. India and China had strong maritime linkages. So it is possible that the Song acquired the compass from the merchants who conducted India–China maritime trade. Naval power and maritime power went hand in hand. In the long run, the sustenance of navies required a state to engage profitably in overseas and river-borne commerce. And this was borne out in case of the Song. The rise in the volume of overseas trade under the Song dynasty in the eleventh century saw increasing monetisation, that is, the increasing use of copper cash. It resulted in a boom of domestic commerce and maritime trade. Adoption and adaptation are two essential qualities of a successful military machine. The Song adopted outriggers from the South-East Asians. In the early Middle Ages, the sails of the Chinese ships were made of textiles and even silk. Cotton became common in China during the Song–Yuan period. Before the twelfth century, the sails were made of woven bamboo mats (Jung-pang 2012: 7, 14, 31). During the thirteenth century, Song ships used rectangular sails (Lien 2017: 10).

For defending the rivers, which functioned as channels for transportation and commerce, the Song navy had flat-decked ships equipped with trebuchets and huge crossbows. Such a ship was known as a mullet ship. Some ships, like the Greek and Roman triremes, had iron rams attached to them for ramming and sinking hostile ships. Unlike the Greeks and Roman ships, however, it was difficult to sink the Chinese ships. This was because each ship had several bulkheads and compartments which were capable of being sealed off if breached by ramming or projectiles. All the planking was fastened with iron clamps and spikes (Waterson 2013: 76–7). Shipbuilding in medieval China relied on the bulkheads to support the hull (Sasaki 2015: 143). So technologically, the Chinese ships at this point of history were more advanced than contemporary Western European ships.

In the mid-thirteenth century, the Mongols emerged as a principal threat to China. The Mongol advance into Central and South China slowed down because of the civil war among the Mongol chieftains, plus the physical geography of the region, which suited riverine war. The Song navy was a powerful one. Long before the advent of the Mongols, in 1161 CE, the Jurchen-Jin navy attacked the Song but was defeated near the coast of Shantung. In 1268, the Mongols planned to attack the Song Empire from the sea. The Mongol plan was to capture Ross Island and Quelpart Island near the coast of Korea and then to launch harassing raids along the littoral of Eastern China. However, the presence of a strong Song navy resulted in rejection of this plan. In 1237, the Southern Song navy was organised in 20 squadrons manned by 52,000 men. Its two main bases were Shanghai, near the entrance of the Yangtze River, and Ting-hai, which guarded the capital Hangzhou (Jung-pang 1955: 490–1). Numerous rivers and paddy fields of South China were not conducive for fast-moving cavalry operations by the Mongols. The Song army's strategy was to hold the strategic fortified centres with infantry equipped with spears and crossbows. Communication among the cities located along the river valleys was maintained with the aid of the riverine navy (Wright 2008a: 51). Taking each of the fortified cities was a time-consuming and costly affair. Besides constructing siege engines with the aid of Persian engineers, the Mongols also had to build a riverine navy. In 1274, the Song deployed approximately 100,000 troops near Yingzhou on both sides of the Han River. About 1,000 Song warships patrolled the river. Moreover, a long iron cable was stretched along the river, and this cable was connected in turn with several large Song warships. The aim was to prevent the entry of Mongol warships downstream (Wright 2008b: 92).

Under Kublai Khan (b. 1215–d. 1294), the Mongols in China partly adapted themselves to the new agro-ecological settings and adopted the Song maritime tradition. The Mongol navy innovated during the conflict with the Song. From the Song navy, the Mongols learnt the art of carrying big trebuchets on their ships. These trebuchets shot both stones and exploding devices. In 1276, the Mongol commander Bayan enlisted two pirates, who joined with 500 ships. In addition, they also brought local knowledge with them. The Mongols were aware that acquisition of topographical knowledge and generation of military power went hand in hand. For smooth conduct of riverine operations, a survey of the Upper Yellow River was completed between 1277 and 1281 (Lien 2017: 9, 12–13).

Vietnam, Java and Japan became the targets of Kublai Khan (a grandson of Genghis Khan), who later founded the Yuan dynasty (1271–1368). Kublai was interested in controlling South-East Asia partly for commercial reasons. The Mongol Khanates west of China were not always on good terms with the Yuan dynasty. So the traditional overland Silk Route through West China, Central Asia and West Asia was not functioning at its maximum capacity. Hence, Kublai attempted to open a maritime Silk Route from China through South-East Asia, South Asia to West Asia. This is also one of the objectives of present day China's Belt and Road Initiative. Then Kublai attempted to impose tributary relationship over China's neighbouring states like Japan, Java and Vietnam (Lien 2017: 8, 12). Further, amphibious expeditions were launched against these countries in order to destroy Southern Song's overseas trade and imperial influence.

Besides seas and rivers, tropical jungles and mountains also obstructed the advance of the Mongol cavalry in South-East Asia. After the conquest of Yunnan, in 1257, Kublai demanded free access for the Mongol army to advance through Dai Viet in order to attack the Southern Song Empire from the south. Dai Viet was the name for Northern Vietnam under the Ly (1009–1225 CE) and Tran (1225–1400 CE) dynasties. Dai Viet refused Kublai's request. In 1258, the Mongols launched their first invasion of Dai Vet under Uriyangkhadai, son of Subedei Bahadur. Emperor Tran Thai Tong decided to offer battle at Bach Hac on the Red River. The Vietnamese force comprised infantry wearing leather helmets and elephants in the style of medieval Hindu armies of India. At that time, the Mongols were not used to fighting war elephants, so the Mongol horses shied away from the war elephants. Further, the Mongols had problems in crossing the river. However, the Vietnamese were skilled in river and coastal navigation. They had long, light boats made of thin planks. Each boat had about 30 to 100 oarsmen (Charney 2004: 38; Lien 2017: 1, 16–17). The Vietnamese retreated into the mountains and jungles and conducted a guerrilla war. Harassed by ambushes and tropical diseases (especially malaria and dysentery), the Mongols retreated.

Upper Yunnan is the source of several major rivers of South-East Asia such as the Salween, Red River, Yangtze and Mekong. A ground force moving from South China towards North Vietnam had to confront diseases, heat, humidity and had to cross several deep rivers. The Mongol horses could not swim over these large and wide rivers. The traditional inflated sheepskin rafts of the Mongol soldiers were also not adequate for crossing these huge water bodies (Lien 2017: 5). Overall, the mountains, marshlands and big rivers in South-East Asia are not conducive for large-scale cavalry operations. But the core of the Mongol ground force was cavalry. For attacking Vietnam successfully, the Mongols realised that they required a navy for conducting amphibious operations to support their land detachments. However, in 1267 Dai Viet had had enough and agreed to send tribute to Kublai Khan.

When the kings of the Pagan dynasty (1044–1287 CE) in Myanmar (Burma) refused to pay tribute and do homage to the Yuan, the Mongols attacked. In 1284, the Mongols captured Upper Myanmar and occupied the capital in 1287. Adverse climatic conditions (heavy tropical showers, heat and high humidity) combined with the guerrilla war conducted by the people of Myanmar forced the Mongols to withdraw in 1303. The forested and mountainous terrain with absence of extensive grasslands was conducive neither to the maintenance of a large cavalry force nor for launching sweeping outflanking cavalry operations. However, Myanmar could not bear the burden of this prolonged conflict and agreed to become a tributary state of the Yuan Empire (Dai 2004: 149–50). Kublai had seriously threatened the kingdoms of South-East Asia but failed to annex them. This was because their geography was not conducive for moving a large ground force from South China to the northern parts of South-East Asia. Only in the mid-twentieth century the People's Liberation Army, using modern technology, was temporarily able to overwhelm Vietnam.

Kublai's attempt to project power across the sea was a failure, and this became evident during the two attempted Mongol invasions of Japan. For the Mongols, Korea was the gateway for invading Japan. Medieval Korea was open to the military influences from Mongolia and China. From China, Korea acquired the crossbow and catapult, and the composite bow from Mongolia. Around the seventh century, the Koreans developed heavy cavalry for attacking the enemy's flanks (Asmolov 1992: 103–16). The Mongols initially attacked Korea to cut the Song's lucrative trade ties with this country. The Koryo dynasty (935–1392) became a vassal of Kublai. The Song's extensive maritime trade with Japan also motivated Kublai to attack the 'land of the rising sun' (Delgado 2008: 88–9).

The military organisation of medieval Japan was influenced by the cavalry traditions of mainland Asia. The Japanese politico-military establishment was decentralised in nature. As regards techniques and technology of warfare, the Japanese were seriously lagging behind the Mongols. During the eighth century CE, the central government lost control over the monopoly of violence. Instead of trying to draft and drill the population directly for induction in the royal infantry formations, the court attempted to co-opt the mounted private forces of the military elites through grants of office and titles. So instead of a centrally raised, equipped and trained infantry-centric military force, the court at Kyoto commanded privately trained and privately equipped mounted professional mercenaries with dubious loyalty. From the late ninth century onwards, the court was dominated by powerful families headed by *kugyo* (senior courtiers) who established complex networks of vertical alliances with low- and middle-ranking nobles. The *kugyo* recruited warriors as their household troops and staffed the officer cadre with their own kinsmen and clients. For tax collection, the court depended on the provincial officers, who in turn delegated this job to the local elites who emerged as the tax farmers. These local elites also controlled private war bands. Besides engaging in government service, many private war bands participated in banditry and also fought with other competing war bands. Towards the end of the twelfth century, a large-scale civil war broke out which is known as the Gempei War. The private military entrepreneurs/warlords were grouped into two opposing groups named Minamoto and Tara. The Kamakura Shogunate emerged out of ashes of this civil war. This shogunate represented a power centre distinct from the imperial court at Kyoto (Friday 2004: 6–13). After 1221, the emperor became a mere figurehead.

Warfare in Japan was dominated by the samurai (mounted knights who were also landowners). The common notion is that the samurai's (*bushi*) code of honour (*bushido*) required death before surrender. Historian Karl Friday's observations about the samurai is more critical. He writes that the early *bushi* were fierce men with a sense of honour and courage, but they did not possess any notion of fair play. The *bushi*/samurai, the military elite of medieval Japan, rode on horses.

They wore armour made of lacquered leather. The mounted warrior aristocracy fought with swords and simple bows made of bamboo and wood. The composite bow came to Japan from the North-East Asian mainland in the twelfth century CE and replaced the simple bow. Besides the composite bow, the most favoured weapon of the samurai was the sword. They used single-edged long swords as well as curved swords. However, the government did not possess a monopoly over weapons production. Instead, a thriving private trade in arms existed in Japan. Unlike in China, in Japan the crossbow never gained popularity. This was because use of the crossbow required drilled and disciplined infantry. However, the private military entrepreneurs of medieval Japan used infantry as an auxiliary branch and depended on hastily recruited untrained peasant militia. This group was unable to use crossbows effectively (Gabriel 2002: 319–26; Friday 2004: 17–18, 65–80). Japan had no navy worth its name in the thirteenth century. If the Mongols could land a significant number of troops with adequate supplies, then Japan will be doomed.

In 1268, Kublai ordered Korea to prepare a fleet of 1,000 warships and an expeditionary army of 10,000 men for his projected invasion of Japan. In 1273, the Koreans were asked to increase rice production for feeding the expeditionary force which would sail for Japan from their country. On 3 October 1274, the Mongol armada sailed from Pusan. This fleet carried 23,000 Mongol, Chinese and Korean troops plus 6,700 sailors. Two days later, the fleet approached the Japanese-held Tsushima Island. After firing a barrage of arrows and stones from the shipborne catapults, the Mongols were able to drive away the defenders and came ashore (Suenaga 2001: 12). Soon samurai resistance collapsed, and on 14 October the Mongol fleet appeared before Iki Island. The Mongol invaders made short work of Japanese resistance on this island. Then the Mongol fleet set sail for Hakata Bay in Northern Kyushu.

On 19 October 1274, when the Mongol fleet approached Hakata Bay, some 6,000 mounted samurai were manning the defence. Mongol crossbows mounted on the ships (a traditional Chinese weapon integrated in Kublai's navy) had a longer range compared to those used by the Japanese. In addition to archery volleys, catapults from the Mongol ships hurled fire bombs (incendiary weapons) which they had adopted from the Song. The Mongols approached the shoreline in small boats from their ships. After landing in Northern Kyushu, the Mongols established camp in the marshy wooded lands of Akasaka and Torikai. The Japanese forces arrived from the eastern direction and set up their base at Hakozaki. Then they attacked in the western direction towards Sumiyoshi and Akasaka. The Japanese mounted bowmen practised their archery skill in shooting dogs and other animals from galloping horses. The Japanese commander ordered his men to take advantage of the difficult terrain to nullify the superior speed of the Mongol cavalry. One Japanese scroll notes: 'As Akasaka has poor terrain, you should pull back here' (quoted from Suenaga 2001: 23). The samurai launched several cavalry charges. The Mongols used fire arrows (another Song weapon) against them. In close-quarter combat, the Japanese *naginata* (a curved blade attached to a long wooden shaft; it was equivalent to a halberd) proved effective (Suenaga 2001: 50, 73). During the night, the samurai retreated inland, but the Mongol commanders, fearing ambushes, traps and nocturnal attacks by the Japanese in unknown territory, recalled their troops to the ships. Furthermore, protracted defence by the Japanese resulted in the Mongols using up all their reserves of arrows (Sasaki 2015: 26). The bow, it must be remembered, was the principal weapon of the Mongol soldiers.

A storm (a reverse easterly wind) in the night caused discomfiture to the Mongol ships. It was also possible that a typhoon had hit the Mongol fleet. The sea between Japan and South China's coast is noted for large cyclonic storms which occur frequently. And the tough resistance by the Japanese plus mobilisation of further reinforcements by the Kamakura regime demoralised the Mongol commanders. They realised that they had lost the chance of surprising the Japanese defenders. The next day, the Mongol ships sailed back to Korea. The failed campaign resulted in the loss of 300 ships and 13,500 men on part of the Mongols. However, Kublai had partly achieved his aim of disrupting overseas commerce (the principal source of revenue for the Southern Song) between South China and Japan (Delgado 2008: 90–9). Disorganisation of the Song–Kamakura commerce resulted in a drop of Song financial reserves, which in turn weakened the Song navy, the empire's principal bulwark of defence. Probably, this naval foray by the Mongols was a raid to test Japanese defence but not really an invasion for conquest of the island nation. What were the lessons learnt by both the combatants from the 1274 campaign? The Mongols concluded that in order to be successful, they needed to deploy more manpower on the beaches. This in turn necessitated deployment of a bigger fleet. The Kamakura regime concluded that in order to prevent another Mongol invasion or any invasion from mainland China, Japan needed to invade Korea (Suenaga 2001: 267). This would become a reality under Toyotomi Hideyoshi (b. 1537–d. 1598) during the Imjin War (1592–1598). Further, the Kamakura *Bakufu* ordered the construction of a stone wall (one may say a prelude to Hitler's Atlantic Wall along the shores of France in 1944) along the western shores of Japan to function as an obstruction against future Mongol landings (Sasaki 2015: 28).

After the collapse of the Southern Song Empire in 1279, Kublai was in a position to send a bigger fleet with a larger expeditionary force against Japan. This was possible because the Mongols had inherited the Song navy and could also use the Chinese and Korean shipbuilders for constructing more warships. Under the Yuan dynasty, South China provided the bulk of the revenue and had the largest concentration of population compared to other regions of the country. For instance, in 1330, out of 58.2 million people in Yuan China, 42.3 million lived in the coastal provinces (Jung-pang 1969: 59). People from the coastal provinces engaged in overseas trade. The Mongols utilised the maritime skills of the people of South China and Korea. In 1279, Kublai ordered the Koryo king to construct another fleet for conquering Japan. The plan was that Korea should build 1,000 vessels and provide an army of 20,000 men. Kublai also ordered his officials in China to manufacture warships. In August 1280, Kublai came up with the campaign plan. Two fleets would sail; one from Korea and another from South China. The fleet from Korea, carrying 40,000 Mongol, Korean and North Chinese troops, and the fleet from South China, carrying 100,000 South Chinese troops, were to rendezvous at Iki Island, and then the joint fleet would sail to Japan (Delgado 2008: 101–2).

Marine archaeology provides us with details about the ships of the Mongol fleet. The Mongol fleet fused two maritime traditions as regards shipbuilding: a Korean one (the Eastern fleet) and a South Chinese one (the Southern fleet). The Chinese and Korean ships (like those of South Asia) used stone anchors. The Eastern fleet used flat-bottomed vessels constructed in Korea, which were used as landing crafts to land the Mongol troops in the beaches of Japan. The Southern fleet from South China comprised V-shaped cargo vessels manufactured at Fujian province. These ships were used for carrying provisions. The Southern fleet also included rounded hull and flat-bottomed vessels constructed along the banks of the Yangtze River, which functioned as reconnaissance crafts, patrol boats and communication ships. Like most of the early medieval Indian shipbuilders, Korean shipbuilders did not use nails but relied on wooden joinery (Rajan 2012: 103; Sasaki 2015: 5–14, 142–3). However, the Chinese shipbuilders used iron nails.

In 1281, the invasion plan was set in motion. However, the two fleets sailing from different regions failed to coordinate their actions. In the absence of wireless technology, this was understandable. The fleet from Korea sailed first, on 3 May 1281, with 900 vessels and more than 30,000 troops. The Japanese islands of Tsushima and Iki were captured easily, the latter on 10 June. The impatient commanders of the fleet from Korea, without waiting for the fleet from South China, sailed from Iki and landed at Hakata Bay. This time the samurai defenders were waiting for the Mongol expeditionary force. The Kamakura regime had accurate intelligence about the second Mongol invasion. The Japanese had built extensive stone walls for protection of their beaches. The samurai took position behind the defensive wall and fought well. The Mongol invaders failed to expand the beachhead. Not only did the Mongol cavalry lacked adequate elbow room for deployment, but the horses also suffered from lack of pastures. Meanwhile, small Japanese boats attacked the stationary ships of the Mongol armada. The darts and huge stones shot by the catapults in the Mongol ships failed to destroy the nimble small Japanese vessels. In modern parlance, this attack would be similar to small and fast torpedo boats attacking stationary cruisers and battleships. During the night, the small Japanese boats came close to the Mongol ships and the crack Japanese warriors boarded the enemy vessels and killed many of the crews with long swords and *naginata* (Suenaga 2001: 119, 213, 268). Then the Japanese sent fire boats towards the Mongol ships. The latter proved to be easy targets, as they were chained together (Delgado 2008: 103–6). Soon a stalemate developed at Hakata Bay.

The fleet from South China, comprising 3,500 ships and 100,000 men, started its journey in June 1281, and after sailing for 480 miles in the open ocean, it did land troops at Imari Bay, 30 miles south of Hakata Bay. Here also the Japanese defenders were waiting for the Mongols. The fighting went on for two months, and then, on 15–16 August, the Mongol fleet was devastated by a storm near the shores of Takashima Island in Imari Bay (Delgado 2008: 106–7; Sasaki 2015: 1). After the Mongol fleet at Imari Bay was destroyed, the Korean fleet from Hakata Bay also sailed back to Korea.

Kublai decided to raise a third fleet for invading Japan, but disaffection in South China resulted in the dislocation of his plan. The Mongols would never dispatch another naval expedition against Japan. Kublai Khan's naval forays against Japan had failed. But the sheer scale of the invasions dwarf any contemporary Western amphibious expeditions like the Norman invasion of Britain in 1066. The 1281 Mongol invasion fleet comprised 4,400 ships (Sasaki 2015: vii). The Norman fleet which invaded Britain under the leadership of William I, the Duke of Normandy in 1066 had 700 ships. The Norman transport ship had oar-holes and was of clinker construction. Each such vessel had a steering oar and a single square sail. Just before William's invasion, the Norwegian fleet under Harald Hardrada comprised 300 ships which carried an army of 15,000 men from Norway to Britain. So the Norman and Scandinavian invasions of Britain were small potatoes compared to Kublai's attempts to invade Japan and Java. Even after the twin disasters in the Sea of Japan, the Yuan navy was not finished. In 1293, Kublai launched the invasion of Java with 1,000 ships (Jung-pang 1955: 493; Lewis and Runyan 1985: 114; Rodger 1997: 33; DeVries 2003: 239–42).

Since the Yuan dynasty faced no maritime threats, it had no rationale to maintain a big fleet. Moreover, the vast agricultural wealth of China provided no incentive to the Mongol rulers to go for overseas trade systematically on a large scale. The Yuan dynasty was right in taking the decision of not maintaining a big fleet, because in the end not any maritime invasion but large-scale peasant rebellion brought down the Mongols in China. And this led to the formation of the Ming Empire (1368–1644).

The first Ming emperor, Hung-Wu (r. 1368–98), used river fleets for operations on the Yangtze River. The river boats were used to fight hostile enemy crafts as well as to transport supplies from the coastal provinces for his ground forces operating inland. His son Yongle (r. 1402–24) strengthened the fleet. The navy under him comprised 3,500 vessels. Of them, 400 were warships stationed near Nanking, 400 ships were used as transport vessels and the rest were patrol boats suited for operation in the coastal areas. In 1403, the Ming fleet was able to deter the Annamese fleet from attacking the capital of Champa (Jung-pang 1958: 149–51). The Ming fleet was equipped with cannons, flame-throwing weapons and fire ships for long-distance combat. For close-quarter combat, the Ming navy also used grappling and boarding techniques (Dreyer 1974: 240).

Between 1405 and 1433, the Yongle emperor launched several maritime expeditions. His ships were commanded by Admiral Zheng He. He sailed across the South China Sea to Indonesia, Australia and India and reached the coast of East Africa. As a point of comparison, Zheng He's treasure ship was about 400 feet long, and Columbus's St Maria was only 85 feet long. The treasure ships were from 1,500 to 3,000 tons in capacity, and da Gama's ships varied between 70 and 300 tons. Zheng He's armada, which sailed across Indian Ocean in 1418, had more than 100 ships. In contrast, in 1498, Vasco da Gama with three small caravels rounded the Cape of Good Hope and sailed along the coast of East Africa and then came to South India (Finlay 1992: 227; Levathes 1996: 20–1).

Luckily for Vasco da Gama, there was no Chola or Ming fleet to confront the imprudent Portuguese mariners. After 1434, Zheng He's ocean-going fleet was decommissioned. This was due to several factors. First, the Ming economy, unlike the Elizabethan British economy, was not geared for acquiring revenue from maritime trade. There was no intimate nexus between the merchants and the strategic managers of China, as was the case in early modern Britain. The Confucian mandarins did not invest heavily in maritime trade. The Confucian mandarins looked down upon the ocean-going navy commanded by a eunuch, Zheng He. In the fourteenth century, the principal maritime threat faced by China was from the Japanese pirates along the Korean coast and in the Gulf of Tonkin. Hence, the focus of the Ming was primarily to build a coastal rather than a high seas navy. Then the rise of nomadic powers in Mongolia forced the Ming Court to pay more attention to the defence of its precarious land frontier along its northern border. In 1449, at Tu Mu, almost 200,000 soldiers of the Ming army were annihilated by the Mongols and the Ming emperor was captured (Mote 1974: 263). The threat which the Ming faced from the Central Eurasian nomads was more intense than the ground-based threats faced by the French or Spanish monarchies of early modern Europe. Britain after the Battle of Hastings (14 October 1066) rarely faced any serious threat of invasion from the continent. And none of these Western European polities were capable of mobilising 200,000 soldiers, like the Ming, for a particular campaign. After the Tu Mu debacle, the Ming Empire went on a strategic defensive and started constructing the Great Wall. Some disjointed walls to keep out the steppe nomads along the northern frontier of China had been built by the earlier regimes. But the Ming took up the task of building the wall systematically in the sixteenth century with a gusto (Waldron 1990).

With the aid of hindsight, we can say that the Ming Empire took the right decision in decommissioning its high sea fleet in the third decade of the fifteenth century. The Ming Empire was brought down not by any phantom maritime threat but by the Manchu nomads from the borders of North-East China. In the European context, also, we must not overemphasise the importance of the maritime prowess of a nation. Little Portugal, despite possessing a maritime empire in the Indian Ocean, was overwhelmed by Spanish land power.

Indian Ocean

Post-Mauryan India had both river boats and sea-going ships. Most naval wars were fought in the dry season right after the end of monsoon. There were ships with one mast each, some had two masts each, and some crafts had three masts each. The maximum number of masts for a ship was four. Ships with cabins towards their prows were used for warfare. We also hear of hundred-oared Indian galleys. Alexander, during his expedition along the Indus River in 326–25 BCE, had introduced 30-oared galleys in the Indian marinescape. Before Alexander, Darius the Great (r. 521–485) sent a fleet under Admiral Scylax which sailed down the Indus and then to Arabia and Egypt. The Achaemenids and the Greeks probably influenced Indian ship construction (Mookerji 1962: 15–21; Arrian 1971: 267; Hourani 1995: 11).

Sean McGrail has argued for a linear progression of marine technology. He writes that the earliest plank boats were fastened by stitching (individual lashings or continuous sewing). The next stage was the use of wooden plank fastenings, and the third stage involved use of iron nails for fastening the planks (McGrail 2012: 231). Iron was not used for constructing ships in medieval India. However, this was not a drawback because iron planking was primarily required for absorbing the shock of cannon fire from the deck of the ship. Ancient Indians cannot be accused of suffering from technological backwardness. They used iron for a variety of purposes. The Indians were using iron-tipped arrows from 450 BCE onwards. They were able to construct specialised ships and marine instruments. The ancient Hindu compass was known as the *macchayantra* (fish machine). It comprised an iron fish that floated in a vessel of oil and pointed to the north (Mookerji 1962: 33, 66). In fact, before the advent of copper sheathing for the hulls of ships, iron planking was vulnerable to rust. Rather, Indians used resin for caulking the ships to increase their longevity. Further, the West Indians continued to use sewn-plank boats even when this technology became outdated in other parts of Eurasia. This was because sewn boats were suitable for the surf and sea conditions of the Western Indian Ocean (Hourani 1995: 96; Ray 2012: 6). Even the Arabs used coconut ropes for sewing the planks of the hulls of the ships rather than using iron nails. Sewn (or stitched) hulls, compared to nailed ships, were less easily broken when the ships accidentally ran onto the coral reefs of the Red Sea and elsewhere. So the use of a particular piece of technology was dependent on local geographical conditions. There is no universal parameter for judging the progress of technology. To sum up, marine geography to a great extent explains the divergence in shipbuilding traditions of the Indian Ocean and the Atlantic.

The Indian ships taking advantage of the monsoon wind sailed from the west coast of India to Arabia, Egypt and East Africa. Ships from Egypt left for India in July and, taking advantage of the south-west monsoon winds sailing through the Gulf of Aden, reached the west coast of India in September and October (Bopearachchi 2012: 69). During the second and third century CE, the Andhras (who ruled the Krishna-Guntur region) maintained a navy along the Coromandel Coast. They carried out maritime trade with West Asia and with South-East Asia. Each ship had two masts and spoon-shaped oars. The deck of the ship was either curved or straight. The Andhra ships had quarter paddles for steering and square-rigged sails. The evidence from Ajanta paintings which could be dated to the sixth century CE portray a seagoing vessel with a high stern and three oblong sails attached to the masts. Each mast was surmounted by a truck and carried a lug sail. The ship had ports and steering oars hung in the sockets. One can argue that there was a gradual evolution from the square-rigged sail (second century CE) to a lug sail (sixth century CE) to fore and aft sails carried on the raked masts during the fourteenth century. The men on the deck of the ships fought with bows. There is no evidence that such ships carried catapults and rams like the Greek and Roman warships. How can these lacunae among the Indian ships be explained? Once rams or heavy catapults were placed on the ships, the vessels became unseaworthy and could sail only with difficulty, hugging the coastline. But the Indian ships, taking advantage of the monsoon wind, sailed across the high sea – the western Indian Ocean – far away from the coastline. Catapult-carrying ships were introduced in the Indian Ocean by the Umayyad Arabs during their invasion of Sind in 712 CE. These ships had sailed from the Gulf of Aden along the Makran coastline up to the coast of Sind. The Arabs learnt the use of placing catapult on the decks of the ships from the Byzantine navy. Top heavy catapult carrying ships were suitable for the inland Mediterranean Sea and for riverine warfare but not for sailing across the turbulent open seas. During the seventh century CE, the mercantile community in Gujarat established a maritime connection with Java. Ships with outriggers, each of which were about 60 feet in length and 15 feet wide, regularly plied between these two regions (Mookerji 1962: 28–36; Deloche 2012: 201–12). Probably such ships were used as transport vessels during wartime. Both the Chinese and the Indians learnt the use of outriggers from the South-East Asians.

In the tenth century CE, the Cholas of South India established a maritime cum land empire. Between the ninth and thirteenth century, the Chola dynasty was able to establish hegemony over South India from Thanjavur. The Cholas were able to project power beyond India because they had an efficient navy and a powerful army. The king was the head of both the army and navy. The army was organised in different regiments. Each such unit had its own distinctive name. Each regiment had its own cantonment, known as a *kadagam*. The men were equipped with bows and swords. The Chola army maintained a strong elephant corps. Soldiers placed on *howdahs* on the back of the elephant fought the hostile troops. The military personnel of the Chola army, like the Maurya army, were of four types: mercenaries, soldiers provided by hereditary military families, personnel provided by guilds of merchants and traders and finally soldiers raised from the various tribes (Sastri 1992b: 745–7).

Chola Emperor Rajaraja I ascended the throne in 985 and immediately started building a navy. In a sea battle, Rajaraja destroyed the Chera navy of King Bhaskara Ravi Varman at Kandalur. Then the Chola fleet attacked the Chera fortress port named Vilnam (Sastri 1992a: 2–3). After establishing control over the Coromandel and Malabar coasts, the Cholas turned their attention to the Arabian Sea. The Chola navy captured the Lakshadweep-Maldives Archipelago as they sat astride the trade routes between South India-West Asia and East Africa (Sakhuja and Sakhuja 2010: 76).

After establishing maritime dominance along the coastline of South India, the Cholas turned their attention towards Ceylon (Sri Lanka). The Cholas were interested in Sri Lanka for several reasons. The Cholas had commercial interest on both the Coromandel and Malabar Coasts, and sailing from the Malabar Coast to the Coromandel Coast and vice versa was much easier and quicker if one controlled the Palk Strait and the Gulf of Mannar, as the Palk Strait and the Gulf of Mannar separate Northern Sri Lanka from the southern tip of the Indian subcontinent. Furthermore, military mercenaries from Tamil Nadu had served with the rulers of Northern Sri Lanka from the second century CE. They frequently rebelled and received support from the South Indian monarchs. Even before the Cholas, powerful kingdoms of South India occasionally intervened in Sri Lanka and established their client rulers. The objective was to secure indirect control over Sri Lanka, and especially over the water bodies surrounding this strategically located island. For instance, in 642 CE, the Pallava monarch sent two naval expeditions to Sri Lanka. The second expedition was successful. As a result, the Pallava client ruler Manavarma was able to kill the monarch of Sri Lanka and captured this island's capital, Anuradhapura (Sastri 1958: 146).

An internal uprising in Sri Lanka was an opportunity for the Cholas to launch an amphibious expedition. We do not have the exact date of the Chola expedition to Sri Lanka. After the accession of King Mahinda V in 981 CE in Sri Lanka, there was a military rebellion against him. The rebellion was led by the Tamil mercenaries in Ceylonese pay. Mahinda had to take refuge in Rohana, which was the hilly region in South-East Sri Lanka. Taking advantage of this, the Cholas launched an amphibious invasion. The Cholas sacked the capital, Anuradhapura, and captured Northern Sri Lanka. Northern Sri Lanka became a Chola province, and a new provincial capital was established at Polonnaruva. This city was garrisoned by the Chola troops in order to maintain order in this part of Sri Lanka and also to deter any attempt of reconquest by the Ceylonese kings from Rohana (Sastri 1992a: 3). Trincomalee, situated near the East Coast of Sri Lanka, was an important harbour (Kulke 2010a: xix). It became the principal port for the Chola naval detachments deployed in Sri Lankan waters.

Besides launching several pillaging raids against Sri Lanka, the Cholas launched two amphibious expeditions against the Kingdom of Sri Vijaya/Kadaram in Southern Thailand. The main expedition came in 1025 CE, with a smaller one in 1070 CE. Most likely, Rajendra Chola (r. 1012–44) launched a minor expedition against Sri Vijaya as early as 1017 CE (Kulke 2010a: xiv–xvi). India imported porcelain and silk from South China (Sen 2010: 62). China and India had extensive overseas trade relations. In fact, a sizeable Tamil community was present in the coastal cities of South China (Karashima 2010: 57). In 1016, a Chola embassy was sent to China (Sastri 1958: 176). The Sunda and the Malacca straits are geographical chokepoints, as all the ships sailing between Arabia and India to China even now have to sail through these narrow water bodies. This is because these two straits offer the shortest and safest sea route connecting the Far East with the Indian Ocean. By using its navy to control the Sunda and Malacca straits, Sri Vijaya forced the foreign ships passing between China and India to stop at their sea ports and pay hefty taxes. Sri Vijaya's ulterior motive was to influence the Song Empire's maritime commerce with Indonesia and South Asia (Sen 2010: 66). The Chola naval raids were probably in retaliation for aggressive Sri Vijayan naval activities along the Malaccan Straits (Kulke 2010b: 2). The Nicobar Islands was used by the Chola fleet as a base for rest and recuperation during their long voyage to South-East Asia (Sakhuja and Sakhuja 2010: 81). Even today, the Indian navy maintains a naval base in the Andaman-Nicobar Island group in order to maintain its long maritime reach in the Bay of Bengal and the Eastern Indian Ocean.

A Chola inscription tells us that in 1025, Rajendra Chola sent an amphibious expedition against Kadaram. The Sri Vijayan ruler was captured, and the Cholas put one of their native clients on the throne (Sastri 1958: 181). This assured that Sri Vijaya's foreign policy would be conducive to Chola maritime commerce. Then the victorious Chola naval detachment raided 12 other port cities in the Malayan Peninsula and Sumatra, and a large amount of treasure was collected (Kulke 2010b: 1). The Chola naval forays into Sri Vijaya were a classic case of power projection across the ocean in the pre-modern era.

Power projection across the seas and ocean required large ships which carried troops, horses and elephants. Bengal built large passenger ships that could carry 800 passengers each. We could surmise that such ships were also used as troop transporters. Such vessels were able to sail from Bengal to Sri Lanka. These ships were constructed with planks sewn together with ropes, and each such craft had a well-rigged lofty mast with a spacious sail. Marco Polo, the Venetian traveller, wrote about the big merchant ships of India in his account. Some of them might have been used by the Cholas for transporting troops. Marco Polo describes:

> ships employed by the merchants which are built of fir-timber. They have a single deck, and below this space is divided into about sixty small cabins. . . . They have four masts, with as many sails. . . . Some ships of the larger class have, besides (the cabins), to the number of thirteen bulk-heads . . . formed of thick planks. . . . The object of these is to guard against accidents.
>
> *(quoted from* Marco Polo *2009: 321–2)*

Both the Chinese and Indian ships used the bulkhead principle for making the ships seaworthy, a technique which was then unknown in the Mediterranean and in the seaboard of the Eastern Atlantic.

Besides soldiers, the Chola fleet also transported many elephants. Elephants were used as battering rams in battles and sieges by the rulers of both precolonial South and South-East Asia. The point to be noted is that medieval European galleys were able to transport horses during the Norman invasion of Britain and during the Crusades. However, elephants were bulkier than horses, so transporting war elephants was quite a feat on the part of the Chola transport ships. We have some evidence of horse transport ships used by the South Indian powers.

The stem and stern of the boat were raked, and the bow was at the same level as the stern. The freeboard was heightened by a bamboo or mat screen. Besides rowers, the ships used quadrangular sails. Some of the transport ships had elevated poop decks for giving the helmsmen a clear view of the front. Each ship had one groom for looking after five horses (Deloche 2009: 551–67).

However, the Cholas could not focus on the maintenance of an ocean-going navy due to presence of existential threats along their northern land frontier. The Western Chalukyas ruling Western Deccan, supported by the Pandyas and the Eastern Gangas of Kalinga (Orissa), attacked the Cholas over the control of the fertile Vengi region. Further, for the Cholas, land revenue remained the principal source of royal income. The Chola political economy was not totally mercantile in nature. So they could not afford to neglect their ground force for maintaining their territorial empire in South India and Deccan. In fact, even today, India continues to follow a continental grand strategy. And among its military services, the navy's budget is lowest compared to its two sister services. In the thirteenth century, a new power known as the Hoysalas, with a strong cavalry force, emerged to threaten the north-western border of the Cholas. Previously, the Cholas were fighting South Indian kingdoms whose armies comprised mainly elephants and infantry. South India was poor in horses. The Hoysala cavalry was dangerous. More deadly were the Islamic mounted archers in Orissa, who threatened the north-eastern frontier of the Cholas (Sastri 1958: 174–213). The Cholas, like most of the Indian subcontinent's Hindu powers, had no counter to the mounted archers. In 1279 CE, the Cholas disintegrated.

After the Cholas, a prince of Kalinga named Magha invaded Sri Lanka in 1214 CE with 24,000 soldiers. Kalinga was indeed a sea power of some repute. Kalidasa, the fifth-century Gupta-era Hindu intellectual, commented that Kalinga ships dominated the Bay of Bengal, so much so that the Bay of Bengal was known as the Kalinga Sea. Kalinga had commercial relations with Sri Lanka. The ships of Kalinga imported pearl and silver from Sri Lanka. From 300 CE onwards, Kalinga established trade connections with Myanmar. The Buddhist merchants from Kalinga established colonies on the coast of the Gulf of Martaban. From the first century CE, Kalinga ships visited Java for commercial purpose. This journey took an average of 90 days. The mariners sailed by observing the position of the sun, moon and stars. Chinese ships came to Kalinga as well as Tamralipti (a port in West Bengal in Midnapur district) for trade during the seventh century CE. Kalinga had commercial connections with China and Arabia. And Kalinga's ships during their western voyages had to visit the ports of Sri Lanka for replenishing. On one occasion, Arab traders in a Kalinga ship transported a giraffe from Africa to this country. So the large transport ships of Kalinga were able to carry horses and elephants in addition to soldiers during naval campaigns (Mishra 2010: 14–46). All these factors enabled Kalinga to establish its maritime hegemony.

The ships used by Kalinga had both rowers and sails. Some of the ships were narrow and heavy, and outriggers were attached for safety. An outrigger was a series of planks or logs joined to the vessel with long poles or spars. Some ships had two masts and had oars at the stern for steering. By the thirteenth century, the ships were double planked and were fastened with wooden nails. The bottoms of the ships were smeared with quick lime and hemp to prevent the growth of sea barnacles. The crew of the ships varied from 60 to 300 men for each vessel. The oars of the big ships were huge, and four men were required for operating each oar (Mishra 2010: 17, 19–20, 58–60). As a point of comparison, in 1417, King Henry V of England (r. 1413–1422) hired cogs from the Low Countries to transport his army to France. The cogs were of 30–50 tons each and the crew of each such vessel comprised merely one master and six mariners (Friel 2008a: 18). Kalinga's naval power declined due to the ground-based threat posed by its neighbours. In the fourteenth century, the Eastern Gangas of Kalinga declined due to an amalgam of external and internal threats. The attack by the Delhi Sultanate's Turkish horse archers combined with internal rebellion by the nayaka chieftains brought about the downfall of the Eastern Ganga dynasty.

Indian ships underwent technological upgrades with the passage of time. For instance, both in the Mediterranean and in the Indian Ocean, the stern post rudder was adopted in the twelfth century (Ray 2012: 6). Again, the Indians also focused on maritime theory. For instance, a naval manual known as *Rahmani* provides nautical details for the mariners who sailed from the Malabar Coast through the Lakshadweep Islands to West Asia and East Africa (*Rahmani* 2004).

Nevertheless, none of the Indian powers had any strong navy when Vasco da Gama reached South India in 1498 CE. King Zamorin of Calicut had a coastal navy, and he was a minor ruler. Why did the big polities of the subcontinent lack an ocean-going navy in the fifteenth century despite the presence of a vibrant maritime tradition in ancient and medieval India? This was because the Rajput principalities were facing threats from the Delhi Sultanate established by the Central Asian Turks and the Afghans. While the Rajputs had lancers and elephants, the core of the Delhi Sultanate's force comprised horse archers. The Portuguese were a minor pinprick for the Delhi Sultanate. More deadly were the Chagatai Turks, who were threatening North-West India and would establish the Mughal Empire in 1526. The rulers of medieval India could not be faulted for not maintaining an ocean-going navy because none of them were destroyed by maritime threat. Rather, invasions by hostile ground forces across the north-west frontier of India resulted in the rise and fall of empires in medieval India.

The Mediterranean: Islam versus Christendom

The characteristic feature of combat in the Mediterranean Sea during the medieval and the early modern era was the confrontation between the Islamic and Christian powers. After the fall of the Western Roman Empire, the Mediterranean became a sort of 'Byzantine Lake'. Initially, the Umayyad and then the Fatimid Caliphate challenged Byzantine naval supremacy in the Mediterranean. After the fall of Constantinople in 1453 CE, the Ottoman navy made a bid for dominating this inland sea. The Ottoman naval tide was contained at the Battle of Lepanto (7 October 1571) and then slowly but steadily pushed back. The Mediterranean maritime culture diverged from the Far Eastern maritime tradition. However, there were some commonalities between the Indian Ocean and Mediterranean maritime traditions. In the Mediterranean maritime space, the Phoenician–Greco-Roman naval tradition was updated and adopted by the Islamic polities and the Byzantine Empire.

Historian John H. Pryor asserts that the geographical constant structured the nature of maritime conflicts in the Mediterranean. The galley remained the prime vessel in the Mediterranean for both commercial and military purpose till the fifteenth century. During this period, the technology of navigation and shipbuilding did not register significant improvement. It was extremely difficult for a galley to sail against the flow of wind for a long time. In the Mediterranean, the wind blows from north-west to north-east. Hence it was easy for Christian ships from Spain and Italy to sail from the Western to Eastern Mediterranean compared to Muslim ships sailing from the Anatolian and Egyptian coast towards the Western Mediterranean. And galleys used for maritime warfare in the Mediterranean were coast-hugging vessels. Since the galleys lacked space for the large-scale storage of food and water for the crew, these vessels had to make landfall frequently. Most of the islands which had adequate population and could produce surplus that can be used to provision the galleys are situated along the northern part of the Mediterranean. Hence, the trunk routes for shipping (both commercial and naval) lay close to the northern shoreline of the Mediterranean. So the Christian powers, which were located along the northern part of the Mediterranean, enjoyed a tremendous geographical advantage over their Muslim opponents, as the Islamic polities were located along the eastern and southern seaboards of the Mediterranean. Thus, naval fleets of the Christian powers could easily control the trunk routes, which lay within their easy reach. Islamic ships sailing along the trunk routes fell prey to the Christian ships, which acquired logistical support from the bases in the islands located along the trunk routes and also from the ports and harbours situated along the shoreline of the Christian states like Spain, France and Italy. If Islamic commercial ships tried to avoid the trunk routes, then the journey became uneconomic due to a longer sailing time. Further, sailing along the Southern Mediterranean coast was dangerous. In contrast to the Christian fleets hunting Islamic shipping sailing along the trunk routes, Islamic corsairs from Algeria and Tunisia had to sail for a longer time to reach the trunk routes to prey upon Christian shipping, and Islamic fleets of Egypt could not seriously threaten the trunk routes because they lacked bases. The bottom line is that the shoreline of the Southern Mediterranean where the Islamic maritime bases were located were far away from the trunk routes, and there are no islands with substantial population generating a large surplus along the southern half of the Mediterranean which could provision the Islamic fleets. End result: Islam could not dominate the Mediterranean; geography saw to it (Pryor 1992).

Geography is definitely important in military matters. But we cannot write out the non-geographical factors while dealing with a complex phenomenon like warfare. Besides geography, certain other factors such as command and control of the naval forces, weapons technology and the strategic threat faced by a polity, etc., also shaped the dynamics of naval warfare. Let us trace the trajectory of the struggle between Islamic and Christian forces before the advent of the early modern era.

In the fifth century CE, naval and land forces of the Western Roman Empire collapsed, and the Western Mediterranean ceased to be a 'Roman Lake'. The Vandals crossed from Spain into North Africa. The Vandals occupied Carthage in 439 CE and established their kingdom in Northern Tunisia. From Carthage, they acquired shipyards and people experienced in constructing and manning ships (Pitassi 2012: 157). The loss of African wheat and revenues further undermined the fiscal-military structure of the disintegrating Roman state. In 466 CE, the Roman fleet of Flavius Basilikos, when anchored at Cape Bon, was destroyed by Vandal fire ships (Pryor and Jeffreys 2006: 9).

Despite being nomads from the desert of the Arabian Peninsula, the Arabs, after establishing themselves in West Asia, adopted sea power as an instrument of their strategy. The shipwrights and other craftsmen of the captured territories of Syria and Egypt were pressed into service by the Arabs for creating a navy for the caliph. The capture of Syria was important for establishing naval dominance in the Eastern Mediterranean because the Lebanon forests provided timber. It was the source of timber for the ancient Egyptians, Phoenicians and the Romans too. The two important Byzantine dockyards in Egypt were Alexandria and Clysma, and the Byzantine naval base at Clysma in Egypt was perilously close to Medina. Further, Clysma was the conduit for maritime trade through the Red Sea. Moreover, Egypt was the source of wheat for the Byzantine capital (Fahmy 1966: 69–72). So Egypt and Syria became targets of Arab attacks.

The Byzantine position in Egypt collapsed in 642 CE. The Arabs captured Alexandria and the coastal towns and established their rule along the Nile Delta. Muslim rule was established up to the First Cataract of the Nile. Initially, the caliphs were hesitant about engaging in naval conflicts with the Christians. In fact, Caliph Umar (r. 634–44 CE) actually forbade Muawiya from engaging in naval battles. Umar stressed the lack of experience of the Arabs in seafaring activities. Umar never considered Alexandria to be the capital of Egypt, though this city had functioned as the provincial capital under the Ptolemaic, Roman and Byzantine empires. The reason was that the Arabs at this stage were not confident of controlling the sea lanes of the Eastern Mediterranean. So Alexandria did not become the provincial capital of Muslim Egypt under the caliphate. Actually, two naval disasters temporarily reduced the confidence of the Arabs in conducting war on the sea. In 638, Al-Hadrami, the Governor of Bahrain, led an amphibious expedition against Persia. He crossed the Persian Gulf and landed his troops for advancing to Persepolis. However, the Persians attacked and cut off his troops from the ships. Al-Hadrami's force was saved when Umar sent a relief army. Again in 641 CE, Umar sent Mujazziz with a naval detachment across the Red Sea to ward off attacks upon the Muslims on the Abyssinian coast. All the ships of this expedition were wrecked. Instead of engaging in naval confrontations, Umar encouraged peaceful navigation. Probably it was the prelude to allowing the Islamic mariners to gather experience of seamanship before establishing themselves as a naval power. The caliph ordered ships to be built in Egypt for the transport of grain from Alexandria to Medina. After capturing Egypt, the caliphate with the aid of forced labour maintained the ancient canal which linked the Nile with the Red Sea (Fahmy 1966: 71, 73–6). This waterway was the forerunner of the Suez Canal. This canal probably allowed the caliphate to transfer maritime assets from the Eastern Mediterranean to the Red Sea, Persian Gulf and Indian Ocean as per the shifting strategic requirements.

The Byzantine rulers made repeated attempts to recover Egypt. In 645, Emperor Constans II ordered construction of a fleet of 300 ships for recovering Alexandria. The Byzantine fleet, carrying marines under the eunuch Manuel, appeared before Alexandria. Eunuchs enjoyed high position in the admiralties of both the Byzantine and Ming empires. Manuel's force captured the town and raided the villages along the Nile Delta. The Arab position in Egypt seemed to be collapsing. Caliph Uthman (r. 644–56 CE) ordered Amar to re-establish control over Egypt. Alexandria was reconquered, and during the course of the attack, the Arabs razed the strong walls of this city. The bulk of the Byzantine land force, including their commander Manuel perished along with most of the ships. Only a small number of Byzantine soldiers were able to reach the few remaining ships, and they were able to pull out of Egypt (Fahmy 1966: 77–8).

Muawiya once again pressed Uthman to expand the scope and objectives of the caliphate's maritime and naval activities. Uthman agreed but ordered that no Muslim should be conscripted for naval duties. Marine service was strictly voluntary (Fahmy 1966: 78–9). This was probably because the Arabs, being landlubbers, had neither interest nor technical skill in maritime affairs. The caliph's navy depended mainly on the captured prisoners of war (especially those who were Christians from the erstwhile territories of the Byzantine Empire). Only the senior officers of the navy were Muslim grandees. After capturing Syria and Egypt, the next step for consolidating maritime control over the Eastern Mediterranean was to capture the islands between Greece and Anatolia. In 648–49, a joint expedition by ships from Egypt and Syria was launched by Muawiya and Abdullah, the Governor of Egypt. The target was Cyprus. This island used to pay an annual tribute of 7,200 pieces of gold to the Byzantine Empire. Cyprus was thoroughly pillaged, but the caliphate's fleet retreated when it heard that a large Byzantine relief armada was approaching (Fahmy 1966: 81–2).

Between 643 and 752, the Umayyad Caliphate had launched more than 50 major naval expeditions in the Mediterranean. Its successor, the Abbasid Caliphate (750–1258), did not launch major naval expeditions in the 'Sea of the Romans'. However, between 779 and 838, the caliphate's fleet conducted amphibious raids along the coastline of Asia Minor and provided logistical support to the Islamic ground forces operating in Anatolia. However, the Iraqi caliphs' navy did not engage in any fleet actions against the Christians (Picard 2018: 70–1).

After the collapse of the Umayyad Caliphate (661–750), the Fatimid Caliphate in Egypt (909–1171) carried on the naval struggle with the Crusaders. During the Crusader invasions, the Fatimid fleet was stationed in a number of major ports in Palestine, the Red Sea and Egypt. The principal ports of Egypt were Cairo, Alexandria and Damietta. The strength of the Fatimid fleet came to about 90 galleys and 10 large transports. Egypt had a large mercantile fleet of private vessels which the sultans commandeered during war for use as transport and supply ships (Hamblin 2008: 215). In 1123, the Fatimid fleet, which was at the roadstead of Ashkelon, was defeated by a Venetian squadron under Domenico Michiel. He divided his ships into two groups. The heaviest crafts were deployed in the van. Some of the largest vessels were big galleys which carried mangonels or ballistas on their bows. Like the Song ships, the ballistas were used to hurl stones and pots of flaming liquids (Greek fire) towards the enemy ships. The Greek fire was a napalm-like substance that burnt in water. It was developed towards the end of the seventh century CE. The Byzantine biremes and dromons were fitted with machinery which threw Greek fire from the bows of the ships. However, the efficacy of this weapon depended on the wind current. Still, one could argue that the Greek fire gave the Byzantine navy a thin margin of superiority over their Muslim opponents. The Venetians attacked at dawn. The surprised Fatimid fleet was assaulted both at the front and rear simultaneously. The Fatimids were defeated completely, and the ship of their admiral was rammed and sunk by Michiel's galley (Rose 2005: 36–7; Roland 2008: 419–43). The fall of Ashkelon in 1153 resulted in the loss of the last Muslim coastal town in Syria. At that time, the Egyptian fleet comprised some 70 ships (Ehrenkreutz 2008: 237).

Saladin came to power in Egypt during March 1169 CE. He earnestly started building a fleet in 1177. A Fatimid galley had 140 oarsmen besides a strong detachment of marines. There were special fire boats which had catapults that threw combustible materials (Greek fire) (Ehrenkreutz 2008: 241). Probably the Fatimids copied the technique of Greek fire from their Byzantine opponents despite the fact that the latter had taken stringent steps to guard this technological marvel. By 1179, Saladin had 60 galleys and 20 transport ships for use against the Franks. Saladin deployed ships in the Eastern Mediterranean as well as in the Gulf of Suez and in the Red Sea. He also fortified Damietta and Suez. Saladin's Fatimid fleet suffered from two serious shortcomings. One, his conscripted crew was inexperienced in naval combat and had no enthusiasm for marine service. Second, the Muslims lacked crossbows used by the Venetians, Genoese and the Aragonese. The light composite bows used by the Muslims were suited for cavalry battles on land but not for long-distance archery required in naval battles (Rose 2005: 38–41).

The Fatimids in Egypt were followed by the Ayyubids (1169–1260). The naval record of the Ayyubids was worse than their predecessors. Besides some occasional spurts of naval activities, the Ayyubids never followed a consistent long-term naval strategy. This was because the military ethos of the mamluks was focused on horses and bows and not on nautical activities. We will see that there was also an additional geopolitical factor which discouraged the Ayyubid government and its successor regime from concentrating on building a navy. In 1182, the Ayyubid ruler Saladin laid siege to Beirut by land and sea. He had 30 ships, but when 33 Christian ships came, the Ayyubid fleet fled without offering battle. In 1187, the same sultan blockaded Tyre with 10 ships. However, a nightly raid by the besiegers resulted in the capture of five ships. The crew of the remaining five ships deserted (Ayalon 2008: 226, 228). In the early thirteenth century, Ala al Din Kayquabad in Alanya on the south-west coast of Anatolia constructed a shipbuilding facility in order to counteract the Crusaders' naval capabilities. This facility had ship sheds for five galleys, and the entrance was fortified (Rose 2005: 7).

In the fourteenth century, the absence of a standing navy on part of the Mamluk Sultanate of Egypt (1250–1517) made the coastal cities of Syria–Palestine susceptible to raids by the Frankish corsairs. After the expulsion of the Crusaders from the Syria–Palestine coast in 1291, the principal aim of the Mamluk Sultanate was to prevent their re-entry in that region. So the Egyptian Sultanate followed a scorched-earth policy along the coastline and destroyed all the harbours in that region. Their main line of resistance was further inland, and the defence was centred around their cavalry army. One could argue that the Mamluk Sultanate had gained a victory of sorts by throwing out the Crusaders from the Syria–Palestine region. They could not afford heavy expenditure on a navy due to the dire threat posed by the Mongols. The bulk of the defence spending went to the army to guard against incursions by the Ilkhanate Mongols of Persia. The Crusaders, unlike the Mongols, had no plan for capturing Egypt. So the mamluks rightly concentrated on maintaining a cavalry force and building fortresses against the Mongol threat rather than for seaward defence (Fuess 2008: 267–9).

Ecological limitations hampered Mongol military operations in West Asia. Nevertheless, they posed a deadly threat to the fledgling Mamluk Sultanate of Egypt. On 3 September 1260, at the Battle of Ayn Jalut, the mamluks of Egypt defeated the invading Mongol force. We can argue that logistics and the prospect of civil war in the Mongol Empire were responsible for the Mongols' defeat. The inadequate pastures of Syria prevented the Mongols from concentrating a huge invasion force for invading Egypt. Further, Halaku Khan (r. 1260–65), founder of the IlKhanate, with his main army, was distracted due to the threat posed by other Mongol *ulus*/appanages, especially the danger posed by the Golden Horde, which wanted to control the Caspian pastures. Halaku Khan mobilised six *tumens* or roughly 60,000 men for the Syrian campaign in September 1259. As a point of comparison, for the 1257–58 campaign against Baghdad, 15 *tumens* (150,000 men-ration strength) were deployed. In comparison, Mongke and Kublai deployed four times more numerous cavalry during their campaigns against the Song Empire. In August 1259, Halaku was gone with the main army to confront the forces of the Chagatai Khanate, which was allied with the Golden Horde, and his commander Kedbuqa was left with 10,000 men in Syria. Baidar, another Mongol commander, locally recruited 20,000 men. However, there was a high probability that Halaku might return to Syria with the main army. Theoretically, the Egyptian mamluk state could muster 24,000 cavalry. Of them, 14,000 were crack cavalrymen (Smith, Jr. 1984: 307–13). In January 1260, Halaku's army descended on Aleppo, and after six days of concentrated fire from the mangonels, the defence of this city cracked. The very possibility that Halaku and his successor Ghazan (r. 1295–1304) could return to Syria with the IlKhanate's Mongol army prevented the Mamluk Sultanate from building a navy for controlling the Mediterranean against the Franks. The Mongol threat was no mere phantom menace. In 1299, the Mongols under Gazan defeated the Mamluk army of Egypt at the Third Battle of Homs. Furthermore, logistical difficulties also hobbled Mamluk Egypt's shipbuilding programme. Medieval Egypt lacked an adequate supply of wood, iron and copper (Hess 1970: 1907; Amitai-Preiss 1995: 223).

Strictly speaking, piracy was not an illegal profession in the medieval world. There was collusion between the pirates, merchants and rulers. At times, the pirates performed sterling services for the rulers. Gudericus, an English pirate, brought King Baldwin I through an Egyptian blockade to Jaffa in 1102 (Unger 2006: 255). Flemish merchants with tacit support from their government preyed on the English merchants. King Edward III (r. 1327–1377) patronised several pirates. And some of the pirates, like John Crabbe, were also merchants by profession. After 1339, Crabbe even served in the Royal Navy (Lucas 2008: 77–93). Henry IV (r. 1399–1413) of England was unwilling to take action against piracy, because during a national emergency pirates provided ships and experienced crews for combating enemy fleets. Realising the monarchs' inability as well as unwillingness to control them, the English pirates engaged in massive plundering of the Flemish merchants, and such piratical activities in turn jeopardised the peaceful relations between England and Burgundy (Pistono 2008: 95–103).

The galley was the principal warship in the Mediterranean. It was a long narrow-beamed vessel which had sails but was mainly driven by oars. The principal battleship for the Byzantine Empire was the dromon, which emerged during the fifth and sixth centuries CE. In the contemporary sources, this ship was described as the 30-oared galley. The dromon was faster than the *liburnae*, the principal Roman warships during the period of the Late Roman Empire. The *liburnae* were developed during the first century BCE from the ship of the Illyrians known as Liburni. In the dromon, the square sail was replaced with a lateen sail. Moreover, the waterline ram of the Classical era was replaced with an above-water spur. The Greek trireme had a straight stempost with a waterline ram that extended from it. In the Hellenic era, the stempost became recurved towards the stern and the ram became three-finned. The bow of the galley of the Roman Republic was modelled on the Hellenistic galley. During the first century CE, the Romans replaced the three-finned ram with a single pointed one, and some of the galleys also acquired forecastles. Many galleys of Roman Emperor Trajan's (r. 98–117 CE) navy had rams which curved upwards from the keel below to a point above the waterline. Such a curved ram was actually not a ram but a spur. The spur was held from the stempost with couplings. The advantage of the upward pointed spur was that it reduced water resistance and increased the speed of the galley (Pryor and Jeffreys 2006: 37–9, 123–37).

According to the sixth-century Byzantine historian Procopius, the early dromon was a monoreme galley with a deck to cover the oarsmen beneath it. The *liburnae* were partly modelled on the Hellenistic galley. And the *liburna*, like the Hellenistic galley, had a lateral part deck. As described by Procopius, the dromons during the reign of Emperor Justinian I were smaller than the Greco-Roman bireme and trireme predecessors. It was partly because the Roman Emperor Constantine I (r. 306–37 CE) defeated Licinius at the naval Battle of Dardanelles (324 CE). Constantine had 200 ships against Licinus's 350 ships collected from Egypt and Phoenicia. Constantine's fleet comprised monoremes, which were smaller than biremes and triremes of the opponent's battlefleet (Pryor and Jeffreys 2006: 7, 128–30). Thus, the smaller ships of Constantine's victorious fleet became the model for the Byzantine rulers. The operating range of a medieval galley in the Mediterranean, like the Classical Greek galley, was about 240 km from its base (Rodger 2008: 160).

By the twelfth century, the dromon in the Byzantine navy was replaced with the galea, a bireme galley developed from ideas and techniques imported from the Latin West. During the twelfth century, the Genoese also used a bireme galley (also known as galea) with two superimposed banks of oars, both rowed through the oar-ports, one above the deck and the other below it (Pryor and Jeffrey 2006: 423–30). The galleys during the thirteenth century had 25 to 28 benches per side with two oarsmen on each bench. So a galley had from 100 to 112 oarsmen. At the end of the thirteenth century, a third oarsman was added to each bench. This in turn led to the restructuring of the hull, which became slightly beamier. The result was 150 oarsmen on a 25-bench galley (Dotson 2006: 65).

Historian Susan Rose asserts that the medieval galley fleets were unable to manoeuvre as a unit. What happened in reality was a series of confused individual combat of a galley versus a galley as in the case of Greco-Persian naval conflicts. In such fluid encounters, the experienced crew had an advantage. The naval battles started with a shower of arrows shot by various engines and bowmen on board the ships. Then the galleys tried to ram each other and the marines boarded the enemy galleys. The Catalan fleet had skilled crossbowmen and light infantry who wore leather jackets. Equipped with javelins, lances and daggers, they engaged in close combat with the enemy crew on the decks of the ships (Rose 2005: 47–9).

The Crusader armies required support ships for logistical purpose. The nave was an Italian ship used in the late medieval era. It was a broad-beamed cargo vessel which carried triangular lateen sails on two to three masts (Dotson 2006: 65). The naves were used for transporting horses. Each nave was able to carry 40 warhorses, 40 knights and foot soldiers and 15 seamen, plus stores for all of them. Each horse required 5 kg of grain, 5 kg of hay and 32 litres of water. The total weight came to about 50 kg per day or 10 times that needed by a single person. The crew and the animals consumed about 2.2 tons of supplies each day.

A nave was able to carry 10 days' weight of food for all these animals and men (Unger 2006: 262). These transport ships were much smaller than those used by the Song navy, Chola navy and Kalinga Kingdom of medieval India. During the Third Crusade (1189–92), there were horse transport galleys with hull ports at the stern known as *taride/chelandre/salandria*. During the night, the convoys sailed with the aid of lanterns (Gertwagen 2006: 95–8). From the twelfth century onwards, English ships while sailing during the night used lanterns and candles (Rose 2008b: 75).

The Atlantic

The diverging shipbuilding traditions of the Mediterranean and the Atlantic was evident from Roman times. In 56 BCE, the Roman legions under Julius Caesar advanced to the coast of Southern Brittany. Here, the Celtic Veneti, who were a seafaring race, had invented sturdy crafts suitable for plying along the stormy Atlantic coast. In contrast, the Roman triremes were suited for the calmer Mediterranean waters. The prows and sterns of the Gallic ships were higher compared to the Roman ships. The hulls were made of oak to withstand the violent shock of the sea waves of stormy Atlantic, and the cross-beams of timbers were a foot thick and were fastened with iron bolts which were as thick as a man's thumb. They used sails made of hide or soft leather, as the use of flax was not known widely. Moreover, the Gallic tribesmen believed that cloth sails would not be able to withstand the force of the violent Atlantic gales or be suitable to steer the heavy ships. Overall, the Gallic ships were better suited to withstand the storms and high waves of the Atlantic and the English Channel compared to the Roman vessels, which were suitable for the calmer inland Mediterranean Sea. Regarding coastal navigation also, the Gallic ships had an advantage over their Roman counterparts. The keel of the ship was flat compared to a Roman ship. As a result, the Gallic ship was able to traverse the shoals and shallow water of the inlets better during a low tide (Pitassi 2012: 153).

Caesar brought Roman warships from the Mediterranean and also manufactured several more in the Roman-controlled Atlantic ports. The tribes of Gaul concentrated 220 ships. The traditional Roman naval tactics involved ramming and then boarding enemy ships. The thick hulls of the Gallic ships were impervious to Roman rams. Moreover, the Gallic ships were too high for the Romans to conduct boarding. Even from the turrets on their ships, the Romans could not easily reach the decks of the Gallic ships. However, the Roman ships powered by oars were faster and more manoeuvrable compared to the Gallic ships, which were completely dependent on the sails. In *The Conquest of Gaul*, Julius Caesar writes about the naval fighting during the First Invasion of Britain in 55 BCE:

> Caesar ordered the warships – which were swifter and easier to handle than the transports . . . to be removed a short distance from the others and then be rowed hard and run ashore on the enemy's right flank, from which

> position slings, bows, and artillery could be used by the men on deck to drive them back. This manoeuvre was highly successful. Scared by the strange shape of the warships, the motion of the oars, and the unfamiliar machines, the natives . . . retreated.
>
> *(quoted from Caesar 1982: 99)*

More importantly, the Roman ships and naval tactics underwent several innovations. The Romans had equipped themselves with long poles tipped with scythe-like blades. When the Roman ships closed with the Gallic crafts, the Romans used these poles to cut the riggings of the Gallic ships. This resulted in the yards and sails of the Gallic ships collapsing and enabled the Roman ships to come close enough for the Romans soldiers to board the hostile ships. In close-quarter combat on the decks of the enemy ships, the heavily armed disciplined Roman infantry had a decisive advantage over the undisciplined lightly armed Gallic infantry. Seeing this the Gallic ships tried to retreat, but the winds dropped, and then they were at the mercy of the oared Roman ships (Pitassi 2012: 153).

Within the Atlantic seaboard area, the shipbuilding tradition of Scandinavia differed from that of North-Western Europe. During the first century CE, the ships of the Suiones of Southern Scandinavia had a prow on each end. The row-locks were moveable, so they could be reversed as circumstances dictated, and these ships could be rowed in either direction. Such boats were also suited for use in the rivers. Borrowings from different military cultures are a characteristic feature in the history of armed conflict. The Scandinavians borrowed the sailing rig from Southern Europe, and sailing ships emerged among them in the seventh century CE (Pitassi 2012: 153).

However, the tribes who inhabited the North Sea Coasts were familiar with the Celtic trading vessels and were acquainted with the Roman ships with sails from Caesar's time onwards. By first century CE, these tribes had copied the sailing mechanism. The rebel leader Civilis mustered a fleet which comprised captured Roman monoremes and biremes, and the indigenous crafts were rigged with improvised sails. However, these tribes never adopted the ram and ship-borne artillery. These two pieces of technology gave the Romans a technical advantage over their opponents' ships (Pitassi 2012: 153–4).

From the late second century CE onwards, the Germanic tribes who inhabited the littoral region of North Sea took to piracy. Since they lacked the power and organisation to engage the Romans in symmetrical warfare, the Germanic tribes waged asymmetric warfare. They used their boats for sailing along the rivers and creeks and raided the Roman coast. In response, the coastal cities under Rome started building walls, and forts were constructed along the strategic points near the coastline. During the early third century CE, the coastal Germanic tribes extended their range of operations to raid along the estuaries of the Rhine and Thames. In fact, the Germanic tribes had a tradition of boat building which extended far back into the misty past. As early as 350 BCE, the North Germanic tribes used to build boats of about 52.5 feet in length with 20 paddles each.

These boats were used for plying along the rivers. With the passage of time, the seamanship of these tribes developed further. By the first century CE, the river boats had both paddles and oars. The German raiders used open clinker-built double-ended boats, and by the fourth century CE they used both sails and oars (Pitassi 2012: 154).

Not only did the seamanship of the Germanic tribes develop over time, but they also formed themselves into larger confederations from the late third century CE onwards. This meant that the raiders could now put to the sea and rivers a larger number of boats with a greater number of armed men, and their capacity to build better and larger boats also increased with time. By the mid-fourth century CE, some of their boats were 79 feet long with 30 rowers each. Each such boat was able to carry a dozen armed men. Raids by dozens of such boats became common. It was the prelude to the Viking raids of the Carolingian Empire (800–88 CE) much later. The Roman Empire responded by building a chain of forts with a string of watchtowers. Some of the big forts, in addition to garrisons, also had naval detachments (Pitassi 2012: 154).

Besides political factors, ecology also played a part in accelerating the trend towards increasing piracy in North-Western European waters. Due to the great inundations of the North Coast of Europe during the mid-third century CE, the level of the North Sea rose. The coastal lands were affected. The Zuider Zee came into existence during this time. Much of the land of Belgium and Flanders were affected. Worse, much more coastal land was transformed into soggy, salty marshland. Further, the climate became colder, which adversely affected cereal cultivation. The disruption of agriculture forced many of the desperate tribesmen to take to piracy. Depopulation of the coastal Roman land between the Rhine and Cap Gris Nez was also accelerated by the end of the warm climate in North-Western Europe (Pitassi 2012: 155). Belgium and Northern Gaul became economically unproductive and military indefensible from the perspective of the Western Roman Empire.

As Roman power declined, conversely, 'barbarian' power increased. In 313 CE, the German raiders reached the coastline of Spain. The decline of Roman maritime power was interrelated with the reduction of the military effectiveness of their ground force. In the case of the Romans, land and sea power were interrelated like a DNA double helix. Continuous ground attacks by the Germanic tribes along the Rhine and Danube frontiers and the arrival of the Hunnic invaders sounded the death knell of the once powerful legion-centric Roman force. By 350 CE, the Roman Rhine defence had almost collapsed, and this allowed the Germanic tribes to move their boats along the channels and inlets newly formed by the great inundations. Now the Germanic tribes could raid even inland towns previously beyond their range of operations (Pitassi 2012: 155). The successor regime of the Western Roman Empire was the Carolingian Empire, which also suffered heavily from Norse maritime attacks.

Conventional sea battles occurred with the rise of France and Britain as regional powers during the Late Middle Ages. The Battle of Sluys, fought on 24

June 1300 between France and England, was a decisive confrontation. The battle was fought in the estuary of the Scheldt near the town of Sluys, in a waterway then known as the Zwyn. The French fleet comprised 200 ships with 30,000 men. When the English attacked, the French ships were chained together. Hence, the French ships lacked mobility. The English ships took advantage of the wind and faced the French with the sun in front of the latter. In the end, the battle was decided by archery. English and Norman/Genoese archers traded heavy fire against each other. In the end, the English longbow proved superior to the Genoese crossbow (DeVries 2008: 131–43; Rose 2008a: xv).

From *c.* 1000 CE, we see continuous development in the technology of sailing ships and development of naval facilities required for repairing and manufacturing the sea vessels in Scandinavia and North-Western Europe. In the thirteenth century, galleys were built at Rouen. The galley yard was enclosed with ditches and a palisade. There were sluices to control the level of water in the ditches as well as the entry and exit of ships from the river to the galley yard. By the end of the fourteenth century, buildings within the enclosures came up. These buildings housed the armoury, workshops and the men who worked for the yard. The principal item of food for the European galley crews was *biscotti* (biscuits), which was a sort of hard-baked sweetened bread that provided calories for the rowers. Biscuit was also an important item of ration for the legionaries. Towards the end of the fourteenth century, the ration for the marines improved. They received beef, pork, wine, salted or dried fish, dried peas, beans, salt, onions and garlic (Rose 2005: 6–16). This improved ration enabled the Western European mariners to undertake long voyages.

During the fourteenth century, the dominant tradition of shipbuilding in the region stretching from Scandinavia to Northern Spain was to build clinker-built crafts. The hulls were formed from overlapping shells built up from the keel. The nails securing the plank were driven in from the outside, and on the inner side of each plank, they were hammered flat ('clenched', hence 'clinker') or riveted over a meal washer. The frames were added at a later stage of construction, and the frames were light. The principal load-bearing element in the structure was the shell planking (Friel 1999: 183). In the fourteenth century, the stern rudder became the principal means for steering large sea-going vessels of North Europe (Friel 2008b: 68).

Cogs emerged as the principal vessel in this region during the Later Middle Ages. The cog had clinker sides and a flat bottom (Friel 1999: 183). Unlike the cog, the Viking longship was not suited for carrying a large volume of cargo. The incapability of the longships to carry substantial amount of cargo and manpower, along with the onset of the Little Ice Age (*c.* 300–*c.* 1700), whose adverse effect was greatest was on the northern part of northern hemisphere, prevented the Vikings from building an economically productive and militarily effective overseas empire stretching from Norway–Iceland–Greenland–Newfoundland. The cog had a deep roomy hull with a low length-to-breadth ratio. The addition of a keel raised the sailing capability of the cog. During naval battles, the high freeboard of the cog made it an efficient platform for hurling missiles on the enemy (Rose 2005: 28). Rose asserts that in Iberia and South-West France, located at the confluence of the Mediterranean and the Atlantic, the clinker-built tradition got fused with the caravel-building tradition of the Mediterranean, and the result was the sea-worthy caravels and galleons of the sixteenth century which gave Western Europe its maritime supremacy (Rose 2008a: xiv).

Initially, it was the Iberian powers who took the lead in conducting voyages along the Atlantic. To an extent, geography aided them. The cross-currents of the Mediterranean proved extremely difficult for the Roman and Venetian galleys to break out from Western Mediterranean to the Eastern Atlantic. In the second half of the thirteenth century, the Portuguese and Castilian mariners sailed down the African shores as far as the Gold Coast, and the Portuguese *volta* allowed the ships to sail north from the West Coast of Africa by going far out in the Atlantic and then allowing the Westerlies to carry them back to the Iberian Peninsula. The Portuguese and Spanish mariners used nature in this profitable manner from the late fourteenth century. Long-distance navigation was further accelerated by marine map-making. The Catalan map of 1375 drawn in Majorca depicted the West Coast of Africa for the benefit of seafarers. By 1500, the Spanish and the Portuguese had started to colonise the offshore islands of the South Atlantic and had sailed to India in the east and Brazil in the west (Lewis and Runyan 1985: 144–9). The path was open for the West's naval dominance of the world's oceans.

Conclusion

During the period under discussion here, there was no clear distinction between cargo and warships and between the crew of regular ships and pirates. Pirates were used as naval crew in the Far East, Mediterranean and the Atlantic. This chapter shows that the Asian ships witnessed continuous innovations with the passage of time. Neither Asian technology nor the Asian mindset was stagnant, as some Eurocentric authors have assumed. In terms of operating range and size, there was no comparison between the gigantic long-range Song, Ming and Chola ships vis-à-vis the puny ships that comprised the Mediterranean fleets of the Byzantine Empire and the Arab Caliphate. This difference was because the Mediterranean was an inland sea and the ports and cities along the coast of the Mediterranean were quite close to each other. In other words, the distance between Alexandria and Constantinople is less than the distance separating Peking and Sri Vijaya by sea. This explains the divergence between the development of navies between Eastern Eurasia and the Mediterranean. Technologically (especially in the use of bulkheads, outriggers, etc.), the Chinese and South Asian ships were ahead of their Western counterparts. Too much has been made about the absence of iron in the ancient and early medieval shipbuilding tradition of the Indian Ocean region. The unique geographical conditions of the Indian Ocean to a great extent explain this absence. Rams and spurs were unique to the ships which sailed in the calm Mediterranean Sea. Such unwieldy weapons cannot be used in ships which plied across the stormy Atlantic with hurricanes and typhoon-ridden seas of the Far East. The Asian ships traversing the marine space between India, East Africa and Arabia took advantage of the monsoon wind (May–October) which travels deep through the Western Indian Ocean, far away from the coastline. Moreover, this marine space is full of coral reefs and surfs. Ram- and spur-equipped ships carrying large trebuchets on their decks made the vessels top heavy and unwieldy and such crafts were just unsuitable for sailing in this region. Compared to close-quarter combat by the Mediterranean ships equipped with rams and spurs, the warships of the Indian Ocean relied on long-distance archery. Probably the use of rams and spurs encouraged the shipbuilders of the Mediterranean to use iron. We find convergence in the use of naphtha/Greek fire. In all likelihood, independent evolution of this weapons technology occurred in China and the Byzantine Empire.

At the end of *c.* 1300 CE, a neutral observer would have thought that the Chinese and Indian navies were going to rule the oceans of the world. But history both on land and sea actually took a different trajectory from *c.* 1500 CE onwards. Just before Vasco da Gama's entry into the Indian Ocean, we see that neither India nor China had any ocean-going navies. The question is, why did such an occurrence take place? It has nothing to do with a cultural mindset but with geopolitics and the respective political-economies of the polities concerned. The Cholas, Yuan and Ming, all these big agrarian bureaucratic empires, faced serious threats from the steppe nomads along their live land frontiers. Internal rebellion, along with the threats posed by the steppe nomadic khanates in Central Asia and Mongolia, brought down the Yuan dynasty. The Chera, Pandya and other kingdoms of South India sounded the death knell of Chola power. The presence of dangerous live land frontiers prevented the rulers of agrarian bureaucratic polities of continental countries like India and China from building ocean-going fleets. Maintenance of an ocean-going fleet seemed to be a luxury fleet for these rulers, especially when the ground-based threats were knocking at their doors. Further, the huge revenues accruing from the agrarian sector also prevented the crying need to extract revenue from maritime commerce.

The Ottoman Sultanate might have been an exception. True, the Ottomans made some forays in the Indian Ocean. Despite the assertion of one scholar that the Ottoman Empire was a merchants' empire (Brummett 1994), the Ottoman political economy was mainly dependent on land revenue. Again, the Ottoman Empire, unlike, let us say, Britain was attempting to become both a land power (in Europe) and sea power simultaneously. On land, it was fighting the East and Central European states and Muscovy. Further, Constantinople's naval attention was directed towards the Mediterranean. In Asia, the principal threat to the Ottomans was posed by the Safavids, and they were a land power.

In contrast to China and India, Britain, protected by the English Channel, had no serious live ground threat after the conquest of Scotland and Ireland. Absence of an extensive agricultural sector also pushed the British merchants and strategic managers to look for fields where maritime commerce could be developed. This in turn resulted in the growth of ocean-going fleets and an active interest displayed by the strategic managers in protecting and extending the commercial activities of the state by deploying aggressive naval forces. In Europe, we see the gradual evolution of two shipbuilding traditions. The one, centred on the inland Mediterranean Sea, focused on rowing ships capable of fighting battles mostly near the coastline. The Mediterranean maritime culture reached a dead end by the fifteenth century. The Mediterranean galleys suited for operating in the inland sea were unable to sail properly either in the Indian Ocean or in the Atlantic. The other tradition, which was an amalgamation of the Scandinavian and North-Western European traditions, produced clinker-built sailing ships. The storms, gales and the wide open Atlantic gave rise to these sorts of ships. This shipbuilding tradition gave rise to ocean-hopping crafts which enabled the Western Europeans to rule the world's oceans from *c.* 1500 onwards.

References

Amitai-Preiss, Reuven, *Mongols and Mamluks: The Mamluk-Ilkhanid War, 1260–1281* (Cambridge: Cambridge University Press, 1995).

Anon, *The Rahmani of M.P. Kunhikunhi Malmi: A Sailing Manual of Lakshadweep*, edited by Lotika Varadarajan (New Delhi: Manohar, 2004).

Arrian, *The Campaigns of Alexander*, translated by Aubrey de Selincourt, Revised with a new Introduction and notes by J.R. Hamilton (1958, reprint, Harmondsworth: Penguin, 1971).

Asmolov, Konstantin V., 'The System of Military Activity of Koguryo', *Korea Journal*, (Summer 1992), pp. 103–16.

Ayalon, David, 'The Mamluks and Naval Power: A Phase of the Struggle between Islam and Christian Europe', in Susan Rose (ed.), *Medieval Ships and Warfare* (Aldershot: Ashgate, 2008), pp. 223–34.

Bopearachchi, Osmund, 'Seafaring in the Indian Ocean: Archaeological Evidence from Sri Lanka', in Himanshu Prabha Ray and Jean-Francois Salles (eds.), *Tradition and Archaeology: Early Maritime Contacts in the Indian Ocean* (1996, reprint, New Delhi: Manohar, 2012), pp. 59–77.

Brummett, Palmira, *Ottoman Seapower and Levantine Diplomacy in the Age of Discovery* (Albany, NY: State University of New York Press, 1994).

Caesar, Julius. *The Conquest of Gaul*, translated by S.A. Handford, revised with a New Introduction by Jane F. Gardner (1951, reprint, London: Penguin, 1982).

Charney, Michael, *Southeast Asian Warfare: 1300–1900* (Leiden/Boston: Brill, 2004).

Dai, Yingcong, 'A Disguised Defeat: The Myanmar Campaign of the Qing Dynasty', *Modern Asian Studies*, vol. 38, Part 1 (2004), pp. 145–89.

Delgado, James P., *Khubilai Khan's Lost Fleet: In Search of a Legendary Armada* (Berkeley/Los Angeles: University of California Press, 2008).

Deloche, Jean, 'Stable-Ships of Tiruppudaimarudur and Tirukkurunkudi (South India)', *Indian Journal of History of Science*, vol. 44, no. 4 (2009), pp. 551–67.

Deloche, Jean, 'Iconographic Evidence on the Development of Boats and Ship Structures in India (2nd Century BC-15th Century AD): A New Approach', in Himanshu Prabha Ray and Jean-Francois Salles (eds.), *Tradition and Archaeology: Early Maritime Contacts in the Indian Ocean* (1996, reprint, New Delhi: Manohar, 2012), pp. 199–214.

DeVries, Kelly, *The Norwegian Invasion of England in 1066* (1999, reprint, Woodbridge: Boydell Press, 2003).

DeVries, Kelly, 'God, Leadership, Flemings, and Archery: Contemporary Perceptions of Victory and Defeat at the Battle of Sluys, 1340', in Susan Rose (ed.), *Medieval Ships and Warfare* (Aldershot: Ashgate, 2008), pp. 131–43.

Dotson, John E., 'Ship Types and Fleet Composition at Genoa and Venice in the Early Thirteenth Century', in John H. Pryor (ed.), *Logistics of Warfare in the Age of Crusades* (Aldershot: Ashgate, 2006), pp. 63–75.

Dreyer, Edward L., 'The Poyang Campaign, 1363: Inland Naval Warfare in the Founding of the Ming Dynasty', in Frank A. Kierman, Jr. and John K. Fairbank (eds.), *Chinese Ways in Warfare* (Cambridge, MA: Harvard University Press, 1974), pp. 202–42.

Ehrenkreutz, A.S., 'The Place of Saladin in the Naval History of the Mediterranean Sea in the Middle Ages', in Susan Rose (ed.), *Medieval Ships and Warfare* (Aldershot: Ashgate, 2008), pp. 235–51.

Fahmy, Aly Mohamed, *Muslim Sea Power in the Eastern Mediterranean from the Seventh to the Tenth Century AD* (1950, reprint, New Delhi: S. Chand, 1966).

Finlay, Robert, 'Portuguese and Chinese Maritime Imperialism: Camoes's Lusiads and Luo Maodeng's Voyage of the San Bao Eunuch', *Comparative Studies in Society and History*, vol. 34, no. 2 (1992), pp. 225–41.

Friday, Karl, *Samurai, Warfare and the State in Early Medieval Japan* (New York/London: Routledge, 2004).

Friel, Ian, 'Winds of Change? Ships and the Hundred Years War', in Anne Curry and Michael Hughes (eds.), *Arms, Armies and Fortifications in the Hundred Years War* (1994, reprint, Woodbridge: Boydell Press, 1999), pp. 183–93.

Friel, Ian, 'Documentary Sources and the Medieval Ship: Some Aspects of the Evidence', in Susan Rose (ed.), *Medieval Ships and Warfare* (Aldershot: Ashgate, 2008a), pp. 3–24.

Friel, Ian, 'The Building of the Lyme Galley, 1294–1296', in Susan Rose (ed.), *Medieval Ships and Warfare* (Aldershot: Ashgate, 2008b), pp. 65–8.

Fuess, Albrecht, 'Rotting Ships and Razed Harbors: The Naval Policy of the Mamluks', in Susan Rose (ed.), *Medieval Ships and Warfare* (Aldershot: Ashgate, 2008), pp. 267–93.

Gabriel, Richard A., *The Great Armies of Antiquity* (Westport, CT: Praeger, 2002).

Gertwagen, Ruthy, 'Harbours and Facilities along the Eastern Mediterranean Sea Lanes to *Outremer*', in John H. Pryor (ed.), *Logistics of Warfare in the Age of Crusades* (Aldershot: Ashgate, 2006), pp. 95–118.

Hamblin, William, 'The Fatimid Navy During the Early Crusades: 1099–1124', in Susan Rose (ed.), *Medieval Ships and Warfare* (Aldershot: Ashgate, 2008), pp. 215–21.

Hess, Andrew C., 'The Evolution of the Ottoman Seaborne Empire in the Age of Oceanic Discoveries', *American Historical Review*, vol. 75, no. 7 (1970), pp. 1892–919.

Hourani, George F., *Arab Seafaring in the Indian Ocean in Ancient and Early Medieval Times*, revised and expanded by John Carswell (1951, reprint, Princeton, NJ: Princeton University Press, 1995).

Jung-pang, Lo, 'The Emergence of China as a Sea Power During the Late Sung and Early Yuan Periods', *Far Eastern Quarterly*, vol. 14, no. 4 (1955), pp. 489–503.

Jung-pang, Lo, 'The Decline of the Early Ming Navy', *Oriens Extremus*, no. 5 (1958), pp. 149–68.

Jung-pang, Lo, 'Maritime Commerce and Its Relation to the Sung Navy', *Journal of the Economic and Social History of the Orient*, vol. 12, no. 1 (1969), pp. 57–101.

Jung-pang, Lo, *China as a Sea Power, 1127–1368: A Preliminary Survey of the Maritime Expansion and Naval Exploits of the Chinese People During the Southern Song and Yuan Periods*, edited and with commentary by Bruce A. Elleman (Singapore: National University Press of Singapore, 2012).

Karashima, Noboru, 'Medieval Commercial Activities in the Indian Ocean as Revealed from Chinese Ceramic-Sherds and South Indian and Sri Lankan Inscriptions', in Hermann Kulke, K. Kesavapany and Vijay Sakhuja (eds.), *Nagapattinam to Suvarnadwipa: Reflections on the Chola Naval Expeditions to Southeast Asia* (2009, reprint, New Delhi: Manohar, 2010), pp. 20–60.

Kulke, Hermann, 'Introduction', in Hermann Kulke, K. Kesavapany and Vijay Sakhuja (eds.), *Nagapattinam to Suvarnadwipa: Reflections on the Chola Naval Expeditions to Southeast Asia* (2009, reprint, New Delhi: Manohar, 2010a), pp. xiii–xx.

Kulke, Hermann, 'The Naval Expeditions of the Cholas in the Context of Asian History', in Hermann Kulke, K. Kesavapany and Vijay Sakhuja (eds.), *Nagapattinam to Suvarnadwipa: Reflections on the Chola Naval Expeditions to Southeast Asia* (2009, reprint, New Delhi: Manohar, 2010b), pp. 1–19.

Levathes, Louise, *When China Ruled the Seas: The Treasure Fleet of the Dragon Throne, 1405–1433* (1994, reprint, New York/Oxford: Oxford University Press, 1996).

Lewis, Archibald R. and Runyan, Timothy J., *European Naval and Maritime History: 300–1500* (Bloomington: Indiana University Press, 1985).

Lien, Vu Hong, *The Mongol Navy: Kublai Khan's Invasions in Dai Viet and Champa*, Nalanda-Sriwijaya Centre Working Paper No. 25 (Singapore: Nalanda-Sriwijaya Centre, 2017).

Lucas, Henry S., 'John Crabbe: Flemish Pirate, Merchant, and Adventurer', in Susan Rose (ed.), *Medieval Ships and Warfare* (Aldershot: Ashgate, 2008), pp. 77–93.

McGrail, Sean, 'The Study of Boats with Stitched Planking', in Himanshu Prabha Ray and Jean-Francois Salles (eds.), *Tradition and Archaeology: Early Maritime Contacts in the Indian Ocean* (1996, reprint, New Delhi: Manohar, 2012), pp. 225–38.

Mishra, Ananta Kumar, *Glimpses of Maritime History of Ancient Odisha* (New Delhi: Kunal Book Publishers & Distributors, 2010).

Mookerji, Radha Kumud, *Indian Shipping: A History of the Sea-Borne Trade and Maritime Activity of the Indians from the Earliest Times* (Calcutta: Kitab Mahal, 1962).

Mote, Frederick W., 'The T'u-mu Incident of 1449', in Frank A. Kierman, Jr. and John K. Fairbank (eds.), *Chinese Ways in Warfare* (Cambridge, MA: Harvard University Press, 1974), pp. 243–72.

Picard, Christophe, *Sea of the Caliphs: The Mediterranean in the Medieval Islamic World* (Cambridge, MA: Belknap Press of Harvard University Press, 2018).

Pistono, Stephen P., 'Henry IV and the English Privateers', in Susan Rose (ed.), *Medieval Ships and Warfare* (Aldershot: Ashgate, 2008), pp. 95–103.

Pitassi, Michael, *The Roman Navy: Ships, Men & Warfare 350 BC-AD 475* (Barnsley: Seaforth Publishing, 2012).

Pryor, John H., *Geography, Technology, and War: Studies in the Maritime History of the Mediterranean, 649–1571* (1988, reprint, Cambridge: Cambridge University Press, 1992).

Pryor, John H., and Jeffreys, Elizabeth M., *The Age of the Apomon: The Byzantine Navy ca 500–1204* (Leiden/Boston: Brill, 2006).

Rajan, K., 'Early Maritime Activities of the Tamils', in Himanshu Prabha Ray and Jean-Francois Salles (eds.), *Tradition and Archaeology: Early Maritime Contacts in the Indian Ocean* (1996, reprint, New Delhi: Manohar, 2012), pp. 97–106.

Ray, Himanshu Prabha, 'Maritime Archaeology of the Indian Ocean: An Overview', in Himanshu Prabha Ray and Jean-Francois Salles (eds.), *Tradition and Archaeology: Early Maritime Contacts in the Indian Ocean* (1996, reprint, New Delhi: Manohar, 2012), pp. 1–10.

Rodger, N.A.M., *The Safeguard of the Sea: A Naval History of Britain*, vol. 1, *660–1649* (London: HarperCollins in association with the National Maritime Museum, 1997).

Rodger, N.A.M., 'The Naval Service of the Cinque Ports', in Susan Rose (ed.), *Medieval Ships and Warfare* (Aldershot: Ashgate, 2008), pp. 159–74.

Roland, Alex, 'Secrecy, Technology, and War: Greek Fire and the Defense of Byzantium, 678–1204', in John France and Kelly DeVries (eds.), *Warfare in the Dark Ages* (Aldershot: Ashgate, 2008), pp. 419–43.

Rose, Susan, *Medieval Naval Warfare: 1000–1500* (2002, reprint, London/New York: Routledge, 2005).

Rose, Susan, 'Introduction', in Susan Rose (ed.), *Medieval Ships and Warfare* (Aldershot: Ashgate, 2008a), pp. xiii–xx.

Rose, Susan, 'Bayonne and the King's Ships, 1204–1420', in Susan Rose (ed.), *Medieval Ships and Warfare* (Aldershot: Ashgate, 2008b), pp. 69–76.

Sakhuja, Vijay and Sakhuja, Sangeeta, 'Rajendra Chola's Naval Expedition to Southeast Asia: A Nautical Perspective', in Hermann Kulke, K. Kesavapany and Vijay Sakhuja (eds.), *Nagapattinam to Suvarnadwipa: Reflections on the Chola Naval Expeditions to Southeast Asia* (2009, reprint, New Delhi: Manohar, 2010), pp. 76–90.

Sasaki, Randall J., *The Origins of the Lost Fleet of the Mongol Empire* (College Station: Texas A&M University Press, 2015).

Sastri, K.A. Nilakanta, *A History of South India from Prehistoric Times to the Fall of Vijayanagar* (1955, reprint, Madras: Oxford University Press, 1958).

Sastri, K.A. Nilakanta, 'The Colas', in R.S. Sharma and K.M. Shrimali (eds.), *A Comprehensive History of India*, vol. 4, Part 1, *The Colas, Calukyas and Rajputs AD 985–1206* (New Delhi: People's Publishing House, 1992a), pp. 1–71.

Sastri, K.A. Nilakanta, 'South Indian Political Organization', in R.S. Sharma and K.M. Shrimali (eds.), *A Comprehensive History of India*, vol. 4, Part 1, *The Colas, Calukyas and Rajputs AD 985–1206* (New Delhi: People's Publishing House, 1992b), pp. 740–72.

Sen, Tansen, 'The Military Campaigns of Rajendra Chola and the Chola-Srivijaya-China Triangle', in Hermann Kulke, K. Kesavapany and Vijay Sakhuja (eds.), *Nagapattinam to Suvarnadwipa: Reflections on the Chola Naval Expeditions to Southeast Asia* (2009, reprint, New Delhi: Manohar, 2010), pp. 61–75.

Smith, Jr., John Masson, 'Ayn Jalut: Mamluk Success or Mongol Failure?', *Harvard Journal of Asiatic Studies*, vol. 44, no. 2 (1984), pp. 307–13.

Suenaga, Takezaki, *In Little Need of Divine Intervention: Takezaki Suenaga's Scrolls of the Mongol Invasions of Japan*, translations with interpretative Essay Thomas D. Conlan (Ithaca/New York: Cornell University East Asia Programme, 2001).

The Travels of Marco Polo the Venetian, with an Introduction by John Masefield (1931, reprint, New Delhi: Asian Educational Services, 2009).

Unger, Richard W., 'The Northern Crusaders: The Logistics of English and Other Northern Crusader Fleets', in John H. Pryor (ed.), *Logistics of Warfare in the Age of Crusades* (Aldershot: Ashgate, 2006), pp. 251–73.

Waldron, Arthur, *The Great Wall of China: From History to Myth* (Cambridge: Cambridge University Press, 1990).

Waterson, James, *Defending Heaven: China's Mongol Wars, 1209–1370* (London: Frontline Books, 2013).

Wright, David Curtis, 'The Northern Frontier', in John Ferris (ed.), *Calgary Papers in Military and Strategic Studies*, vol. 2 (Calgary: Centre for Military and Strategic Studies, University of Calgary, 2008a), pp. 51–72.

Wright, David Curtis, 'Artillery Is Not Needed to Cross a River: Bayan's Swift Riparian Campaigns Against Southern Song Chinese, 1274–1276', in John Ferris (ed.), *Calgary Papers in Military and Strategic Studies*, vol. 2 (Calgary: Centre for Military and Strategic Studies, University of Calgary, 2008b), pp. 91–104.

8

EVOLUTION OF WARFARE BEYOND EURASIA

Introduction

Environmental conditions generated a different military trajectory in the continents outside Eurasia. Most of the precolonial societies of North America, Central America (Mesoamerica), South America, Africa (minus its Mediterranean coastline) and Oceania (Australasia, Melanesia, Micronesia and Polynesia) remained at the level of Neolithic culture before the European conquest. This was partly due to absence of intensive agriculture. The 'agricultural revolution' which swept across Eurasia some 8,000 years ago did not occur in these continents partly due to ecological conditions. We will touch upon this issue, but a detailed examination of this question remains outside the purview of this book. The traditional societies of these four continents outside Eurasia have left us with almost no written records. We have to rely on the accounts generated by European observers and observations by the anthropologists. For medieval societies in the interior of Africa, we have some accounts by the Islamic scholars. The detailed narratives of the 'outside' agents are affected by their biases. Further, these authors were often ignorant of local customs and conditions. Moreover, these traditional societies changed considerably due to contact with the European settlers. It is questionable how far observations by the twentieth-century anthropologists of the Stone Age cultures in the continents outside Eurasia represent these traditional societies as they existed one to two thousand years ago. Besides the modernising influence of the Europeans, these 'native' societies might have undergone changes by themselves with time. In addition, we have art objects (paintings on the sides of the temples, caves, etc.) generated by some of these societies. Analysing warfare from iconography is problematic. It is questionable whether iconography depicts warfare in the idealised form or combat as it had really occurred in reality. Nevertheless, a historian has to work

DOI: 10.4324/9780429284069-9

with the available materials, regardless of how defective they are. We will start with the evolution of warfare in North America.

North America

By 9000 BCE, the migrants from North-East Asia who had crossed into North America had moved down to the tip of South America. By 11,000 BCE, the inhabitants of the Kamchatka Peninsula equipped with stone blade technology were constructing double-walled houses. By 8500 BCE, the people of Alaska, due to intensive hunting, had decimated the bison, horse and elk species. Around this date, big game animals including mastodons had been wiped out by the hunter-gatherers, even in South America (Snow 1996: 130–5). The absence of draught animals prevented the emergence of wagons and carts with wooden wheels, so no chariots came into existence in the 'New World'. Warfare occurred among the Asiatic Eskimos who inhabited Northern Alaska. Their armour and shields were manufactured from wood, ivory, seal skin etc. Ambush and raiding enemy localities rather than battles were their preferred modes of combat. Ambushes often involved a dozen men equipped with bows and arrows. Conflicts among the various groups within the Eskimos took place, but violence was more frequent, especially at the boundary lands between Eskimos and other groups (LeBlanc 1999: 17). Around 200 CE, bow and arrow technology emerged in Alaska. One author states that terrestrial fauna that might be hunted with a bow was limited in the Lower Alaskan Peninsula. The atlatl was generally used in this area for hunting, so the bow and arrow technology represented a proliferation of combat. Slender arrow points fashioned of basalt were used for warfare (Lambert 2002: 212).

Along the North Pacific Coast of North America, extensive warfare started in around 1000 BCE. The Aleuts, Alutiiq and other tribes and clans fought with bone and stone clubs, slate daggers, etc. Warfare occurred due to the acquisition of slaves and food stores and occasionally for territorial extension and control. Moreover, these people constructed quite sophisticated fortified centres. So besides battles and raids, siege warfare also occurred. Most of the North Pacific forts were situated on elevated landforms like rocky headlands or small islands with sheer rock walls. The natural defensive features of the forts were further improved by building log ramparts, palisades and causeways. Some of the forts in the Vancouver Islands and along the Gulf of Georgia also had trenches for protection. Some of these trenches were 6 feet deep and about 15 feet wide. The dirt of the trench was piled and used to support the base of the wooden palisade (Moss and Erlandson 1992: 73–90).

Around 2500 BCE, conflicts occurred between the Laurentian and Lamoka tribes in the modern day New York area (Levy and Thompson 2011: 22). Prehistoric conflicts in the Eastern Woodlands of North America were products of revenge and status enhancement as well as interest in acquiring land, meat and hides. Increased competition resulted in an escalation of conflicts, which in turn gave birth to the Mississippian chiefdoms. By 1000 CE, ambushes by small parties were replaced with a large number of warriors battling with each other for conquest of land (Milner 1999: 108). It is to be noted that compared to groups of hunter-gatherers, chiefdoms were able to mobilise a larger number of fighters.

The indigenous people of the pre-Columbian south-west part of North America were organised in tribes and chiefdoms. Since this region was more densely populated compared to North Alaska, organised violence in the south-western areas of North America was more frequent and intense. The objectives of warfare were not merely blood feuds and vengeance but also the conquest of land. Warfare in the prehistoric south-west region of North America took the forms of raiding, ambushes, pillaging, scalping and captive taking. At times, war became so intense that certain groups ceased to exist as a result of their defeat in combat. Celebrated war leaders not only acquired higher status in their societies but also gained access to resources, including women. Military leadership was important because when a war chief died, Pueblo resistance generally faltered. The term 'Pueblo' is Spanish for 'village'. Pueblo Indians shared a number of cultural characteristics in addition to their farming and settlement pattern (Cordell and Smith 1996: 202).

Steven A. LeBlanc asserts that among the precolonial North Americans, compared to battles, raids were frequently more destructive. Battles were formal affairs involving manoeuvres and resulted in few deaths. In contrast, raiding targeted even the non-combatants. Women collecting firewood and engaged in food procurement were often wiped out by the hostile raiding parties. Further, the threats of attack on the villages forced a group to guard their settlements day and night, and the fear of ambuscades reduced the frequency of hunting and harvesting crops, which in turn caused nutritional insecurity among the group concerned (LeBlanc 1999: 1–47).

Between 2000 and 1000 BCE, indigenous North American crop plant species were domesticated in Eastern North America. Between 250 BCE and 200 CE, agriculture became the principal mode for supplying food vis-à-vis hunting, gathering and foraging. Barley was also grown during this time period. Between 800 and 1100 CE, maize cultivation dominated farming (Smith 1989: 1566–71). Between the beginning of the Common Era and 900 CE, agriculture spread in the south-western part of North America. However, lack of draught animals and iron implements prevented agriculture in precolonial North America from crossing the threshold, as in Eurasia. Further, the arid climate and lack of navigable rivers also obstructed the development of intensive agriculture, trade and commerce in the south-western parts of North America (Cordell and Smith 1996: 203). All these factors prevented the rise of a state system in this region.

Permanent villages with rudimentary fortifications came into existence with the spread of agriculture. Warfare increased in scope and intensity due to disputes over ownership of land and general population pressure. The burning of the hostile settlements, massacring the prisoners, taking of body parts of the dead enemies as trophies and rounding up the women and children were part and parcel of warfare. Between 900 and 1200 CE, the frequency of warfare declined due to two reasons. First, the climate was very good. Second, the Chaco tribe (a branch of the Pueblos who inhabited the Colorado region) established a sort of peace among the subordinate tribes. However, peaceful conditions and favourable climate resulted in the expansion of agriculture, which in turn generated more surplus. Hence, the population tripled during this time period (LeBlanc 1999: 36–44).

From 1250 CE onwards, in the south-west areas of North America, small habitations were replaced with larger sites. Probably this was a by-product of political centralisation. What role did warfare play in this process? The 50-year period between 1250 and 1300 saw an increase in organised violence and a massive displacement of population. Deteriorating weather, the decline of agriculture and population pressure probably caused these conflicts. Warfare declined between 1300 and 1400 CE because the climate was warmer and favoured agriculture. However, colder weather starting from *c.* 1400 resulted in the collapse of agriculture and the decline of population, increasing warfare and abandonment of many settlements. The threat to the declining sedentary groups escalated due to the incursions of hunter-gatherers from the margins of agricultural settlements in the south-western parts of North America (LeBlanc 1999: 31, 38, 52). Probably, the hunter-gatherers were forced to attack the agricultural settlements due to adverse climatic conditions.

Besides climate, technology also played an important role in intensifying conflicts in this region. The indigenous people of the south-western regions of North America fought with shields, clubs, spears, slings, broad knives and bows. The sinew-backed recurve bow came into operation during *c.* 1300 CE, and this deadly weapon raised the level of casualties in warfare. This weapon was common among the steppe nomads and probably spread from Central Eurasia through Alaska to North America (LeBlanc 1999: 39, 46–7). However, the indigenous people of North America lacked horses, and their arrows were not iron tipped due to the absence of iron technology in their culture. So militarily they were not as dangerous as the Eurasian steppe nomads.

Overall, we can say that warfare was an integral part of life among the Pueblos. Scalp dances, war dances and the presence of war chiefs were common. The war priests organised the community for defence and at times took command of the fighting force. Scalps of the bravest enemies were taken and brought back to the villages. By securing the scalp of the dead, the Pueblo captors tried to gain the mental and physical abilities of the enemy for themselves and their tribe (Haas and Creamer 1997: 235–61).

Increasing intergroup conflicts to a great extent were responsible for the emergence of the North-Eastern League of Iroquois and the mid-Atlantic chiefdoms like the Powhatan during the middle of the second millennium CE. So war functioned as a motor of political growth. The Iroquoian people's homeland, known as Iroquoia, was in the north-eastern parts of North America. These people's homeland sat astride several important waterways. A large Iroquois village contained more than 100 longhouses and had a population of about 2,000 people. Smaller villages of 1,000 people each were also common. In addition to the villages, there were smaller settlements called hamlets. There were also seasonal camps for hunting, fishing and for purposes of conducting war. In the early seventeenth century, there were 10–13 large villages in Iroquoia. The Mohawks had three to four of them, the Oneidas had one, the Onondagas had one or two, the Cayugas had two or three and the Senecas had three. The total population of the Iroquois came to about 20,000 people. The five groups of Iroquoian speaking people who inhabited the region constituting modern New York state combined to form the Iroquois League. However, it was not a monolithic centralised organisation. All the five Iroquois people did not jointly wage war against the common enemy. There were always factions and divisions which prevented 100 per cent Iroquois participation in any conflict (Barr 2006: xvii, 1).

These five groups of Iroquois people constantly raided each other, resulting in a sort of blood feud. The death of each warrior demanded retaliation, which in turn created more cries for vengeance. This came to be known as mourning war. The women of the villages carved the body of the deceased into pieces and, after boiling it, served it to the warriors, who consumed it. The warriors believed that thus they were consuming the dead enemy's power. Historian Daniel P. Barr writes that ritual torture and cannibalism were designed to restore the spiritual power lost when Iroquois people died (Barr 2006: 2–9).

Though there were some distinctions between mourning wars and wars called for by village governing councils, the two forms of Iroquoian warfare were not mutually exclusive. In the case of these two types of wars, revenge was a powerful motivation, and the principal aim was the expansion of Iroquois power expressed through the taking of captives. Either through adoption or ritual execution, the Iroquois believed that they obtained the power of the captured enemy, thereby adding it to their own. Killing an enemy demonstrated bravery and skill in combat, but it did not bring equivalent forms of power to the community (Barr 2006: 14–15). Here we find some sort of similarity with the Aztec's 'flower wars', which are discussed later.

Just before the coming of the Europeans, about 75,000 North American Indians were living in the region that later came to be called New England. Most of them occupied the coastal plains and river valleys, especially in the southern half of this region. While the men hunted, the women grew maize, squash and beans. These people were organised in bands led by the *sachem*. Since the *sachems* had no hereditary rights, many of the bands broke up to form new bands. So political alliances were in a flux, which resulted in incessant warfare. The *sachems* kept their followers in military readiness. During military expeditions, the men carried grounded dry corn in their bags. When hungry, they ate it mixed with water. They wore moccasins manufactured from the skins of deer or hides of moose. The North American Indians strapped snowshoes to their moccasined feet for ease of travel across difficult terrains. They had 50-foot-long canoes, each of which carried 20 men. While launching an amphibious raid, a *sachem* could gather a fleet of 30–40 canoes filled with warriors. Further, these people were able to construct forts (square or rectangular in shape) with wood and earth. One of their principal weapons was the bow, each of which was 5 feet long, with a flat belly and a convex face. The maximum range of such a bow was 500 feet (Malone 2000: 1–14). Compared to pre-Columbian North America, where only tribal chieftains existed, in Mesoamerica and parts of South America, more advanced polities with a different military culture came into existence. The next section turns the spotlight on Mesoamerica.

Mesoamerica

Mountains run along the west and east coasts of North Mexico. At the centre of Mexico, another mountain chain runs east to west, almost from coast to coast. The union of these three mountain ranges creates a vast highland region in Central Mexico characterised by mountains, well-watered valleys and a temperate climate. This region, known as the Mexico Basin, is a broad interior valley at an elevation of 7,218 feet, and tall mountains surround it. North of the Mexico Basin is the great desert where low rainfall and sandy soil make agriculture impossible. To the north-east of the Mexico Basin is the Teotihuacan Valley. To the south of this region are extensive river valleys at lower altitudes, with a subtropical climate and fertile soil (Grove 2000: 122–3).

In the Maya Lowlands (the Yucatan Peninsula, Belize, Honduras, El Salvador, Guatemala etc.), forest clearance for agriculture started with the beginning of the Preclassic Era (i.e. 2500 BCE), which lasted till 250 CE. Around the same time, agriculturally based societies living in small, settled villages and utilising a ceramic technology emerged in the central highlands of Mexico. Semi-sedentary societies based on wild plants and animal resources had emerged in the Mexico Basin around 4000 BCE. Maize had been domesticated in this region roughly around 5000 BCE (Grove 2000: 123, 125). Around 1150 BCE, warfare started around the Gulf Coast of Mexico with the rise of the Olmecs. Urban centres, complex religions and interregional exchange networks emerged under the Olmecs. They introduced clubs, maces and spears as tools of war. By 550 BCE, the Olmec Empire (if the term could be used at all) started contracting, and independent city states emerged in Mesoamerica (Hassig 1988: 7, 1999: 361; Webster 1999: 335).

During the Middle Preclassic Period (1000–400 BCE), conflict among the smaller group of people occurred in the Maya Lowlands. Decapitation and the use of trophy heads became common during this period (Inomata 2014: 38, 40). The Late Pre-classic Period (500 BCE-350 CE) settlements of the Western Lowlands were defensively situated on or near the hilltops, and some of them had protective walls. The spread of warfare probably forced the relocation of the scattered population inside the fortified centres. A population expansion during the Late Classic Period (250–950 CE) in the Maya Lowlands and increased concerns of the polities with territoriality gave warfare a new dimension (Scherer and Golden 2014: 61, 63).

Warfare was casualty prone. Yaxuna in North-Eastern Yucatan at the end of Late Classic Period was in a weakened state. Around 600 CE, the city of Coba took advantage of Yaxuna's vulnerability and conquered it. Between 760 and 830 CE, warfare in the Pasion Valley of Guatemala resulted in a demographic collapse in this region. A unique aspect of warfare in this region is that the victor introduced a new style of ceramic production in the conquered areas to signify as well as to legitimise its victory (Bey III 2003: 19–29).

The objective of Maya warfare during the Classic Period was mainly captive taking rather than killing the enemies. Most of the combatants wore light costumes, and some of them wore only loincloths. However, the war leaders wore jaguar-skin tunics and capes. However, even the latter's arms and legs were mostly exposed. Only the rulers and top commanders wore large feather headdresses. Some of the elite warriors also wore animal headdresses, and most of the warriors fought barefoot. Their weapons were darts and spears (Inomata 2014: 27).

During the eighth century CE, bigger Maya polities in Guatemala attempted to establish territorial control over their smaller neighbours. Localities along the trade routes or those positioned along strategically vital areas from where both offensive raids as well as a positional defence against hostile parties could be conducted were ruled by subordinate nobility, and the latter were appointed by the Maya ruler of the big (hegemonic) polity. After being conquered, the existing buildings were ritually destroyed, and the local lord swore an oath to the Maya ruler of the big polity and continued as his vassal. A loyal and efficient local lord from a *sajal* (small locality/principality) might become an *ajaw* but never *kuhul ajaw*, the chief title held by the ruler of the hegemonic polity. One example was the ruler Bird Jaguar of Yaxchilan, the *kuhul ajaw* in the mid-eighth century CE who conquered La Pasadita in Northern Guatemala and appointed the local lord as his *sajal*. La Pasadita, a naturally fortified hilltop centre, guarded the trade routes and also functioned as a defensive barrier against the Piedras Negras polity (Golden 2003: 31–48).

During the Terminal Formative Period (100 BCE–200 CE), the Monte Alban polity controlled a territorial empire which extended over 20,000 km. The core of the Monte Alban polity was the Valley of Oaxaca in Southern Mexico. In the Aztec Empire, males between the ages of 20 and 49 became soldiers, and they comprised 43 per cent of the total male population. During the Late Formative Period, the population of the Valley of Oaxaca was 50,920, with 70 per cent of them residing in the Central and Etla sections of the valley, which were under Monte Alban's direct control. This estimate yields a total population figure for Monte Alban of 35,644, with 7,663 men aged between 20 and 49. Not all of them participated in warfare. So we can conclude tentatively that for military expansion beyond the Valley of Oaxaca, some 5,000 men were available (Joyce 2014: 118, 128). The Monte Alban polity did not go for territorial conquest and direct administration. Instead, their policy was to establish indirect hegemonic control through co-opting the local elites in the peripheral regions. Cooperation of the local elites was achieved through military threats backed by occasional raids. Once the local elites submitted, then they were integrated into imperial control through marriage ties and gift exchanges. Later, the Aztecs also followed this model of imperial expansion (Joyce 2003: 49–72).

Between 200 and 900 CE, Teotihuacan, situated in the north-eastern portion of the Mexico Basin, emerged as a Mesoamerican Empire. This polity dominated Central Mexico and Southern Guatemala. Teotihuacan was under theocratic rule. The basic armaments were atlatls, darts, spears, clubs, cotton body armour, etc. These weapons continued to be used by their successors, including the Aztecs. Aztec power was established in Mexico roughly around 1298 CE. Tenochtitlan, the centre of Aztec power, emerged in 1345. The Aztec military organisation and technology was a development of the Toltecs centred at Tollan. The Toltecs dominated Central Mexico from 900 to 1200 CE. During this period, power passed from the theocratic priesthood to the secular warriors. This trend of the rise of secular power holders continued under the Aztecs (Hassig 1988: 5).

The view that the Aztecs practised flowery war merely for acquiring sacrificial victims is now debatable. Frederic Hicks asserts that acquisition of sacrificial victims was incidental and occasional by-products of the flowery wars, and the primary function of these wars was a practical one. Participation in the flowery wars provided training for the nobles and the commoners. For the nobles, military training was obligatory, and the commoners were encouraged to acquire military training. To be effective, military training had to be realistic in order to be effective in actual combat. The aim of flowery war was not to conquer the enemy. Those who were courageous and skilful enough to take captives received special recognition. And from them (both commoners and nobles) were chosen the leaders for manning the high posts. The realistic military training provided in the course of the flowery wars aided the Aztecs during wars of conquest (Hicks 2007: 1–8). The premier historian of Aztec warfare, Ross Hassig, writes that before going to war, the priests played an important role in communicating with the gods. Human sacrifices were made and supernatural aid was sought. Priests bearing images of the gods accompanied the army during campaigns and encouraged the soldiers by performing rituals. Senior priests also fought in the battlefields. The religious element was interwoven with Aztec warfare. But this practice was not unique to the Aztecs. The Christianising Spaniards also gave due importance to religion during combat (Hassig 1988: 9). Hicks and Hassig make an argument for reducing the perceived cultural gulf between the Aztecs and Western European war making. Nevertheless, the cultural gulf did exist, which in turn had adverse consequences for the Aztecs when the Spanish confronted them in 1519.

The primary objective of Aztec warfare was to establish tributary relationships rather than annexation of enemy territories. One author argues that rather than acquisition of captives for sacrifice, the principal motive of Aztec imperialism was acquisition of tribute (both human and material goods). Acquisition of captives by the warriors resulted in their upward mobility. Heroic warriors received fame, wealth, land and power. And not all the prisoners were sacrificed. Some of them became slaves. Some captives were, however, sacrificed. Human sacrifice in the public domain demonstrated power and domination by the Aztecs. Human sacrifice was not unique among the Aztecs but quite common in Mesoamerican culture. In the Mesoamerican worldview, sacrifice was a mechanism to convert a living being into sacred matter. Through death, the victims became part of the divine world. Thus, death became a fertilising element in which bones and seeds were equivalent (Balderas 2014: 171–3).

Local allies were important in the expansion of the Aztec Empire. For instance, the Aztecs, with the local Nahua groups, were engaged in a long attritional war from 1460 to 1487 in the province of Tlapa, where the Tlapa-Tlachinollan polity was functioning. The Aztec warriors were equipped with obsidian swords and wore cotton armour. Rarely did the Aztecs leave an occupation force or bureaucrats in the territories of groups or communities defeated by them. This was because the Aztec Empire was hegemonic and based on indirect control. The conquered territories were not annexed, and the defeated local chiefs were kept in place after taking an oath of loyalty to the Aztec ruler. They had to provide tributes (human beings and exotic trade goods) to the Aztec ruler. Occasionally, the vassal territories rebelled and refused to pay the tributes, obstructed the trade relationships and murdered the merchants and emissaries of the Aztecs (Gutierrez 2014: 143–52). Then the Aztecs would send a military force comprising Aztecs and their loyal local warriors. Two serious weaknesses of the Aztecs which later the Spaniards would take advantage of was the fact that the empire lacked a permanent standing army comprised of professional soldiers (during emergencies, the Aztec men were ordered to join the expeditionary force) and the presence of disgruntled local chiefs in the periphery of the empire (Hassig 1988: 11).

All the precolonial Mesoamerican empires were hegemonic and based on indirect control rather than cohesive territorial entities based on direct control exercised by a permanent bureaucracy and a standing army. This was due to the difficult terrain of Mesoamerica. Dense forest, mountains and absence of navigable rivers obstructed quick communications and easy movement of the infantry forces. Absence of wheel technology and draught animals were further obstacles to political centralisation. Next, we turn further south of Mesoamerica, into South America.

South America

We have evidence of occasional conflicts among the foraging bands in Peru during 4500 BCE. Around 3000 BCE, chiefdoms emerged, and the men were equipped with clubs, spears and slings (Levy and Thompson 2011: 42–3). The Moche lasted from the beginning of the Common Era till 800 CE in the North

Coast of Peru. There was no single superstate but a loose confederation of decentralised polities, and they often fought against each other and at times also cooperated among themselves. The Moche economy was based on cultivation of maize, sweet potatoes, avocados, beans, peanuts and fruits. In addition, cotton and gourds were also produced for commercial transactions. The principal domesticated animals were pigs, ducks and dogs. Llamas were found in the high altitudes but probably not in the coastal region (Quilter 2002: 145–58). Further, the Moche constructed large canals for irrigation purposes. The Moche warriors liked to fight from a long distance with slings rather than engaging in close-quarter hand-to-hand combat with maces. Moche art celebrated militarism. About 60 per cent of the existing Moche art portrayed warriors, combat, captive sacrifice, weapons, etc. Portable arts like ceramics, metal works, carved woods, etc., which depicted military activities had a limited audience. However, displays of ritual killing of the captives on the sides of the temples were designed for public consumption on a grand scale. The locations of the latter type of art forms had a high visibility as they were permanently displayed before a large number of people. The Moche sacrificed the captives before a large audience, and repeated slashing of the victims before finally killing them was designed to create a dramatic impact on the onlookers (Butters 2014: 259, 268; Verano 2014: 284–94). The rise of the Wari Empire weakened the Moche.

To an extent, the decline of Moche culture was also due to climatic changes. They were heavily dependent on large-scale irrigation agriculture. With canal water, they grew mainly beans and maize. The climate changed from around 500 CE. Successive droughts occurred during the sixth and seventh centuries. Environmental degradation due to the expansion of human settlements made the droughts more intense. The water table was reduced, and salinisation occurred. Torrential rains due to the El Niño effect resulted in flash flooding and soil erosion in the hyper-arid north coast. Sand dunes emerged in the mouths of the rivers. The Chancay River is the principal river in this region. Flash floods and expansion of the sand dunes in the southern portion of Moche culture destroyed many irrigation channels, so the amount of water reaching the cultivated land was significantly reduced. Hence, the Moche abandoned their erstwhile cultivated land in the southern portion of their domain and concentrated in the northern portion of their occupation zone. Due to the shortage of water reaching the cultivated areas in the coast from the highlands, the Moche shifted to calabash cultivation, as this crop has a shorter growing season and requires less water. The Moche gradually abandoned their coastal cultivation, including the marginal and lower valley farmlands, and concentrated along the Rimac Valley neck due to availability of water and good land. By 750 CE, as the agro-climatic crisis gripped the Moche, the Wari expansion started (Shimada et al. 1991: 247–70).

Between 600 CE and 1000 CE, the Wari Empire existed in Peru. The core of the Wari power was the Ayacucho Basin in the central highlands of the Andes. Huari was the capital city of the Wari Empire, and the walled site encompassed 2 sq km. The Wari warriors used bows which shot projectiles with obsidian points, and they also used stone maces and copper bronze maces shaped like a pine cone (Tung 2014: 227–44). Between 400 and 800 CE, circular and quadrangular settlements emerged in the hilltops, and these settlements were protected with defensive trenches and walls (Lau 2014: 331–2).

In the Late Intermediate Period (1100 CE–1450 CE) in the Andean Highlands due to rising military threats, the families concentrated into defensive towns and villages. These communities then organised themselves into local and regional confederations. Then these local political groupings engaged in small-scale conflicts amongst themselves. Around 1000 CE, the Tiwanaku superstate in the Southern Titicaca Basin in the South-Central Andean Highlands (Bolivia) collapsed. This resulted in the disruption of the exchange routes and existing social hierarchy. Again, the colder climate with highly variable precipitation and long severe droughts resulted in the decline of agriculture in the Titicaca Basin. The decentralised Colla state with its capital Hatuncolla emerged during this time (Arkush 2014: 199–207).

Between the Maranon River in the west and the Huallaga River in the east, over 150,000 km of the eastern slopes of the Andean cordillera, the Chachapoya people were organised as chiefdoms. The Chachapoya power was partly based as being the trading intermediaries between the highland societies and the lowland tropical region. The latter region supplied the highland societies with medicinal plants, bird feathers, animal pelts, etc. Keulap is centrally located in the Chachapoya region, about 9,842 feet above sea level in the Luya Province. Keulap was a fortified city surrounded by a wall 62 feet in height. The defenders were equipped with slings, star-shaped maces, and copper or bronze spears, among other weapons. The Incas conquered the Chachapoya in 1490 CE. They constantly revolted till the coming of the Spaniards. Continuous resistance to the Incas generated an identity among the various Chachapoya groups (Toyne and Vargas 2014: 342–53). Mass production of bronze tools started in the Andes just around this time (Diamond 2005: 16). By comparison, the Bronze Revolution had occurred in Eurasia almost 4,000 years ago. We can speculate that the spurt in bronze production might have aided Inca expansion.

The Incas depended on soldiers conscripted from the subjugated territories, who were commanded by the Inca nobles and skilled leaders from the provincial groups. Their weapons were slings, stone maces, clubs, axes, spears and bows and arrows. In a big campaign, an invasion force mustered at Cuzco (the capital of the Inca Empire) numbered to tens of thousands. Numerical superiority was the principal reason behind repeated Inca victories. The Incas gave their potential victims the options of either voluntary submission or war. In the former case, the local leaders were allowed to retain their positions and given high quality clothes, golden drinking vessels and women in marriages. If the enemies refused to submit voluntarily, then the Inca soldiers engaged their opponents either in pitched battles or assaulted the hostile hilltop fortified sites. The groups defeated in combat had their leaders replaced, and individuals chosen by the Incas were appointed as their leaders. In many cases, defeated groups were broken up and resettled in other parts of the empire. Their land was appropriated by the Incas, and the subjugated people were awarded burdensome labour obligations (Ogburn 2014: 365).

John Haldon, a historian of Byzantine Empire, writes that pre-Columbian warfare was characterised by the culture of 'visual warfare'. Representation of organised violence was present in the ceremonies associated with the rulers and also in the games. The visual violence legitimised the status quo. Technological limitations, he claims, encouraged the polities to focus on symbolic violence. Ritualised sacrifice of the prisoners mobilised spiritual and ideological resources for reinforcing subservience and cooperation among the defeated groups. For instance, the Incas, unlike the Aztecs, at least had llamas which were used to transport goods. Logistical and communication difficulties in Mesoamerica were more difficult for the Aztecs in order to achieve long-distance hegemony like the Incas did in the Andes. No polities in Latin America had wheels and siege machinery. So the focus was on symbolic or ritual aspects of fighting. Haldon goes on to say that technological innovations did not occur because in the 'static' mental universe of pre-Columbian society, any radical innovative shifts were not acceptable. Why pre-Columbian mentality was static, unlike the dynamic West Eurasian mentality, is of course not explained. Then Haldon goes on to compare the precolonial Latin American states with the Chola Empire (849 CE–1279 CE) of medieval South India. Following the anthropologist Burton Stein, he argues that all these polities were segmentary states. And in such segmentary polities, temples functioned as centres for redistribution of surplus and tributes, and such temple centred functions were crucial in maintaining royal prerogatives and authority. In other words, segmentary states lacked hierarchical bureaucracies and standing armies. Political authority was maintained among the fluid, ever-shifting local units through symbolic religious activities. It is interesting to wonder why the Cholas, who had access to iron technology, unlike the precolonial Latin American polities, remained a segmentary state. Haldon points to the cosmological universe of Hinduism, which was also resistant to change (Haldon 2014: 385–401).

However, the same Hinduism produced centralised bureaucratic polities like the Maurya and the Gupta empires. And the Chola polity, despite being labelled as a 'segmentary state', was able to project power, as we have seen in the previous chapter, across the vast Indian Ocean with its oceangoing navy and disciplined standing army. The static mentalities of the pre-Columbian South Americans and the pre-modern Hindus, as put forward by the culturalist scholars, is somewhat similar to the argument of the nineteenth century British scholar-officials that the non-European races, being racially inferior, were resistant to modernising changes.

Jared Diamond comes up with an alternative argument. He writes that 17,000 years ago, the two Americas were filled with large mammals like camels, giant ground sloths, elephants, horses, woolly mammoths, lions and cheetahs. But the humans who entered the Americas between 17,000 and 12,000 years ago killed all of them by 11,000 BCE. As a result of this overkill, the humans who settled in the Americas were left with no indigenous animals which they could domesticate. Since the animals in the Americas had not coevolved with the humans, as in the case of Eurasia, the latter found the former easy to kill and exterminate them. And this seriously impeded their further development, especially as regards the development of agriculture. Domesticated animals not only supplied milk and flesh to the humans but also made agriculture more fertile. Animal manures acted as natural fertilisers, which in turn increased the yield of crops. Further, domesticated animals were used as draught animals for pulling the plough and also for carrying goods. Hence, domestication of livestock had a multiplier effect on intensification of productive agriculture and aided the growth of trade and commerce. In other words, domestication of animals, animal husbandry, intensive agriculture and the resultant economic and population growth were all interrelated. An acre of land in which a community practised animal husbandry and agriculture was able to sustain 10 to 100 times more population than could be sustained through hunting and gathering. The birth rate among the sedentary population practising agriculture and animal husbandry was at least two times higher than that of the hunter-gatherers. Moreover, the latter being mobile, hunter-gatherer women faced problems with child birth. In contrast, the sedentary lifestyle of agricultural societies enabled their women to produce more children. Infant mortality was also lower in the agricultural societies due to the availability of more food. Dense population clusters due to agriculture and domesticated animals gave birth to epidemics. Most of the epidemic diseases (smallpox, measles, flu, etc.) evolved from the germs of the domestic animals themselves, and humans in response developed such disease-resistant genes. As a result, both epidemic disease and such disease-resistant capacity were present among the Eurasians but absent among the indigenous people of the New World (Diamond 2005: 46–89), and this had a devastating impact when the Spaniards came to the New World and spread smallpox among the indigenous Americans. This process was repeated in Australia and New Zealand too.

As a contrasting example, cattle were domesticated in India and Western Eurasia separately by around 8,000 BCE. In around 8500 BCE, plant and animal domestication occurred in the Fertile Crescent. In China, both these processes occurred in around 7500 BCE (Diamond 2005: 98–100). The agricultural revolution following close on heels of the Neolithic Revolution put Eurasian history on a different track. Agriculture, by increasing the food surplus, enabled the specialists to thrive and gave birth to stratification in society. Demographically rich stratified societies supported by agriculture gave birth to political hierarchies which in turn ultimately mutated into nation-states in Eurasia. Densely populated regions backed by effective political leadership and iron technology then produced the next stage of historical evolution in Western Eurasia: the Industrial Revolution and rise of capitalism. Now we turn to see the evolution of warfare in the 'dark' continent.

Africa

North Africa is in fact part of the Mediterranean cultural system. From antiquity, North African political, economic and military history was interrelated with the social, political and cultural changes in the Near East and Mediterranean Europe. However, for the history of precolonial Sub-Saharan Africa, we have almost no written sources. The Roman historians and the Arab geographers only touched upon Sub-Saharan Africa in their accounts. We have to depend mainly on ethnography and archaeology for constructing the history of the regions in the interior of Africa. In the previous chapters, while dealing with the rise and fall of the Achaemenid Empire, Hellenistic monarchies, the Roman Empire, and the caliphate, we have touched on North Africa. This section mainly focuses on West, Central and East Africa.

Africa's coastline, except along the Mediterranean Sea, is very regular and is not conducive for the growth of harbours. This, in turn, writes historian Richard J. Reid, obstructed the growth of ocean-going navies among the precolonial polities along Eastern and Western Africa. The regularity of India's coastline also discouraged the development of ports and harbours and somewhat decelerated the growth of sea-going navies among the traditional Indian polities. The equatorial forest which covers most of West Africa's hinterland and the Congo Basin, along with most of the Atlantic coastline, accounts for slightly more than 10 per cent of the 'dark' continent's geography. The tropical zone of the continent is infested with the tsetse fly. It attacks animals and creates trypanosomiasis (and its human variant is known as sleeping sickness). The presence of the tsetse fly prevented the use of horses and pack animals on a large scale in the region between the Sahara and Kalahari deserts. The presence of this fly also prevented the spread of horse-drawn chariot culture in this region when Eurasia experienced chariot warfare. Cavalry was limited to the Sahel Belt and in the southern tip of the continent. In the rest of the region, mainly infantry operated. So large parts of the interior of Africa did not witness the rise of large cavalry states in the medieval period. The inability to use pack animals resulted in the use of human porters till the First World War (Reid 2012: 2), and this reduced the mobility and striking capability of the pre-modern infantry armies in the interior of the continent. Let us first start with East Africa.

The C-Group or C-Horizon people inhabited Lower Nubia (Northern Sudan) around 2500 BCE. Their origins are uncertain. Either they were descendants of an earlier indigenous group or migrants driven from the region east or maybe west of the Nile due to the gradual desiccation of the Sahara. It is to be noted that in Northern Mauretania a humid climate supported an extensive Neolithic population of cultivators and herders. Most of them were black. This agricultural civilisation lasted from 2000 to 300 BCE and then vanished. The C-Group people cultivated cereals and domesticated cattle, and these techniques by 2000 BCE spread from them into Ethiopia. Southern Uganda was at that time highly forested (Oliver and Fagan 2002: 349; McDougall 2007: 71). So cattle and cereal cultivation did not spread in this area. In Eastern Africa, the migration of the Cushitic speakers reached Kenya by 2000 BCE, and they introduced domesticated cattle in this region (Reid 2012: 39). The Egyptians entered Northern Sudan (old name Kush) in about 1500 BCE and conquered the region up to Kurgus near Abu Hamed. Egyptian administration was introduced, and the rich chiefly families of Kush moved into the cities and adopted Egyptian customs. By 750 BCE, an independent Kushite state came into existence. The Kushites even invaded Egypt but withdrew in 654 BCE. The introduction of the *shaduf* from Egypt (originally introduced in the Nile Delta by the Hyksos) in around 300 BCE and the ox-driven water wheel during 200 BCE substantially increased food production in Nubia (Shinnie 2002a: 210, 219).

We have seen in Chapter 2 that the Hyksos introduced chariots in Egypt in 1600 BCE. Soon the Egyptians adopted the war chariots in their military organisation. The Egyptian war chariot with two wheels was light and made of wood and leather and weighed about 34 kilos. The horse-drawn war chariot was adopted by the Libyans from the Sea Peoples in around 1300 BCE and by the Nubians around 1000 BCE. The Libyans with their war chariots carrying archers in fact attacked Ethiopia. The chariots were maintained for both military and prestige factors. However, it was difficult to maintain chariots and horses due to the scarcity of water in the Sub-Saharan region. Though horses were kept as markers of social prestige, it was the camels which proved to be most effective for crossing the desert and semi-desert conditions. The camel came into Egypt in around 1600 BCE through the Hyksos and then slowly spread into Libya, the Maghreb and the Sahara (Mauny 2002: 276–86). The camels played an important role both in commerce and warfare in Africa as well as in Arabia. They are the best beasts of burden for desert conditions. And besides carrying the baggage of the armies, the camels in many cases were also ridden by soldiers in battles and campaigns. In fact, the spread of Islam in the initial phase to a large extent was possible due to the use of camels by the Arab tribes.

From the fifth century CE, Christianity spread from Roman Egypt into Nubia. After the collapse of Byzantine rule in Egypt in 626 CE, the Arabs occupied Egypt. In 641 CE, an Arab cavalry army under Uqba Nafi advanced south of the First Cataract. The Arab cavalrymen were equipped with swords. Before the Arab mounted soldiers could come in contact with the Nubians for sword play, the latter with their bows played havoc among the former. In 651 CE, Abd Allah led a force of 5,000 men equipped with catapults into Nubia. Old Dongola, the capital of Makuria and Nobatia were captured. The Fatimids, like Xerxes, recruited 'black' troops from Nubia. In 1171 CE, the Nubians attacked Aswan and the next year advanced into Upper Egypt in order to aid their ally the Fatimids. However, Saladin (Ayyubid dynasty) sent his brother Turan Shah, who drove back the Nubians. In thirteenth century under the Mamluk Sultanate of Egypt, Islam replaced Christianity in Nubia. During the fourteenth and the first half of the fifteenth century, mass migration of the Arab tribes from Arabia and Upper Egypt in the southern direction occurred under the aegis of the

Egyptian Mamluk Sultanate (Shinnie 2002b: 559–86). The process of Islamisation of Nubia was complete by then.

By the fifth century CE, the whole of Sub-Saharan Africa moved towards iron-based farming (Reid 2012: 39–40). The introduction of Islam and the greater use of the Saharan trade routes which linked the coastline of North Africa with Sub-Saharan Africa somewhat transformed the African political and military landscape. In 663 CE, Uqba Nafi, the leader of Arab troops in Tripoli, led a force of 400 horsemen and 400 camel riders south into the Sahara and reached Sudan (Levtzion 2002: 637–8). Unlike the Romans, the Arabs were adept in using camels in their homeland in Arabia. Anthropologist Brian Fagan writes that use of improved saddles on the camels facilitated Arab expansion across the Sahara. This saddle lies at the camel's shoulder, forward of the hump, and thus is positioned to maximise the load-carrying capacity, endurance and control. Hence, the camels were able to carry an adequate amount of water and provisions, thus allowing the humans to cross from North Africa into Western Sudan. And the Muslim armies generally followed the trade routes of the merchants across the desert (Fagan 2009: 72). Hence, the Arabs were able to penetrate deeper into the interior of Africa compared to the Romans.

The volume of traffic carried along the trans-Saharan trade increased with the advent of Islam due to the introduction of the new camel saddle. Moreover, there was a demand for Central African commodities among the Islamic polities of North Africa and Near East. Gold dinars minted by the caliph circulated in the Maghreb and in Spain. By the eighth century CE, West African gold was in high demand in the Islamic World. The metal was acquired from the miners in the Bambuk region of the Senegal River. This region was to the south of the Kingdom of Ghana in the Sahel region. Ghana became an important trading post in the gold trade. Control over the gold route was one of the principal factors behind the rise of the Kingdom of Ghana. The Arab geographers described Ghana as the 'land of gold'. From the ninth century CE onwards, the rulers of the Maghreb started using Sudanese gold in minting their dinars (Fagan 2009: 72).

Infantry remained the striking force of the precolonial African polities. For instance, the Almoravids who ruled Mauretania in the eleventh century CE maintained mostly an infantry force. This is despite the fact that the Spanish Muslims copied Frankish heavy feudal cavalry and the Western style tournaments were adopted by the Syrian and Egyptian Muslim cavalrymen (Fisher 2007: 32–3). Bows and arrows diffused into East Africa from the Nile Valley, and it is also possible that bows and arrows evolved independently in West Africa as part of hunting techniques. Rock paintings show that sub-Saharan African people used bows at a very early time. The West African infantry used bows, each of which was between 2.5 feet and 5 feet in height. The English longbow in comparison was about 6 feet. The African bow was made from various types of hardwood, and the bowstring consisted of twisted rawhide (antelope hide was the favourite). The effective range of such a bow was 225 feet compared to 750 feet for the longbow. The bowmen using short bows often wore leather bracelets on their left arms for protection. The arrow was about 1 foot in length. The shaft was made of palm wood or savanna grass or reed and was sometime fletched with feathers. The arrows were carried in a tubular quiver made of bamboo or leather. Each such quiver contained 50 arrows (Smith 1989: 70–4).

In West Africa, the shields were made of hide, wood and basketwork. In the Gold Coast, the shields were often covered with a thin copper plate. By the end of first millennium CE, hilltops in Central Angola were fortified with stone walls. In the equatorial region, the principal weapons used were stabbing spears, shields and curved scimitars. In Eastern Africa, spearmen and bowmen constituted the armies. During battle, the spearmen of Eastern Africa fought several ranks deep behind a wall of shields. Only with the introduction of muskets, was this closed order replaced with an open order. Most of the soldiers were not full-time professional military men but cattle keepers, farmers etc. who fought as part-time soldiers. Overall, we see that in the savanna region, infantry forces fought from a distance with spears, javelins, bows and arrows. And in the equatorial forest, movement along the narrow, twisted pathways prevented large-scale deployment of forces. In such regions, close fighting and ambushes were prevalent, and the main weapons used for close-quarter combat were swords and clubs. The infantry soldiers moved barefoot, which enabled them to advance silently through the forest tracks, and this was advantageous for launching sudden ambuscades (Smith 1989: 4, 58, 79; Reid 2012: 42–4).

In Africa, south of the Sahara, the role of agriculture in food production increased after 500 BCE, when use of metal tools spread in this region. By 1000 CE, agriculture and then stock raising spread rapidly across the continent (except in some large enclaves) (Oliver and Fagan 2002: 342). The introduction of horses and slave raiding also occurred in this time period. For example, the Arab rulers of Tripoli imposed a tribute of slaves on the Libyan chiefs. The Libyan chiefs were raiding the neighbouring black peoples from the fifth century BCE. If anything, the frequency and intensity of slave raiders increased in the medieval period. In the tenth century CE, the Arab governors of North Africa pressured the rulers of Sudan to export slaves. Black slaves, both male and female, were sold to Egypt and the Muslim countries of the Middle East. Thus, slave raiding was in operation long before the Western Europeans came into West Africa. The advent of the Europeans resulted in the replacement of trans-Saharan slave trade with the trans-Atlantic slave trade. Before the intrusion of the Western Europeans, in return for slaves, African polities like Sudan and Chad received horses from the Arabs (Levtzion 2002: 638–9). The Europeans would replace horses with firearms in exchange for slaves. The military effectiveness of horse in the interiors of Africa was somewhat limited due to terrain and climate. To a great extent, the rulers of the 'black' African polities used horses for conducting slave raids.

Between 1000 and 1500 CE, a series of state systems emerged in certain regions, like the Great Lakes, the coastal forest of Western Africa, the west-central savanna and South-East Africa. These states were products of superior iron weaponry and iron implements (Reid 2012: 77). Big polities in fact emerged in the Middle Niger Bend. The central floodplain stretching from Djenne far downstream made agriculture profitable. The desert, in the mid-fifteenth century, was much further north than it is today. Important cities like Gao, Timbuktu and Awdaghust emerged in this region. These cities derived food from the surrounding fertile hinterlands and became trading emporiums. Urban growth and political centralisation went hand in hand. By the tenth century, Gao in the north-eastern corner of the Middle Niger Bend became the centre of Songhay, then a small kingdom. Songhay's power was based on its control of the trade routes which ran through the capital. It brought gold and silver from the south in exchange for Saharan salt and North African goods. Salt was not available in Central Africa. Camel caravans travelled southward from Sijilmassa south to Taghaza, where they picked up salt cakes from the nearby mines. From Taghaza, the camel caravans went to Walata, Ghana and Jenne on the middle Niger River. Many caravans also followed the route, which entailed crossing over the dunes of Central Sahara to Taghaza and Awdaghust on the borders of the desert. The latter was an important salt trading centre. Salt, sheep and honey from the Sahel region were also available in this town for commerce. The nomads provided guides who used the stars and rocky outcrops for navigation across the desert and also sold camels to the merchants. The Berbers, who were inhabitants of the desert, bred camels and also accompanied the caravans. The Berber pastoralists peopled the region south of Mauretania between the fourth and eighth centuries CE. The camels were loaded at very early hours, and then the caravan moved till the sun appeared over the horizon. Then it became too hot to move further. The caravans rested till late afternoon and then travelled in the night (McDougall 2007: 74; Fagan 2009: 72–6).

Songhay became part of the Mali Empire. The Mali Empire's principal income came from exporting gold. By the end of the fourteenth century, two-thirds of Europe's gold came from Mali. The gold was transported across the Sahara by caravans (Fagan 2009: 66). However, towards the end of the fourteenth century, Mali was weakened due to succession struggles. The Berbers and the Fulani exploited this weakness. The desert dwellers conquered Timbuktu in 1433 CE along with other important cities like Walata and Nema, and the Fulani increased their power in the west (Holst 2015: 46–7).

In the fifteenth century, the collapse of Mali resulted in the rise of Songhay which was previously a vassal state of Mali. Songhay maintained cavalry, infantry and a fleet of canoes which operated along a large stretch of Central Niger. Saleiman Dandi/Sunni Ali (r. 1464–92) was responsible for the establishment of the Songhay Empire (Reid 2012: 59). He pushed the Berbers back into the desert and conquered many of the areas occupied by the Fulani and Bozo. He extended Songhay's power from Gao in the east, running along the Niger, towards Djenne in the south-west, encompassing its inland delta and the Bandiagara Uplands. Songhay supplied black slaves from the surrounding servile tribes to the Muslim polities of North Africa in return for horses. Songhay was located at the borderland between the Sahara and the savanna. The economic power of Songhay was dependent on long-distance foreign trade, which occurred both in the east–west and in the north–south directions. For instance, the town of Djenne was at the south-western corner of the Middle Niger Delta. This area was transformed annually into a vast shallow floodplain which provided opportunities for good harvests and fishery. Simultaneously, Djenne was part of the east–west trading routes of the savanna. Thus, this city became an important commercial entrepot. Since it was situated on the Niger, it tied these trade routes into the north–south trade which shipped goods along the river, connecting the east–west trade to the northern outpost of Timbuktu (Fisher 2007: 37; Holst 2015: 47, 49).

For maintaining political power, Sunni Ali had a well-trained army. Slaves played an important role in the army of Sudan. While the cavalry of the savanna states was made up of freemen, most of the infantry personnel were unfree. For instance, the archers of the Mali army were slaves. At the core of the Mali army were the archers, who constituted the longstanding permanent troops under the authority of the central government. In West Africa, military service was obligatory for all the freemen, and this in turn enabled the rulers to raise large armies. For example, the Empire of Ghana, which lasted from the eighth century to the first half of the thirteenth century CE, was able to mobilise theoretically 200,000 men. Generally, the freemen provided their own arms, clothing and weapons. But from *c.* 1400 CE, as weapons became more sophisticated and differentiated from hunting tools, most of the weapons were provided from the arsenals of the kings (Smith 1989: 43, 55).

Robert S. Smith asserts that from the thirteenth century onwards, in the savanna of West and Central Sudan, cavalry became the principal striking force. This was possible because of the introduction of improved saddles, stirrups and larger breed of horses from North Africa. Till the fifteenth century, the mounted warriors of West Africa used small swords (like scimitars, adopted from the North African Muslims) and light spears for throwing. A horseman carried several spears with wooden shafts and leaf shaped blades. Small swords were also used by the infantry. Only from sixteenth century onwards did lances, chain mail armour and long broadswords become important. For Songhay, we can say that cavalry and heavy infantry remained of equal importance. Songhay was not a cavalry state per se. For transporting its army and for provisioning of the ground troops, Songhay maintained a large fleet of canoes. It is to be noted that most of the battles which Songhay fought took place along the banks of the Niger River. Before Songhay, the rulers of Kanem (Chad and Nigeria) used canoes in Lake Chad. Each such canoe was able to carry 100 soldiers (Smith 1989: 48–70).

In addition, Sunni Ali also initiated social changes to buttress his rule. The big towns like Timbuktu and Djenne were dominated by the Muslim learned class. In order to gain legitimacy, Sunni Ali converted into Islam. Though the Muslims were numerically smaller, due to their expertise in trade and learning, they were valuable in running the empire's political and economic apparatus. Nevertheless, not all the Muslims proved loyal to Sunni Ali. For instance, the *ulemas* (Islamic theologians) of Timbuktu maintained connections with the

Tuaregs, who were enemies of the Songhay Empire. In response, between 1469 and 1486, Sunni Ali carried out large-scale purges of the Muslim clerics, expelling many of them from his domain (Holst 2015: 49–52).

After Sunni Ali's death in 1492, his son Abu Bakr Dao (Sunni Barou) became the ruler of Songhay. In order to establish his authority, he rejected Islam. Instead, he utilised the non-Islamic model of magician-king. Then he sidelined the Muslim scholars, clerics and merchants. This rift between the Muslim stakeholders and the ruler was exploited by Abu's political enemies. Muhammad Tune, who was one of the famous commanders of Sunni Ali, recruited troops from North-West Songhay where Muslim influence (scholarly and commercial) was strongly entrenched, and he challenged the 'heretic' Abu. On 19 February 1493, Tune's force clashed with Abu Bakr near Gao. Tune emerged victorious and ascended to the throne as Askaya Muhammad, thus establishing the Askyas dynasty. We could speculate that Tune's victory was probably because his Muslim soldiers were better skilled with bows and arrows and were motivated by Islamic religious zeal. He further extended the empire through military conquest, and under him the Songhay Empire became the largest precolonial African empire, controlling about 1,400,000 sq km (Holst 2015: 53–9).

North-West Africa, due to its close linkage with the Mediterranean Basin, had developed an advanced state system from the time of Carthage onwards. By the fifteenth century, Morocco was a powerful kingdom. Due to its interaction with Christian Europe, it readily adopted gunpowder warfare. However, it had to import firearms and powder from Europe, and this required more money. So it decided to attack Songhay in order to tap the latter's gold resources (Holst 2015: 63–4).

The Moroccan expedition across the Sahara in 1591 resulted in a pitched battle in West Africa at Tondibi on the north-east bend of the Niger. The Moroccans deployed cavalry and musketeers, while Songhay had cavalry and infantry but no firearms. Morocco's ruler, Sultan Ahmad al-Mansur, in order to control the trans-Saharan trade, sent 5,000 men under Judar Pasha across the Sahara. Most of the personnel were Iberian mercenaries, including a number of recent Muslim converts. Judar Pasha's force had horses, muskets and some light cannons. However, dehydration and starvation killed 2,000 of the personnel. Meanwhile, Songhay put into the field 12,500 cavalry and 10,000 infantry. So on 12 March 1591, the Moroccan troops were outnumbered when the Songhay force attacked them. The Songhay army deployed infantry (spearmen and archers) in the centre and cavalry at both the flanks. Songhay infantry advanced behind cover of the cattle. It was a typical Central African battle tactic. The Moroccan musketeers fired at the cattle, which ran back among the Songhay infantry, causing disorganisation among the latter. Then the Songhay cavalry charge also failed. This was probably because the Moroccans had bigger horses and were more adept in fighting from horseback compared to the mounted men of Songhay. The Moroccans remained master of the battlefield (Reid 2012: 60–1). This was probably the first victory of gunpowder weapons in Africa. Cavalry played a minor role in deciding this battle.

Till the sixteenth century, the Angolan army was deployed in open order. The soldiers spread out fairly widely in order to allow the individual soldiers to use their skills effectively while still being mutually supportive. The army was not a disorderly mass but was organised in regular units with a formal command structure. Each unit had its own standard, and flags and orders were issued through drums and trumpets of various sizes. The basic tactics was to use light infantry to make frontal attacks in order to pin down the hostiles and at the same time attempting to envelop or outflank the enemy army by attacking its flanks or rear. The elite heavy infantry was used as a reserve either to pierce the enemy line or to prevent eruption of the hostile force through one's main battle line. In Central Africa, for ecological reasons, infantry remained the key force (Thornton 1988: 360–78). Sleeping sickness disease, jungles and swamps prevented large-scale use of cavalry.

For studying warfare in pre-modern South Africa, our source is the rock art. The Bushmen (San), who were hunter-gatherers, inhabited the region known in modern times as the Orange Free State and the Kalahari Desert. Rock art, especially in red, black, white and ochre, was the creation of *shaman* (religious magicians) artists. These pictures depict bows, arrows and quivers, broad-bladed spears, short stabbing spears, battle axes, etc. The arrowheads were made of bone, and some of the arrows were barbed. The argument by some scholars that the drawings with arrows on the rock faces did not represent combat but metaphorical scenes of fighting diseases is just not tenable. The Bushmen were in contact with the Bantu-speaking agro-pastoralists. In some areas, the San entered into a sort of serf relation with the Bantu-speaking mixed farmers. The former got cattle and crops from the latter, especially if the *shamans* succeeded in bringing rainfall. At times, some Bush women were also married to the Bantu males (Dowson 1994: 332–45). We can speculate that warfare occurred among these two communities for cattle, crops and women. Occasionally the San raided the Bantu settlements for capturing cattle (Campbell 1986: 255–7). Now we turn the spotlight to see the evolution of warfare in the world's smallest continent and its neighbouring islands.

Oceania

Humans reached New Guinea and Australia about 35,000 years ago and soon killed the megafauna. This megafauna included giant kangaroos, rhino-like marsupials called diprotodonts, ostrich-like flightless birds, each weighing 181 kilos, and big reptiles like a one-ton lizard, land-dwelling crocodiles, etc. Diamond speculates that the giant mammals of Australia and New Guinea died but the big mammals of Africa and Eurasia survived into modern times because the later had coevolved with proto-humans for hundreds of thousands of years. Hence, the mammals of Africa and Eurasia had lot of time to evolve a fear of the humans as the proto-humans' hunting skill slowly improved. In contrast, the giant mammals of Australia and New Guinea, without any evolutionary preparation, suddenly confronted the humans with developed hunting skills and were soon wiped out. For the giant mammals of New Guinea and Australia, it was a sort of 'culture clash'. The humans in New Guinea and Australia killed these giant mammals for food, and some of these animals also died indirectly due to fire and habitat modification by the humans. The net result was that the humans in New Guinea and Australia, as in the case of North and South America, were left with no indigenous animals which could be domesticated (Diamond 2005: 42–4).

So the destruction of the inmates of the 'quasi-Jurassic Park' in New Guinea and Australia seriously impeded the economic and technological evolution of the humans in these regions. We could speculate about a counterfactual scenario. If the one-ton lizards were domesticated and used for agriculture and riding, then the course of history would have been different. Such big animals would have greater traction power than the cattle and horses. Thus, agriculture along the coastal regions of Australasia then would have been much more intensive than the pre-modern agriculture of North India, South China, Fertile Crescent, and Western Eurasia. And in the medieval era, the Australasian Aborigines, riding giant lizards which were more powerful than horses and elephants, would have overwhelmed the sedentary states of Eurasia. In such a setting, the pre-modern world would have been dominated by lizard armies of Australasia rather than cavalry armies which swept out from the Central Eurasian steppes.

New Guinea

The indigenous people of New Guinea were not organised in any centralised political unit. Farmers in Eurasia from about 5000 BCE, by generating a surplus, were able to support full-time craft specialists who were not engaged in producing their own food but in developing techniques and technologies. Due to the absence of an agricultural revolution in New Guinea, this specialisation did not occur among the indigenous people. In contrast, sedentary food-producing communities based on agriculture in Eurasia from around 3000 BCE were able to support chiefs, who later became kings supported by bureaucrats and professional soldiers. Stored food made available by agriculture fed the priests who developed religion and provided a justification for royal power and wars of expansion. Again, surplus food allowed the emergence of scribes (necessary for accounting and administering) and artisans such as metalworkers who developed sophisticated weapons. The emergence of writing formed the seedbed of information explosion. This not only allowed the emergence of complex administrative infrastructures but also the development of complex knowledge systems. Intensive farming did not develop in Australasia due to the absence of domesticated animals. Hence, political centralisation and subsequent progress in theoretical and practical knowledge of warfare did not occur in New Guinea and Australia (Diamond 2005: 30, 90–1).

Here we will focus on the Dugum Dani tribe, who are Neolithic warrior farmers inhabiting the Grand Valley of the Baliem River in the central highlands of Western New Guinea. The Grand Valley is 40 miles long and 10 miles wide and 5,000 feet above the tropical jungles of New Guinea. The region is filled with wildlife like kangaroos, rats, foxes and several types of birds. Large animals which could be domesticated and harnessed for agricultural purpose are absent. Hence, most of the agricultural work has to be done by human muscle power rather than by draught animals. Moreover, the hilly swampy region is also not suited for large-scale intensive cultivation and transportation. All these factors to an extent explain why the agricultural revolution missed New Guinea. And this in turn resulted in the tribes of New Guinea being stuck in the Stone Age culture. In addition to scattered farming, the Dani people raise pigs. Besides pigs, some of them also keep dogs which aid them in hunting birds and furry marsupials in the forests. Their food consists of sweet potatoes and pigs' flesh (Gardner and Heider 1969: xiii, 4–41).

The tools of the Dani people are made from stones, bones, wood and bamboo. Metal and even pottery were not known among them till recent times. They made fire by sawing a bamboo thong rapidly back and forth under a split hardwood stick. The friction creates embers which fall into the tinder and ignite it. The whole process takes one or two minutes. They use stone adzes to sharpen the blades of their digging sticks. Adzes are used to chop down trees, and stone axes are used to split the wood. Limestone rocks and sand in the stream beds are used to shape and smooth spear shafts. The Dani buildings are held together with vines and bamboo thong lashings (Gardner and Heider 1969: 28–9).

Generally, there exists a division of labour among the Danis. War is a gendered activity. While women work on the fields, men take turns as sentries and perform guard duties. Robert Gardner and Karl G. Heider, after observing the Dugum Dani warfare in 1961, concluded that they fought ritualised battle. The warriors were equipped with stone axes, bows and arrows. They decorated themselves with pig tusks in their noses and wore feathered headdresses and ornaments of various shells across their chests before participating in combat. The Dani warriors engaged in battle not merely to save their families and settlements but also to defend and exhibit their masculine identities (Gardner and Heider 1969: 8–30).

Anthropologist Paul Roscoe challenges the assertion that the Dugum Dani in the central highlands of New Guinea engaged in ritualised battle. Roscoe claims that rather than culture, the terrain shaped the dynamics of battle. For instance, the waterlogged terrain of the mid-Baliem Valley obstructed the capacity of the Dani force to deploy and carry out large-scale manoeuvres. Moreover, the surrounding swamps also discouraged pursuit of the defeated enemy. In fact, tribal warfare could be bloody. According to one calculation, 16 per cent of all the deaths in two generations occurred due to organised violence. Generally, the death rate is 0.4 per battle and 0.9 per raid (Roscoe 2011: 56–61). So raids and ambushes rather than battles were more dangerous in the case of indigenous combat in New Guinea and also in North America. For protection against raids, the villages of Dugum Dani have watchtowers (made with poles and vines) which are always manned (Gardner and Heider 1969: 4, 8, 30). Next, we study the evolution of warfare in the continent of Australia.

Australia

The nineteenth-century settlers portrayed the Aborigines as violent savages in order to justify their subjugation and expropriation of land from them. Emphasis on the violent savagery of the Aborigines legitimised the draconian measures of the colonial government towards the former and also appropriation of their land by the industrious 'white' people. The current view that the Aborigines lived in a peaceful society without war before the European intrusion is equally erroneous (Connor 2013: 8–9).

We can recreate the nature of combat from the rock art available at the Arnhem Land Region of the Northern Territory of Australia. The oldest of them are 10,000 years old. At that time, the climate became arid. Drier climate and a lower sea level characterised the environment at that time. Between 20,000 and 10,000 years ago, a period of extreme aridity resulted in the shrinkage of food and water resources. Ecological stress might have resulted in competition over these resources. Some of the rock art specimens depict men fighting with spears and boomerangs. Quite a few figures depict men dodging spears and others chasing enemies with weapons raised. A number of figures show men down with spears embedded in their torsos. Some of the high-ranking war leaders wore ceremonial headdresses. The climate changed about 6,000 years ago. This period, known as the Intermediate Phase, saw the rise of the world's oceans, which cut off Australia from the surrounding islands. Some of the rock figures which are 3,000 years old show the use of barbed spears. During this time, battles between tribes occurred (Tacon and Chippindale 1994: 211–48). Historian Azar Gat claims that the Australian Aborigines did not invent bows and arrows because, due to their isolation from mainland Asia, no further technological development occurred among them. The only long-range weapon they used was the boomerang, which is depicted in their rock art (Gat 2006: 18).

Just before the European arrival in Australia in 1788, there were 300,000 hunter-gatherers in this continent organised in 400–700 regional groups. Each band had about 500–600 people (Gat 2006: 17). Aboriginal society was organised in non-hierarchical kinship groups, which in turn limited the scale and scope of warfare among them. Aborigines fought to protect their kinship group and to uphold their group's status vis-à-vis the neighbouring groups. Many groups presented young males with weapons as part of the initiation ceremony. For instance, Darug youths were given spears, while in North Queensland youths were given shields after their initiation (Connor 2013: 11).

There were about 4,000 indigenous people in Tasmania before the arrival of the Europeans. This island got separated from Australia about 10,000 years ago. As a result, their technology remained poorer than the tribes of mainland Australia. They did not possess even the boomerang. Nevertheless, raiding and counter-raiding occurred among them (Gat 2006: 19).

New Zealand

On the Chatham Islands, 500 miles east of New Zealand, the Moriori people were attacked by the Maoris on 19 November 1835. A ship carrying 500 Maori warriors arrived, equipped with guns acquired from the Europeans, clubs and axes. On 5 December 1835, another ship came with 400 more Maoris. Since the Morioris were numerically superior, an organised resistance could have defeated the Maoris. But the Morioris had a tradition of resolving disputes amicably. They decided for peace and a division of resources. However, the Maoris attacked and killed many Morioris, and many of their dead bodies were cooked and eaten by the victors. The rest of the Morioris were enslaved (Diamond 2005: 53). It was a case of successful amphibious expedition by the Maoris.

The Morioris were a small, isolated group of hunter-gatherers, inexperienced in war and lacking organised leadership. In contrast, the Maori invaders from New Zealand's North Island came from a dense population of farmers who frequently engaged in intense warfare and were assembled under effective leadership. In fact, both the Maoris and the Moriori were Polynesian farmers. The Maoris colonised New Zealand around 1000 CE. Another group of Maoris who colonised the Chatham Islands became Morioris. The Morioris became hunter-gatherers, while the North Island Maori turned to intensive farming. The Morioris had to revert to being hunter-gatherers because Maori tropical crops could not be grown in the cold climate of Chatham Island. The Moriori depended on shellfish, seabirds and fish which were captured with hands and clubs and required no further sophisticated technology. The Chatham Islands were able to support only about 2,000 hunter-gatherers. With no other accessible islands to colonise, writes Diamond, the Morioris renounced war. They further controlled population by castrating some male infants (Diamond 2005: 54–6). Hence, they proved to be 'easy meat' for the aggressive Maori farmers.

Malaita

A large volcanic island in the Solomons, before the coming of the Europeans, Malaita had a population of 65,000. Mountain ridges divide the island into several small habitable areas, some containing fewer than 100 people. A big chief was able to control 200 households. Taro farming and pig raising were the principal economic activities of the people. The presence of so many small chiefships resulted in intense competition, and violence was endemic. Boys were trained for war in mock battles and joined raiding expeditions armed with a club. No women married them unless they had shed enemy blood. At times, raids by 200–300 men occurred, and the target group could be annihilated completely. If there were survivors among the defeated parties, then revenge killings lasted for generations (Younger 2014: 1–14).

Conclusion

In precolonial America, Oceania and Africa (except the Maghreb, Egypt and the mid-bend of the Niger), high intensity wars waged by bureaucratic standing armies supported by a hierarchical political setup were absent. From the perspectives of deep history, we have two structuralist arguments: a cultural one by Haldon and a geographical one by Diamond. At least to me, the geographical one seems to carry more traction. Agriculture was more developed in Mesoamerica and in parts of South America compared to the Eastern and south-western North America. Moreover, in South America, llamas were available for carrying baggage. Agricultural surplus generated trade which could be carried on the back of the llamas. So big empires like the Aztec, Maya and Inca emerged in Latin America. Camels, gold and salt resulted in long-distance overland trade, which in turn accelerated the formation of big polities in parts of North and West Africa, and such polities were able to muster a large number of soldiers. But in precolonial North America, the political unit remained in the form of tribes and bands. Again, warfare was not static. The introduction of recurve bows among the North American tribes and horses and firearms in certain regions of Africa raised the intensity of conflict. However, the absence of iron technology, wheels and draught animals prevented conflicts waged by the precolonial Latin American empires from becoming as deadly as those waged by the Eurasian polities since the time of the Neo-Assyrians.

But warfare was not always a ritualised cosmetic affair among the tribes and chiefdoms and in the case of Latin American empires like the Maya, Aztec and the Inca. Aztec warfare was certainly more than being flower wars. However, formal battles both in New Guinea and in south-west regions of North America resulted in few casualties. As we have seen, raids and ambushes often targeted the non-combatants So, prehistoric warfare involved an element of what modern theorists call 'total war'. Both among the Aztecs and the Dugum Danis, a large percentage of the adult males participated in warfare. So this was another similarity with total war which emerged in the industrial societies of the West during the first half of the twentieth century. However, the absence of iron weapons and baggage-carrying animals severely limited the casualties, at least in absolute terms, in the case of the precolonial societies of North America, Latin America and Australasia.

One author claims that the characteristics of precolonial Latin American warfare were similar to hunting (Lau 2014: 313). Even the Mongol way of war has certain characteristics of hunting. All sorts of warfare emerged from hunting. Captive taking was an important if not the primary aspect of pre-Columbian Latin American warfare. Prehistoric North American warfare involved scalping of the enemies. In Latin America, this process became more refined and elaborated in mass ritual killings of the captives in the public domain. In the case of pre-Columbian warfare in Mesoamerica and Andes, torture of captives, taking of trophies and dismembering the bodies of the prisoners of war were important mechanisms of celebrating victory beyond the battlefield. Decapitation of the captives and trophy taking functioned as symbols of dominance of the victors over the vanquished. The act of obtaining and wearing trophies made from the body parts of defeated enemies reflected the linkages between combat, display and emergence of masculine warrior identities. Broadly, ritualised deaths of the captives were part of a cultural system of Mesoamerica that linked warfare, religion and power into a holistic worldview. And this characteristic was missing in Eurasian history.

Absence of draught animals and intensive agriculture in the two Americas and in Australasia coupled with the ecological conditions of Africa resulted in the dominance of infantry warriors till the coming of colonialism. These four continents missed the chariot-centric war of the Bronze Age and cavalry-centric forces during the medieval era. This explains the military divergence of these four continents from Eurasia. And the militaries of these continents were no match against the firepower-effective military machine which developed in Western Eurasia during the early modern era.

References

Arkush, Elizabeth, '"I Against My Brother": Conflict and Confederation in the South-Central Andes in Late Prehistory', in Andrew K. Scherer and John W. Verano (eds.), *Embattled Bodies, Embattled Places: War in Pre-Columbian Mesoamerica and the Andes* (Washington, DC: Dumbarton Oaks, 2014), pp. 199–226.

Balderas, Ximena Chavez, 'Sacrifice at the Templo Mayor of Tenochtitlan and Its Role in Regard to Warfare', in Andrew K. Scherer and John W. Verano (eds.), *Embattled Bodies, Embattled Places: War in Pre-Columbian Mesoamerica and the Andes* (Washington, DC: Dumbarton Oaks, 2014), pp. 171–97.

Barr, Daniel P., *Unconquered: The Iroquois League at War in Colonial America* (Westport, CT/London: Praeger, 2006).

Bey III, George J., 'The Role of Ceramics in the Study of Conflict in Maya Archaeology', in M. Kathryn Brown and Travis W. Stanton (eds.), *Ancient Mesoamerican Warfare* (Walnut Creek, CA: Altamira Press, 2003), pp. 19–29.

Butters, Luis Jaime Castillo, 'Taming the Moche', in Andrew K. Scherer and John W. Verano (eds.), *Embattled Bodies, Embattled Places: War in Pre-Columbian Mesoamerica and the Andes* (Washington, DC: Dumbarton Oaks, 2014), pp. 257–82.

Campbell, C., 'Images of War: A Problem in San Rock Art Research', *World Archaeology*, vol. 18, no. 2 (1986), pp. 255–68.

Connor, John, 'Traditional Indigenous Warfare', in Craig Stockings and John Connor (eds.), *Before the Anzac Dawn: A Military History of Australia to 1915* (Sydney: University of New South Wales Press, 2013), pp. 8–20.

Cordell, Linda S. and Smith, Bruce D., 'Indigenous Farmers', in Bruce G. Trigger and Wilcomb E. Washburn (eds.), *The Cambridge History of the Native Peoples of the Americas*, vol. 1, *North America*, Part 1 (Cambridge: Cambridge University Press, 1996), pp. 201–66.

Diamond, Jared, *Guns, Germs and Steel: A Short History of Everybody for the Last 13,000 Years* (1997, reprint, London: Vintage, 2005).

Dowson, Thomas A., 'Reading Art, Writing History: Rock Art and Social Change in Southern Africa', *World Archaeology*, vol. 25, no. 3 (1994), pp. 332–45.

Fagan, Brian, *The Great Warming: Climate Change and the Rise and Fall of Civilizations* (2008, reprint, New York/London: Bloomsbury Press, 2009).

Fisher, Humphrey J., '"He Swalloweth the Ground with Fierceness and Rage": The Horse in the Central Sudan, II. Its Use', in John Lamphear (ed.), *African Military History* (Aldershot: Ashgate, 2007), pp. 27–51.

Gardner, Robert and Heider, Karl G., *Gardens of War: Life and Death in the New Guinea Stone Age* (London: Andre Deutsch, 1969).

Gat, Azar, *War in Human Civilization* (Oxford: Oxford University Press, 2006).

Golden, Charles W., 'The Politics of Warfare in the Usumacinta Basin: La Pasadita and the Realm of Bird Jaguar', in M. Kathryn Brown and Travis W. Stanton (eds.), *Ancient Mesoamerican Warfare* (Walnut Creek, CA: Altamira Press, 2003), pp. 31–48.

Grove, David C., 'The Preclassic Societies of the Central Highlands of Mesoamerica', in Richard E.W. Adams and Murdo J. Macleod (eds.), *The Cambridge History of the Native Peoples of the Americas*, vol. 2, *Mesoamerica*, Part 1 (Cambridge: Cambridge University Press, 2000), pp. 122–55.

Gutierrez, Gerardo, 'Aztec Battlefields of Eastern Guerrero: An Archeological and Ethnohistorical Analysis of the Operational Theater of the Tlapanec War', in Andrew K. Scherer and John W. Verano (eds.), *Embattled Bodies, Embattled Places: War in Pre-Columbian Mesoamerica and the Andes* (Washington, DC: Dumbarton Oaks, 2014), pp. 143–70.

Haas, Jonathan and Creamer, Winifred, 'Warfare Among the Pueblos: Myth, History, and Ethnography', *Ethnohistory*, vol. 44, no. 2 (1997), pp. 235–61.

Haldon, John, 'Some Concluding Remarks', in Andrew K. Scherer and John W. Verano (eds.), *Embattled Bodies, Embattled Places: War in Pre-Columbian Mesoamerica and the Andes* (Washington, DC: Dumbarton Oaks, 2014), pp. 385–401.

Hassig, Ross, *Aztec Warfare: Imperial Expansion and Political Control* (Norman: University of Oklahoma Press, 1988).

Hassig, Ross, 'The Aztec World', in Kurt Raaflaub and Nathan Rosenstein (eds.), *War and Society in the Ancient and Medieval Worlds: Asia, the Mediterranean, Europe and Mesoamerica* (Washington, DC: Center for Hellenic Studies, distributed by Harvard University Press, 1999), pp. 361–87.

Hicks, Frederic, '"Flowery War" in Aztec History', in Miguel A. Centeno (ed.), *Warfare in Latin America*, vol. 1 (Hampshire: Ashgate, 2007), pp. 1–8.

Holst, Christian, *Muslim Traders, Songhay Warriors and the Arma: The Social Destruction of the Middle Niger Bend from 1549 to 1660* (Kassel: Kassel University Press, 2015).

Inomata, Takeshi, 'War, Violence, and Society in the Maya Lowlands', in Andrew K. Scherer and John W. Verano (eds.), *Embattled Bodies, Embattled Places: War in Pre-Columbian Mesoamerica and the Andes* (Washington, DC: Dumbarton Oaks, 2014), pp. 25–56.

Joyce, Arthur A., 'Imperialism in Pre-Aztec Mesoamerica: Monte Alban, Teotihuacan, and the Lower Rio Varde Valley', in M. Kathryn Brown and Travis W. Stanton (eds.), *Ancient Mesoamerican Warfare* (Walnut Creek, CA: Altamira Press, 2003), pp. 49–72.

Joyce, Arthur A., 'Warfare in Late/Terminal Formative-Period Oaxaca', in Andrew K. Scherer and John W. Verano (eds.), *Embattled Bodies, Embattled Places: War in Pre-Columbian Mesoamerica and the Andes* (Washington, DC: Dumbarton Oaks, 2014), pp. 117–41.

Lambert, Patricia M., 'The Archaeology of War: A North American Perspective', *Journal of Archaeological Perspective*, vol. 10, no. 3 (2002), pp. 207–41.

Lau, George F., 'A Materiality of Opposition: On Ancient Andean Conflict and Organization in Northern Peru', in Andrew K. Scherer and John W. Verano (eds.), *Embattled Bodies, Embattled Places: War in Pre-Columbian Mesoamerica and the Andes* (Washington, DC: Dumbarton Oaks, 2014), pp. 311–40.

LeBlanc, Steven A., *Prehistoric Warfare in the American Southwest* (Salt Lake City: University of Utah Press, 1999).

Levtzion, Nehemia, 'The Sahara and the Sudan from the Arab Conquest of the Maghrib to the Rise of the Almoravids', in J.D. Fage and Roland Oliver (eds.), *The Cambridge History of Africa*, vol. 2, *From 500 BC to AD 1050* (1978, reprint, Cambridge: Cambridge University Press, 2002), pp. 637–84.

Levy, Jack S. and Thompson, William R., *The Arc of War: Origins, Escalation and Transformation* (Chicago: University of Chicago Press, 2011).

Malone, Patrick M., *The Skulking Way of War: Technology and Tactics among the New England Indians* (1991, reprint, Lanham: Madison Books, 2000).

Mauny, Raymond, 'Trans-Saharan Contacts and the Iron Age in West Africa', in J.D. Fage and Roland Oliver (eds.), *The Cambridge History of Africa*, vol. 2, *From 500 BC to AD 1050* (1978, reprint, Cambridge: Cambridge University Press, 2002), pp. 272–341.

McDougall, E. Ann, 'The View from Awdaghust: War, Trade and Social Change in the Southwestern Sahara from the Eighth to the Fifteenth Century', in John Lamphear (ed.), *African Military History* (Aldershot: Ashgate, 2007), pp. 69–99.

Milner, George R., 'Warfare in Prehistoric and Early Historic Eastern North America', *Journal of Archaeological Research*, vol. 7, no. 2 (1999), pp. 105–51.

Moss, Madonna L. and Erlandson, Jon M., 'Forts, Refuge Rocks, and Defensive Sites: The Antiquity of Warfare Along the North Pacific Coast of North America', *Arctic Anthropology*, vol. 29, no. 2 (1992), pp. 73–90.

Ogburn, Dennis E., 'The Scope of Inca Warfare as an Imperial Strategy of Conquest and Control', in Andrew K. Scherer and John W. Verano (eds.), *Embattled Bodies, Embattled Places: War in Pre-Columbian Mesoamerica and the Andes* (Washington, DC: Dumbarton Oaks, 2014), pp. 365–83.

Oliver, Roland and Fagan, Brian M., 'The Emergence of Bantu Africa', in J.D. Fage and Roland Oliver (eds.), *The Cambridge History of Africa*, vol. 2, *From 500 BC to AD 1050* (1978, reprint, Cambridge: Cambridge University Press, 2002), pp. 342–409.

Quilter, Jeffrey, 'Moche Politics, Religion, and Warfare', *Journal of World Prehistory*, vol. 16, no. 2 (2002), pp. 145–95.

Reid, Richard J., *Warfare in African History* (Cambridge: Cambridge University Press, 2012).

Roscoe, Paul, 'Dead Birds: The "Theater" of War Among the Dugum Dani', *American Anthropologist*, vol. 113, no. 1 (2011), pp. 56–70.

Scherer, Andrew K. and Golden, Charles, 'War in the West: History, Landscape, and Classic Maya Conflict', in Andrew K. Scherer and John W. Verano (eds.), *Embattled Bodies, Embattled Places: War in Pre-Columbian Mesoamerica and the Andes* (Washington, DC: Dumbarton Oaks, 2014), pp. 57–92.

Shimada, Izumi, Schaaf, Crystal Barker, Thompson, Lonnie G., and Ellen, Mosley-Thompson, 'Cultural Impacts of Severe Droughts in the Prehistoric Andes: Application of a 1,500-Year Ice Core Precipitation Record', *World Archaeology*, vol. 22, no. 3 (1991), pp. 247–70.

Shinnie, P.L., 'The Nilotic Sudan and Ethiopia, c. 660 BC to c. AD 600', in J.D. Fage and Roland Oliver (eds.), *The Cambridge History of Africa*, vol. 2, *From 500 BC to AD 1050* (1978, reprint, Cambridge: Cambridge University Press, 2002a), pp. 210–71.

Shinnie, P.L., 'Christian Nubia', in J.D. Fage and Roland Oliver (eds.), *The Cambridge History of Africa*, vol. 2, *From 500 BC to AD 1050* (1978, reprint, Cambridge: Cambridge University Press, 2002b), pp. 556–88.

Smith, Bruce D., 'Origins of Agriculture in Eastern North America', *Science*, New Series, vol. 246, no. 4973 (1989), pp. 1566–71.

Smith, Robert, *Warfare & Diplomacy in Pre-colonial West Africa* (1977, reprint, London: James Curry, 1989).

Snow, Dean R., 'The First Americans and the Differentiation of Hunter-Gatherer Cultures', in Bruce G. Trigger and Wilcomb E. Washburn (eds.), *The Cambridge History of the Native Peoples of the Americas*, vol. 1, *North America*, Part 1 (Cambridge: Cambridge University Press, 1996), pp. 125–99.

Tacon, Paul and Chippindale, Christopher, 'Australia's Ancient Warriors: Changing Depictions of Fighting in the Rock Art of Arnhem land, NT', *Cambridge Archaeological Journal*, vol. 4, no. 2 (1994), pp. 211–48.

Thornton, John K., 'The Art of War in Angola, 1575–1680', *Comparative Studies in Society and History*, vol. 30 (1988), pp. 360–78.

Toyne, J. Marla and Vargas, L. Alfredo Narvaez, 'The Fall of Kuelap: Bioarchaeological Analysis of Death and Destruction on the Eastern Slopes of the Andes', in Andrew K. Scherer and John W. Verano (eds.), *Embattled Bodies, Embattled Places: War in Pre-Columbian Mesoamerica and the Andes* (Washington, DC: Dumbarton Oaks, 2014), pp. 341–64.

Tung, Tiffiny A., 'Making Warriors, Making War: Violence and Militarism in the Wari Empire', in Andrew K. Scherer and John W. Verano (eds.), *Embattled Bodies, Embattled Places: War in Pre-Columbian Mesoamerica and the Andes* (Washington, DC: Dumbarton Oaks, 2014), pp. 227–56.

Verano, John W., 'Warfare and Captive Sacrifice in the Moche Culture: The Battle Continues', in Andrew K. Scherer and John W. Verano (eds.), *Embattled Bodies, Embattled Places: War in Pre-Columbian Mesoamerica and the Andes* (Washington, DC: Dumbarton Oaks, 2014), pp. 283–309.

Webster, David, 'Ancient Maya Warfare', in Kurt Raaflaub and Nathan Rosenstein (eds.), *War and Society in the Ancient and Medieval Worlds: Asia, The Mediterranean, Europe, and Mesoamerica* (Washington, DC: Center for Hellenic Studies, distributed by Harvard University Press, 1999), pp. 333–60.

Younger, Stephen P., 'Violence and Warfare in Precontact Melanesia', *Journal of Anthropology* (2014), pp. 1–14, http://doi.org/10.1155/2014/658597.

CONCLUSION

> Wars and different kinds of fighting have always occurred in the world since God created it.
>
> – Ibn Khaldun (b. 1332–d. 1406; quoted from Chaliand 1994: 420)

What was the nature of warfare across the world before the rise of the West in the early modern era? A monolithic master-narrative of infantry-based warfare centred around decisive battles originating in Classical Greece, then continuing through Roman Europe and the medieval period till the onset of gunpowder revolution in Western Europe during the early modern era, is faulty. Similarly, a linear story of heavy infantry armies of Greece and Rome succeeded by medieval Western European knights mounted on armoured cavalry till the advent of gunpowder infantry in the sixteenth century is also too simplistic to explain pre-modern global military history. This book has shown that multiple and varied military traditions (read cultures) existed across various parts of the world till *c.* 1500 CE. And these military cultures borrowed from each other. Just as variety is the spice of life, charting out multiple histories of the different regions is also essential to gaining a more nuanced understanding of our collective past.

The convergence of certain economic and demographic factors resulted in the emergence of warfare in Neolithic times. Hunting practices which evolved almost a million years ago gave birth to the techniques of attacking the target at both its flanks and rear while holding its attention by making a frontal feint. This technique later metastasised into a double envelopment battle, long before the German General Staff conceptualised cauldron/kettle battle during the nineteenth century. Hunting techniques were used in combating other human

DOI: 10.4324/9780429284069-10

groups from about 22,000 BCE, when the population of *Homo sapiens* increased and the number of big game species decreased. Economic inequality among the foragers, bands and tribes also accelerated the propensity towards organised violence (Scheidel 2018: 15–22). The spread of agriculture from 8000 BCE onwards allowed the humans not only to survive but also to grow their numbers. Our unique genes, the use of tools and fire and the propensity of humans to form groups with a strong identity enabled *Homo sapiens* to fight with an intensity and organisational sophistication unknown to other species. The spread of bronze and iron further accelerated these trends.

Then emerged the diversity in war in different regions due to geographical variations. The use of bronze and iron tools supported by intensive agriculture along the river valleys of Eurasia facilitated the rise of states. Due to its managerial capabilities, the state system gave birth to expanding armies supported by elaborate logistical setups. This in turn made combat more intense and casualty prone. So agriculture and the state on one hand allowed humanity to expand and survive but on the other hand also forced *Homo sapiens* to pay a heavier blood price for their progress. Within this macro setting, human agency played an important role in shaping the trajectory of violence. The rise of politico-military entrepreneurs such as Alexander, Julius Caesar, Chandragupta Maurya, Attila and Genghis Khan, among others, decisively shaped the evolution of military organisations in certain regions. This factor further added specificities to the emerging military divergence in different regions of Eurasia at various moments in history.

However, it is necessary to mention a caveat at this juncture. In the two Americas and in Australasia, the absence of iron, bronze and extensive agriculture prevented the rise of a centralised state and the subsequent emergence of intensive large-scale combat. In such a setting, even a potential Julius Caesar or Attila would not have been able attain much success. They probably would have ended their careers as minor tribal chieftains, and their activities would not have been worth even a footnote in world military history. In my analysis, no great importance of culture was found, despite the recent hype of the cultural studies approach. To a great extent, environment and economy shaped the ideas, ideologies and techniques of warfare.

To a person looking at the world, in *c.* 1500 CE, it was not evident that the West was going to achieve global military dominance. From Korea in the east to France in the west, the cavalry tradition held sway. However, there were significant variations within this military tradition. While the Central Eurasian nomads excelled in mounted archery, the sedentary agrarian societies came up with lance-wielding heavy cavalry as the second-best solution. However, the winds of change were already evident. In North-West Europe, seeds were sown for the early modern era military divergence. Western Europe was experiencing the rise of drilled and disciplined infantry which in the near future would be equipped with handguns and supported by light field artillery. This marked the beginning of early modern military divergence from the sixteenth century onwards.

Military divergence in early modern Eurasia occurred not only on land but also in the maritime space. Prior to *c.* 1400 CE, there were four maritime traditions present in Eurasia. The Far Eastern maritime tradition was the product of Chinese and Korean shipbuilding activities. Partly shaped by Arab and Indian influence, it retained its dominance till *c.* 1450 CE. Next in importance was the Indian Ocean maritime tradition. The reach of the Mediterranean maritime tradition (due to the availability of sources, most of the scholarly attention has gone to this tradition), which was mainly the product of Egypt, the Near Eastern polities, Greece, Rome and the Balkans, remained limited outside the inland Mediterranean Sea. Till *c.* 1200 CE, the Atlantic maritime tradition was marginal. But geographical, political and economic factors resulted in the dominance of the Atlantic maritime tradition from *c.* 1500 CE onwards. This maritime tradition generated ocean-going ships equipped with guns. And these ships transported the Western European gunpowder infantry-centric armies to the extra-European world. The Atlantic maritime tradition could have met serious challenges from the Indian Ocean and especially the Far Eastern maritime cultures. However, due to geopolitical and geo-economic factors, the navies of China and South Asia were mothballed at that critical juncture.

In fact, this monograph shows that a military culture based on different forms of cavalry dominated medieval Eurasia, and the military advantage lay with the powers in Eastern Eurasia. In the thirteenth century, a world economic system also emerged, and the quasi-hegemon of this system was China. The disruptive military activities of the steppe nomads (Mongols) and the Black Death partly resulted in the breakdown of this commercial world system (Abu-Lughod 1991). Steppe nomadic activities along with the breakdown of the east-oriented Eurasian commercial system provided Western Eurasia the time and opportunity to generate a gunpowder-centric military system and make a bid for world power. Ironically, the dominance of the Chinese and the Arabs in the Indian Ocean trade and blockade of the overland 'Silk Route' by Timur forced the Western Europeans to search for an alternate route to reach Asia for trading purposes. This imperative motivated the Western European states to explore the Atlantic route. This resulted in the rise of the Atlantic maritime tradition, which set the stage for global supremacy by the Westerners.

However, it was not evident in the sixteenth century that the Western European military paradigm comprising of gunpowder infantry supported by field artillery, siege artillery and ocean-going cannon-armed battleships would dominate the globe. Military divergence was not unique, and as this book has shown, had occurred several times before the sixteenth century in Eurasian history. The technology-centric military revolution concept has become a cliché. After Geoffrey Parker's introduction of this concept to explain the rise of early modern Europe, historians of medieval and ancient history as well as contemporary history use these concepts on and off. We need to think afresh to explain radical changes in the militaries of the various regions of the world at different times. Nevertheless, the role of technology cannot be negated completely. Hence, this book gives adequate space to show the gradual development of military hardware in various regions of the world at different moments in time. For instance, composite bows played an important role in allowing the mounted Eurasian steppe nomads to run roughshod over the sedentary states' infantry armies. In recent times, the scholarly trend is to undermine the importance of military hardware in explaining the rise of the West during the early modern era. True, firepower-intensive weapons by themselves were not adequate in enabling the West to subdue the 'rest'. But superior weapons gave the Western Europeans an advantage which enabled them to recruit local allies in Afro-Asia and in the New World. Superior political management resulting in logistical supremacy and the spread of diseases completed the Western conquest in the non-Western world.

A Global History of Pre-Modern Warfare: Before the Rise of the West, 10,000 BCE–*1500* CE has shown that the traditional notion of inhabitants of the prehistoric era being peaceful and medieval battles being highly ritualised requires a re-evaluation. In the pre-modern era, military mercenaries and especially invading armies were the principal medium for transfer of advanced military technologies from one area to another. For instance, the Hyksos transferred the technology and techniques of war chariots from the Near East to Egypt, and the Mongols transferred gunpowder technology from China to South Asia, West Asia and Western Europe. Long-distance mercenary military service was an established profession in ancient Eurasia. To give an example, Indians from Punjab served as infantry in the Achaemenid army and as elephant drivers and trainers in the Carthaginian and Seleucid armies. And military service functioned as a vehicle for upward mobility. Again, logistics from ancient to modern times remain crucial for conducting military operations. Be it the Neo-Assyrian army or the Mongol cavalry, supply determined the pace of military operations. Logistics is linked with physical geography. We have seen that the issue of pastures qualified the strategic and operational effectiveness of the steppe nomadic cavalry armies.

One way to increase the range of combat effectiveness of a military organisation was to blend elements and ideas from diverse military systems, thus generating a hybrid military organisation. Lieutenant-General Carl von Clausewitz, the nineteenth-century Prussian military philosopher, rightly conceptualises war as a duel. Not only does the duellist have to innovate new techniques, but he also is required to copy some of the practices of his opponent. By integrating innovative techniques with those copied from the enemy, the successful duellist creates a hybrid military organisation. When hybrid military formations emerged, the countries possessing such war machines were able to dominate large parts of Eurasia. Before the emergence of the Mongols, the Macedonian military machine created by Philip II and his son, Alexander, was an example of a hybrid military system. The Mongols were more adaptive than the Huns. South China lacks extensive grasslands, but the Mongol military machine adapted to changing circumstances by adopting selected military elements such as siege engines, techniques of riverine warfare and large numbers of infantry soldiers. Thus, the transformed Mongol army (a hybrid military system) was able to conquer China.

Why did the Mongols fail to exhibit the same level of adaptation and adoption techniques to annex Central and Western Europe? This was probably because China was nearer to the base of their power: Mongolia. Second, the Chinese sedentary empires had been traditional enemies of the tribes of Mongolia. Strong sedentary Chinese empires had played divide and conquer games to keep the tribes of Mongolia disunited and occasionally had launched armies inside the steppe zone. However, Europe was too far away from Mongolia to expend so much energy and resources. Had the Mongols annexed Central and Western Europe, then the gunpowder revolution in Western Eurasia might not have occurred and the course of world history probably would have followed a different trajectory. However, there is a point of caution. Kublai Khan's Mongol regime had changed itself so extensively in attempting to conquer South China that the Yuan Empire became sedentary and was soon overthrown by Chinese indigenous elements. A similar fate might have befallen the Golden Horde Empire if the latter had tried to annex Western and Central Europe. However, Batu Khan did not attempt a wholesale transformation of his appanage in an attempt to control Europe. As a result, the Golden Hordes outlived the Yuan dynasty of Kublai. So there was a limit to the operational range of the hybrid military establishment, and this limit was set by the physical environment. Thus, a hybrid military system could partly but not totally overcome ecological constraints.

The absence of draught animals coupled with ecological conditions in the two Americas, Australasia and Africa (which prevented extensive and intensive agriculture) resulted in the dominance of infantry warriors in these continents till the coming of colonialism. If we take a 'deep history' perspective, then the absence of certain strains of grains in these four continents obstructed the rise of lucrative agriculture. Eurasia was just lucky in this perspective, possessing the right sorts of strains of cereals which enabled the agricultural revolution to flower from 8000 BCE onwards. These four continents largely missed the agricultural revolution and the related iron-wielding chariot and cavalry-centric forces that had evolved in ancient and medieval Eurasia, and this explains the military divergence of these four continents from Eurasia. Thus, the militaries of two Americas and Africa were no match against the firepower-effective military machine which developed in Western Eurasia during the early modern era. Only the steppe nomads of Central Eurasia and the quasi-nomadic sedentary societies (agricultural civilisations ruled by steppe nomads) of China, South Asia and the Middle East were in a position to seriously challenge the military dominance of Western Eurasia.

One part of Eurasia, that is, South-East Asia, was not seriously threatened by the horse-riding nomads due to its tropical forests, hot and humid climate, mountains and rivers. Such physical geography was not conducive for the emergence of horse-drawn chariotry and cavalry armies. Why did South-East Asia fail to develop centralised polities and gunpowder infantry? We could hazard a guess. The very environment which protected South-East Asia from the Islamic Turks and the Mongols also obstructed the growth of centralised bureaucratic states. Intensive agriculture only developed along the coastal plains of South-East Asia. Difficulties in transportation and the absence of intensive agriculture in the hinterlands due to the presence of mountains and dense forests prevented the rise of big agrarian states. Pre-modern agrarian bureaucratic polities depended on extracting surplus from large hinterlands with extensive agriculture and on easy transportation for establishing administrative and military tentacles. Occasionally big states like Sri Vijaya, etc., dependent on revenues from overseas trade, emerged in South-East Asia. However, these South-East Asian polities depended on taxing the maritime trade which was carried on between China, India and West Asia. Thanks to their unique geopolitical location, by dint of controlling the Malacca and Sunda straits, the South-East Asian maritime polities (which were partly rentier states) were able to skim off revenue from maritime commercial ventures between the Far East, South Asia and West Asia. A rentier state is not dependent on extraction of revenue from the domestic base but from the revenues which accrue to it due to its unique geopolitical position. However, when China declined, this trade dropped, and South-East Asian maritime polities' revenue declined, and then these states gradually withered away.

But then, even the most dreaded steppe nomads who dominated the sedentary states of China, India, Persia and South Russia during the early modern era failed to adopt the techniques of naval power projection efficiently. Horse ethos just does not go with seamanship. The infantry-centric gunpowder revolution and the rise of ocean-going ships, which in turn facilitated maritime commerce and the rise of nation states, in the course of time displaced the horse warriors and the quasi-nomadic agrarian bureaucratic polities from their dominant position. And this military transformation occurred in the Western part of Eurasia partly because this region lacked the skill of mounted archery. Further, Western Eurasia, unlike China, India and Persia, was not threatened by the possibility of repeated Central Eurasian nomadic incursions. This allowed the Western European polities to experiment with slow-moving infantry squares and smoothbore muskets whose range, rate of fire and accuracy was inferior compared to a recurve/composite bow even in *c.* 1700 CE. It is to be noted that in the mid-seventeenth century, the rate of fire of a steppe nomadic composite bow-equipped soldier was three times higher compared to the rate of fire by a musketeer (Pant 1978: 21). For instance, the English longbowmen of the fifteenth century and Spanish tercios of the seventeenth century would have been crushed by nomadic mounted archers. Hence, despite possessing knowledge about gunpowder weapons, neither Mughal India nor Safavid Persia and certainly not Ming China poured significant amounts of energy and resources (for research and development) on slow-moving drilled and disciplined infantry equipped with slow-firing, inaccurate muskets. The Spanish, Dutch and the French monarchs could afford to experiment with such gunpowder-equipped 'toy infantry' because no steppe nomadic cavalry force was hovering along their military horizon. One noted historian of the ancient military history probably overstates his case when he asserts that from Alexander to Napoleon, the military systems of the Western world did not register much change in terms of tactics, hardware and organisation (Ferrill 1986: 215–23). But one thing is sure: that Genghis Khan's horse archers in areas with adequate grasslands would have been able to decimate all types of armies which Western Europe gave birth to till at least *c.* 1750. Counterfactuals are not always Cleopatra's nose.

The superiority of gunpowder weaponry was episodic and not intrinsic. In the thirteenth century, the Song Empire had the most advanced gunpowder weaponry. But they proved to be ineffective against the fast-moving Mongol cavalry. Only in *c.* 1750 were gunpowder weapons able to overwhelm the mounted archers. The military dominance of Western Eurasia became apparent only in the eighteenth century. Only after *c.* 1800, when the Western European military and economic dominance of the world was more or less complete, could the concept of a paradigmatic military system be used. Every power worth its salt in the nineteenth century attempted to copy the Western military system to survive. The nineteenth and twentieth centuries witnessed a convergence in matters military when the trend setter became the capital-intensive gunpowder-centric armed forces of Western Europe, and the industrial economy supported such military apparatus.

The military convergence of the modern era (the nineteenth and twentieth centuries) is not a permanent feature of global military history. In the beginning of new millennium, the military and economic rise of China and India and also the emergence of new forms of asymmetric warfare seem to spell the doom of the capital-intensive Western way of warfare. Against the skirmishing light infantry-centric insurgencies conducted by Hamas, Al-Qaeda, the Islamic State of Iraq-Syria (now Islamic State) and the Taliban in Palestine, Syria, Iraq and Afghanistan, the missiles, artillery, aircraft, armour, capital ships and heavy drilled and disciplined infantry of the Western military model are repeatedly failing. Nor do the regular Western-modelled capital-intensive armed forces have any easy answer to the urban terrorism practised by the transcontinental criminals and drug cartels operating in the urban centres of Latin America. Just as the Sea Peoples during the late Bronze Age emerged from their hideouts of Caucasus and attacked the urban river valley civilisations of Eurasia – pillaging, plundering, raping and killing – the Islamic insurgents and transnational criminals are moving out from the mountains and coming to the coastal plains and causing mayhem in the urban megalopolises.

The twenty-first century is also witnessing climatic disturbances (rise in global temperature, shortages of sub-soil water reserve, decline of the ozone layer, etc.) as the earth witnessed in around *c.* 1200 BCE. Radical shifts in climate patterns, the rise of novel technologies (especially information technology in the new millennium), economic redistribution and the emergence of new social groups are triggering military divergence. The point is that the military convergence–divergence model is not redundant but is still operating. We will end by asserting that the twenty-first century is witnessing the beginning of another cycle of military divergence.

References

Abu-Lughod, Janet, *Before European Hegemony: The World System AD 1250–1350* (1989, reprint, New York/Oxford: Oxford University Press, 1991).

Chaliand, Gerard (ed.), *The Art of War in World History from Antiquity to the Nuclear Age* (Berkeley/Los Angeles: University of California Press, 1994).

Ferrill, Arther, *The Origins of War: From the Stone Age to Alexander the Great* (1985, reprint, London: Thames and Hudson, 1986).

Pant, G.N., *Indian Arms and Armour*, vol. 1 (New Delhi: Army Educational Stores, 1978).

Scheidel, Walter, *The Great Leveler: Violence and the History of Inequality from the Stone Age to the Twenty-First Century* (Princeton/Oxford: Princeton University Press, 2018).

INDEX